Tutorials, Schools, and Workshops in the Mathematical Sciences

This series will serve as a resource for the publication of results and developments presented at summer or winter schools, workshops, tutorials, and seminars. Written in an informal and accessible style, they present important and emerging topics in scientific research for PhD students and researchers. Filling a gap between traditional lecture notes, proceedings, and standard textbooks, the titles included in TSWMS present material from the forefront of research.

Sergio Luigi Cacciatori • Alexander Kamenshchik
Editors

Einstein Equations: Local Energy, Self-Force, and Fields in General Relativity

Domoschool 2019

Editors
Sergio Luigi Cacciatori
Department of Science and High Technology
University of Insubria
Como, Italy

Alexander Kamenshchik
Department of Physics and Astronomy
University of Bologna
Bologna, Italy

ISSN 2522-0969 ISSN 2522-0977 (electronic)
Tutorials, Schools, and Workshops in the Mathematical Sciences
ISBN 978-3-031-21847-7 ISBN 978-3-031-21845-3 (eBook)
https://doi.org/10.1007/978-3-031-21845-3

Mathematics Subject Classification: 83C05, 83F05, 83C20, 83C25, 83C35, 83C40, 83C47, 83C57, 83B05

This book is published under the imprint Birkhäuser, www.birkhauser-science.com by the registered company Springer Nature Switzerland AG
The registered company address is: Gewerbestrasse 11, 6330 Cham, Switzerland

Preface

This book presents four lecture courses and in addition eight talks, given at the second edition of the *Domoschool*, the *International Alpine School in Mathematics and Physics*, held in Domodossola in July 2019 with the title "Einstein Equations: Physical and Mathematical aspects of General Relativity." Unfortunately, the COVID-19 pandemic delayed the writing of the present book and prevented the realization of the 2020 and 2021 editions.

Domoschool is a 5-day summer school aiming to bring together young mathematicians and physicists working in the same field of physics. In 2019, general relativity was chosen as the topic, the same as the previous edition, but with a very different focus. Four days were devoted to four lecture courses given by prominent experts, while an additional day of Domoschool was devoted to social activities (such as a hiking tour and a dinner), aiming to show the participants the town of Domodossola and its surroundings. The school also included a public meeting with the local community in order to explain to the citizens of Domodossola the contents of the summer school and the scientific relevance. Public lectures were given by **Andrea Accomazzo** and **Marco Giammarchi**.

The book is divided into two parts. The first part is devoted to the courses given at the school by outstanding world experts.

The first course, given by **Pengzi Miao**, had the primary scope of introducing graduate students to the mathematical problem of defining local energy in general relativity. Indeed, while global energy is well defined, the definition of local energy is still an open problem. Therefore, the notion of quasi-local energy is a basic subject of study in general relativity. In a four-dimensional spacetime, there are different ways of defining the quasi-local energy for a spacelike, closed twosurface. In particular, the "surface Hamiltonian approach" is naturally tied to the classic problem of isometric embeddings in differential geometry. The lectures by Pengzi Miao provide a short introduction to the Wang-Yau quasi-local energy of closed spacelike surfaces in spacetimes. After introducing the subject by discussing a geometric problem of isometrically embedding a surface into the Minkowski spacetime, the lecture proceeds with a review of the formula of surface Hamiltonian and its properties. Then, the definition of the Wang-Yau energy is explained, focusing on its physical and variational features, and relating it to some previously known quasi-local quantities. In the end, the ideas behind the proof of its positivity are outlined.

The second course was given by **Donato Bini**, and concerns the method of gravitational self-force in curved background. The most important application of gravitational self-force concerns metric and curvature perturbations in black hole spacetimes due to moving particles or evolving fields. In these notes, for teaching purposes, the method is illustrated and discussed at the level of a (massless) scalar field. Indeed, in this simple case one can look at the various steps implicit in any self-force computation without facing the additional difficulties of implementing them in a more involved tensorial background.

The third course was given by **Lars Andersson** and is devoted to geometry and analysis in black hole spacetimes. Black holes play a central role in general relativity and astrophysics. The problem of proving the dynamical stability of the Kerr black hole spacetime, which describes a rotating black hole in vacuum, is one of the most important open problems in general relativity. This course presents features of the geometry of the Kerr spacetime, including its algebraically special nature and consequences thereof. Then it introduces the analysis of some aspects of the black hole stability problem and presents the main steps in the recent proof of linearized stability of the Kerr black hole spacetime.

The fourth course was given by **Marco Giammarchi** and it provides a quick introduction to the experimental search for antimatter gravity. Gravitational properties of antimatter are related to both the validity of the CPT theorem in particle physics and the weak equivalence Principle of General Relativity. The course presents the motivation and techniques of the main approaches to this topic.

The school also included a further course, not transposed in the present book, given by **Alexei A. Starobinsky**, "Inflation and pre-inflation: The present status and expected discoveries."

The second part of the book presents the talks given by participants in a conference proceeding type manner. In Domoschool 2019, eight of the participants presented a short talk:

Lennart Brocki et al. (University of Wroclaw, Poland): *Quantum ergosphere and brick wall entropy*;

Francesco Cremona (University of Milan, Italy): *Geodesic structure and linear instability of some wormholes*

Vittorio De Falco (University in Opava, Czech Republic): *New trends in the general relativistic Poynting-Robertson effect modeling*;

Mario L. Gutierrez Abed (Newcastle University, UK): *Brief Overview of numerical relativity*

Colin MacLaurin (University of Queensland Brisbane, Australia): *Length-contraction in curved spacetime*

Jïrí Ryzner et al. (Charles University, Czech Republic): *Exact solutions of Einstein-Maxwell(-dilaton) equations with discrete translational symmetry*

Tereza Vardanyan et al. (University of Bologna, Italy): *Exact solutions of the Einstein equations for an infinite slab with constant energy density*

Adamantia Zampeli (Charles University in Prague): *Emergence of classicality from an inhomogeneous universe*

Domoschool was a project initiated by Sergio Cacciatori, but its realization was made possible thanks to the openminded people and the enthusiastic support of the **municipality of Domodossola**, which supported the school both financially and concretely in its organization. For this, we are indebted to the mayor, **Lucio Pizzi**; the councilor for culture, **Daniele Folino**, who was the first one welcoming our proposal and connecting us to the right people; and, especially, the deputy mayor, **Angelo Tandurella**, who was fully available to us, assisting step-by-step through the evolution, the logistics, the public event, and so on.

Additionally, we want to acknowledge the nonprofit association **ARS.UNI.VCO** (Associazione per lo Sviluppo della Cultura di Studi Universitari e della Ricerca nel Verbano Cusio Ossola); its vice president, **Giulio Gasparini**; and the president, **Stefania Cerutti**. We are particularly indebted to the secretary **Andrea Cottini** and the communication manager **Federica Fili** for their continuous support, organizing the school, assisting all participants, and, in particular, together with the Rosminian Fathers (especially **padre Fausto**), making it possible to gain Collegio Rosmini as the location of the School.

Furthermore, we are particularly grateful to the following speakers of Domoschool who kindly accepted our invitations and because of their participation brought high prestige to the school:

Lars Andersson from Albert Einstein Institute (Max-Planck Institute for Gravitational Physics), Potsdam, Germany

Donato Bini from Istituto per le Applicazioni del Calcolo "M. Picone," CNR, Rome (Italy)

Marco Giammarchi from Istituto Nazionale di Fisica Nucleare – Sezione di Milano, Italy

Pengzi Miao from the University of Miami, Miami, Florida

Alexei A. Starobinsky from Landau Institute for Theoretical Physics, Moscow, Russia

We acknowledge the speakers who accepted our invitation for the public lectures given at Rovereto Square just outside the city hall of Domodossola. Thanks to them, the public event was a complete success. They are:

Andrea Accomazzo from ESA, Germany

Marco Giammarchi from Istituto Nazionale di Fisica Nucleare – Sezione di Milano, Italy

Moreover, we extend our acknowledgments to the participants of the School, including two high school teachers, who, with their active participation, contributed to making the atmosphere of the school welcoming and pleasant. And, to the other members of the scientific board, **Francesco Belgiorno**, **Alessandro Carlotto**, **Simone Noja**, **Batu Güneysu**, **Stefano Pigola**, **Riccardo Re**, **Mauro Giudici**, and **Pietro Antonio Grassi**, as well as the members of the organizing committee, **Andrea Cottini** and **Giorgio Mantica**, we thank you for your support.

Last but not least, we are grateful to the sponsors of Domoschool: **Città di Domodossola**, *Università dell'Insubria*, **INFN**, **FONDAZIONE CRT**, and **CONSIGLIO REGIONALE DEL PIEMONTE**.

Como, Italy
Bologna, Italy
April 13 2022

Sergio Luigi Cacciatori
Alexander Kamenshchik

Contents

Part I
Main Lectures

Introduction to the Wang–Yau Quasi-local Energy

Pengzi Miao

2010 Mathematics Subject Classification Primary 83C99; Secondary 53A05

Contents

1 Introduction

In [36, 37], Wang and Yau gave a new definition of quasi-local energy for closed, spacelike surfaces in a spacetime. The Wang–Yau definition is based on a Hamiltonian approach relevant to the Einstein equation. Starting from the surface Hamiltonian, which is the boundary term in the Hamiltonian formulation, one

P. Miao (✉)
Department of Mathematics, University of Miami, Coral Gables, FL, USA
e-mail: pengzim@math.miami.edu

S. L. Cacciatori, A. Kamenshchik (eds.), *Einstein Equations: Local Energy, Self-Force, and Fields in General Relativity*, Tutorials, Schools, and Workshops in the Mathematical Sciences, https://doi.org/10.1007/978-3-031-21845-3_1

isometrically embeds the physical surface into a reference spacetime, and the quasi-local energy is taken as the difference between the two surface Hamiltonians.

The Wang–Yau definition involves elegant use of classic results on isometric embedding from differential geometry, and it makes a prudent analysis of a boundary value problem of elliptic differential equations and applies fundamental results on manifolds with nonnegative scalar curvature that include the positive mass theorem. The construction leading to the Wang–Yau energy demonstrates the intimate link among geometry, PDEs, and physics.

In this chapter, we want to give a succinct introduction to the Wang–Yau energy definition. The note is by no means a survey on the topic of quasi-local mass. It is not a survey on the current development of the Wang–Yau energy or mass neither. Instead, it aims to quickly introduce audiences whose background is mainly in a geometric analysis to the subject of the Wang–Yau quasi-local energy. The interested readers are then encouraged to further explore many of the intriguing and challenging problems related to this topic.

2 Isometric Embedding of Spheres into $\mathbb{R}^{3,1}$

Let Σ be a closed surface that is topologically a sphere. Let σ denote a smooth Riemannian metric on Σ. In this section, we discuss some results concerning the existence of isometric embedding of (Σ, σ) into the Minkowski spacetime $\mathbb{R}^{3,1}$.

The following fact is a direct consequence of a theorem of Pogorelov [30] on the existence of isometric embedding into hyperbolic spaces.

Theorem 2.1 *There exist infinitely many isometric embeddings $\iota : (\Sigma, \sigma) \to \mathbb{R}^{3,1}$ such that $\iota(\Sigma)$, the image of Σ, is spacelike in $\mathbb{R}^{3,1}$ and is a graph over some 2-surface in $\mathbb{R}^3 = \{t = 0\}$.*

Proof Let K denote the Gauss curvature of σ. Choose $\kappa > 0$ to be any constant so that $K > -\kappa^2$. By the result of Pogorelov [30], (Σ, σ) can be isometrically embedded into the hyperbolic space $\mathbb{H}^3_{-\kappa^2}$ with constant sectional curvature $-\kappa^2$. Using the hyperboloid model of the hyperbolic space, one can take

$$\mathbb{H}^3_{-\kappa^2} = \{(x, t) \mid t^2 - |x|^2 = \kappa^{-2}, \ t > 0, \ x \in \mathbb{R}^3\} \subset \mathbb{R}^{3,1}.$$

As a result, (Σ, σ) isometrically embeds into $\mathbb{R}^{3,1}$. Since the hyperboloid $\mathbb{H}^3_{-\kappa^2}$ is spacelike and is a graph over $\{t = 0\}$ in $\mathbb{R}^{3,1}$, so is the image of Σ under the embedding. As κ can be arbitrarily chosen to satisfy $K > -\kappa^2$, such embeddings are infinitely many. □

Remark 2.1 The image $\iota(\Sigma)$ above is also convex in $\mathbb{H}^3_{-\kappa^2}$. It may be useful to explore the implication of this convexity on the projection of $\iota(\Sigma)$ in $\mathbb{R}^3$.

In [37], Wang–Yau provided an embedding theorem that allows one to prescribe the time function of the embedding. It makes use of a result of Nirenberg [26] (and also Pogorelov [29]) on the existence of isometric embeddings into the Euclidean space.

Theorem 2.2 ([37]) *Given a Riemannian metric σ on Σ, suppose τ is a function on Σ such that $\sigma + d\tau \otimes d\tau$ is a Riemannian metric with positive Gauss curvature. Then, there exists an isometric embedding $\iota : \Sigma \to \mathbb{R}^{3,1}$, unique up to congruence in $\mathbb{R}^{3,1}$, such that $\iota(\Sigma)$ is the graph of $t = \tau$ over a convex surface $\hat{\Sigma}$ in $\mathbb{R}^3 = \{t = 0\}$.*

Proof Let $\hat{\sigma} = \sigma + d\tau \otimes d\tau$. Since $\hat{\sigma}$ is a metric with positive Gauss curvature, by the result in [26, 29], there exists an isometric embedding $\hat{\iota} : (\Sigma, \hat{\sigma}) \to \mathbb{R}^3$, unique up to congruence in $\mathbb{R}^3$, and $\hat{\iota}(\Sigma)$ is convex. Define $\iota : \Sigma \to \mathbb{R}^{3,1}$ by $\iota(q) = (\hat{\iota}(q), \tau(q))$, $q \in \Sigma$, and then ι is an isometric embedding of (Σ, σ) in $\mathbb{R}^{3,1}$ with the required property. The uniqueness of ι follows from that of $\hat{\iota}$. □

Remark 2.2 Given a metric σ on Σ, it is interesting to know if there exist functions τ so that $\sigma + d\tau \otimes d\tau$ is a metric with positive Gauss curvature. By Theorem 2.1, one does know there exist many functions τ so that $(\Sigma, \sigma + d\tau \otimes d\tau)$ isometrically embeds in $\mathbb{R}^3$.

3 Surface Hamiltonian and Its Properties

Let $\mathcal{S}$ denote a four-dimensional spacetime with a Lorentz metric $\bar{g}$. For convenience, the metric will also be denoted by $\langle \cdot, \cdot \rangle$.

Boundary terms in a Hamilton–Jacobi analysis of the Einstein–Hilbert action were considered by Brown–York [8, 9], Hawking–Horowitz [18], and Kijowski [20]. The collection of such terms leads to a surface integral known as the surface Hamiltonian. In this section, we review its definition and geometric properties.

3.1 Surface Hamiltonian

Let Σ be a closed spacelike 2-surface in a spacetime $\mathcal{S}$. Suppose $\Sigma = \partial\Omega$ for some compact spacelike hypersurface Ω. Let n_Ω denote the future-directed, timelike unit normal to Ω. Along Σ, let ν_Ω denote the outward unit normal to Σ in Ω.

Given a future-directed, timelike unit vector field T along Σ, one can decompose T as

$$T = X + N n_\Omega, \tag{1}$$

where X is tangential to Ω and N is a scalar function on Σ. The surface Hamiltonian of Σ in $\mathcal{S}$, with respect to T and n_Ω, is the integral

$$\mathcal{H}(T, n_\Omega) = -\frac{1}{8\pi}\int_\Sigma \left[NH - (K - (\mathrm{tr}K)g)(X, \nu_\Omega)\right] d\sigma. \tag{2}$$

Here,

- H is the mean curvature of Σ in Ω with respect to ν_Ω.
- K is the second fundamental form of Ω in $\mathcal{S}$ with respect to n_Ω, and $\mathrm{tr}K$ denotes the trace of K on Ω.
- g is the induced Riemannian metric on Ω, and $d\sigma$ is the area form on Σ.

(An exposition of the derivation of (2) can be found in [24] and the references therein.)

To see the role of T and Ω more explicitly in (2), one can further decompose X along Σ as

$$X = \langle X, \nu_\Omega\rangle \nu_\Omega + T^{\|}. \tag{3}$$

Here, $T^{\|}$ is the projection of X (hence of T) to the tangent space of Σ. Then,

$$(K - (\mathrm{tr}K)g)(X, \nu_\Omega) = K(T^{\|}, \nu_\Omega) - (\mathrm{tr}_\Sigma K)\langle X, \nu_\Omega\rangle, \tag{4}$$

where $\mathrm{tr}_\Sigma K$ is the trace of K restricted to Σ. The mean curvature vector $\vec{H}$ of Σ in $\mathcal{S}$ is given by

$$\vec{H} = (\mathrm{tr}_\Sigma K) n_\Omega - H\nu_\Omega. \tag{5}$$

Let $\vec{J}$ denote the dual of $\vec{H}$ in $(T\Sigma)^\perp$, the normal bundle of Σ, obtained by reflecting $\vec{H}$ across the inward future null direction. Then,

$$\vec{J} = -(\mathrm{tr}_\Sigma K)\nu_\Omega + Hn_\Omega, \tag{6}$$

and

$$\langle \vec{J}, T\rangle = -NH - \langle X, \nu_\Omega\rangle \mathrm{tr}_\Sigma K. \tag{7}$$

In terms of $\vec{J}$, $\mathcal{H}(T, n_\Omega)$ takes the form of

$$\mathcal{H}(T, n_\Omega) = \frac{1}{8\pi}\int_\Sigma \langle \vec{J}, T\rangle + \langle \overline{\nabla}_{T^{\|}} n_\Omega, \nu_\Omega\rangle \, d\sigma. \tag{8}$$

The right side of (8) indicates that $\mathcal{H}(T, n_\Omega)$ depends on Ω only through the 1-form $\alpha_{\nu_\Omega}(\cdot)$ on Σ, given by

$$\alpha_{\nu_\Omega}(\cdot) = \langle \overline{\nabla}_{(\cdot)} n_\Omega, \nu_\Omega\rangle, \tag{9}$$

where $\overline{\nabla}$ denotes the spacetime connection.

Remark 3.1 The term $-\int_\Sigma \langle \vec{J}, T\rangle\, d\sigma$ relates to the total future null expansion of Σ. Suppose $\{l, k\} \subset (T\Sigma)^\perp$ is a future-directed null frame along Σ, chosen so that l is outward-pointing and $\langle l, k\rangle = -2$. It can be checked

$$-\int_\Sigma \langle \vec{J}, T\rangle\, d\sigma = \int_\Sigma \langle \vec{H}, l\rangle \langle k, T\rangle\, d\sigma + \int_\Sigma \langle \vec{H}, T\rangle\, d\sigma. \tag{10}$$

As $\langle k, T\rangle < 0$, one can scale k so that $\langle k, T\rangle = -1$. Let $\theta_+ = -\langle \vec{H}, l\rangle$ for the corresponding future-directed null l, and then one has

$$-\int_\Sigma \langle \vec{J}, T\rangle\, d\sigma = \int_\Sigma \theta_+\, d\sigma + \int_\Sigma \langle \vec{H}, T\rangle\, d\sigma. \tag{11}$$

In particular, $-\int_\Sigma \langle \vec{J}, T\rangle\, d\sigma = \int_\Sigma \theta_+\, d\sigma$ if T is a timelike Killing vector field on $\mathcal{S}$.

The connection term in (8) transforms suitably under a change of choice of Ω. Suppose $\{\nu, n\}$ and $\{\tilde{\nu}, \tilde{n}\}$ are two orthonormal frames in $(T\Sigma)^\perp$ so that n and $\tilde{n}$ are future-pointing and ν and $\tilde{\nu}$ are outward-pointing. Let

$$\nu = \cosh\phi\, \tilde{\nu} + \sinh\phi\, \tilde{n} \quad \text{and} \quad n = \sinh\phi\, \tilde{\nu} + \cosh\phi\, \tilde{n}$$

for some function $\phi : \Sigma \to \mathbb{R}$. Then, $\forall\, v \in T\Sigma$, and it is readily checked

$$\langle \overline{\nabla}_v \nu, n\rangle = \langle \overline{\nabla}_v \tilde{\nu}, \tilde{n}\rangle - \langle v, \nabla\phi\rangle. \tag{12}$$

In terms of $\alpha_\nu(\cdot) = \langle \overline{\nabla}_{(\cdot)} n, \nu\rangle$, one has

$$\alpha_\nu(\cdot) = \alpha_{\tilde{\nu}}(\cdot) + d\phi. \tag{13}$$

As a result, by (8) and (13),

$$\mathcal{H}(T, n) = \mathcal{H}(T, \tilde{n}) + \int_\Sigma \langle T^\parallel, \nabla\phi\rangle\, d\sigma. \tag{14}$$

Here, ∇ denotes the gradient on Σ.

Remark 3.2 The timelike vector field T in $\mathcal{H}(T, n)$ serves as a physical observer. In the case of $\langle T, \nu\rangle = 0$, i.e., $T = Nn + T^\parallel$, one has $\langle \vec{J}, T\rangle = N\langle \vec{H}, \nu\rangle$ and

$$\mathcal{H}(T, n) = \frac{1}{8\pi}\int_\Sigma N\langle \vec{H}, \nu\rangle + \langle \overline{\nabla}_{T^\parallel} n, \nu\rangle\, d\sigma. \tag{15}$$

3.2 *Graphical Surfaces in* $\mathbb{R}^{3,1}$

Take $\mathcal{S} = \mathbb{R}^{3,1}$ and suppose $\Sigma \subset \mathbb{R}^{3,1}$. Given any constant, future timelike, unit vector field T_0, one can decompose

$$T_0 = T_0^{\perp} + T_0^{\parallel} \text{ along } \Sigma,$$

where $T_0^{\perp} \in (T\Sigma)^{\perp}$ and $T_0^{\parallel} \in T\Sigma$. As $T_0^{\perp} \neq 0$, one can normalize $T_0^{\perp}$ to define

$$n_0 = N^{-1} T_0^{\perp}, \tag{16}$$

where $N = \sqrt{-|T_0^{\perp}|^2}$.

Suppose Σ is a graph over a closed surface $\hat{\Sigma}$ in a hyperplane ζ that is orthogonal to T_0. Let $\hat{\nu}$ denote the outward unit normal to $\hat{\Sigma}$ in ζ. Let ν_0 denote the lift of $\hat{\nu}$ along T_0 in $\mathbb{R}^{3,1}$. Then, ν_0 is orthogonal to $C(\hat{\Sigma})$, where $C(\hat{\Sigma})$ is the cylinder over $\hat{\Sigma}$ in $\mathbb{R}^{3,1}$. As $\Sigma \subset C(\hat{\Sigma})$, one has

$$\{\nu_0, n_0\} \subset (T\Sigma)^{\perp} \text{ and } \nu_0 \perp n_0. \tag{17}$$

Lemma 3.1 ([37]) *With the above choice of* $\{\nu_0, n_0\}$, *along* $\Sigma \subset \mathbb{R}^{3,1}$,

$$N \langle \vec{H}_0, \nu_0 \rangle (q) + \langle \overline{\nabla}^0_{T_0^{\parallel}} n_0, \nu_0 \rangle (q) = -N \hat{H}(\hat{q}). \tag{18}$$

Here, $\vec{H}_0$ *is the mean curvature vector of* Σ *in* $\mathbb{R}^{3,1}$, $\overline{\nabla}^0$ *denotes the connection in* $\mathbb{R}^{3,1}$, q *denotes a point in* Σ, $\hat{q}$ *is the projection of* q *on* ζ, *and* $\hat{H}$ *is the mean curvature of* $\hat{\Sigma}$ *in* ζ *with respect to* $\hat{\nu}$.

Proof $C(\hat{\Sigma})$ is a timelike hypersurface in $\mathbb{R}^{3,1}$. Let $\tilde{\Pi}(\cdot, \cdot)$ denote its second fundamental form with respect to ν_0, i.e., $\tilde{\Pi}(v, w) = \langle \overline{\nabla}^0_v \nu_0, w \rangle$ for v, w tangential to $C(\hat{\Sigma})$. The trace of $\tilde{\Pi}$ on $C(\hat{\Sigma})$ satisfies

$$\mathrm{tr}\tilde{\Pi}(q) = \mathrm{tr}\tilde{\Pi}(\hat{q}). \tag{19}$$

By definition, $\mathrm{tr}\tilde{\Pi}(\hat{q}) = \hat{H}(\hat{q})$ and

$$\mathrm{tr}\tilde{\Pi}(q) = \sum_{\alpha} \langle \overline{\nabla}^0_{e_\alpha} \nu_0, e_\alpha \rangle - \langle \overline{\nabla}^0_{n_0} \nu_0, n_0 \rangle, \tag{20}$$

where $\{e_\alpha\}_{\alpha=1,2}$ is an orthonormal frame in $T\Sigma$ at q. Plug in $T_0 = N n_0 + T^{\parallel}$, and then one has

$$\hat{H}(\hat{q}) = -\langle \vec{H}_0, \nu_0 \rangle - N^{-1} \langle \overline{\nabla}_{(T - T^{\parallel})} \nu_0, n_0 \rangle, \tag{21}$$

which proves (18) as $\overline{\nabla}^0_T \nu_0 = 0$. □

Next, suppose $\tau : \Sigma \to \mathbb{R}$ is the time function on Σ with respect to T_0 and ζ. For instance, one may choose standard coordinates (x_1, x_2, x_3, t) on $\mathbb{R}^{3,1}$ so that

$$T_0 = \partial_t = -\overline{\nabla} t, \quad \zeta = \{t = 0\}$$

and Σ is the graph of $t = \tau$ over $\hat{\Sigma}$. By abuse of nation, τ can be viewed as a function on both $\hat{\Sigma}$ and Σ.

Let $\hat{\sigma}$ and σ denote the induced Riemannian metrics on $\hat{\Sigma}$ and Σ, respectively. Then,

$$\hat{\sigma} = \sigma + d\tau \otimes d\tau.$$

Let ∇ denote the gradient on (Σ, σ). Then $T^{\|} = -\nabla\tau$, $N = \sqrt{1 + |\nabla\tau|^2}$, and $d\hat{\sigma} = \sqrt{1 + |\nabla\tau|^2} d\sigma$. Thus, by (18),

$$\int_\Sigma -N\langle \vec{H}_0, \nu_0\rangle - \langle \overline{\nabla}_{T_0^{\|}} n_0, \nu_0\rangle \, d\sigma = \int_\Sigma N\hat{H} \, d\sigma = \int_{\hat{\Sigma}} \hat{H} \, d\hat{\sigma}. \tag{22}$$

It follows from (15) and (22) that

$$-\mathcal{H}(T_0, n_0) = \int_{\hat{\Sigma}} \hat{H} \, d\hat{\sigma}. \tag{23}$$

(See equation (3.4) in [37].)

3.3 Some Related Inequalities

Suppose Σ is a graph of $t = \tau$ over a convex surface $\hat{\Sigma} \subset \mathbb{R}^3 = \{t = 0\}$. Let $|\Sigma|$ and $\hat{\Sigma}|$ denote the area of Σ and $\hat{\Sigma}$, respectively. The Minkowski inequality in $\mathbb{R}^3$ gives

$$\frac{1}{8\pi} \int_{\hat{\Sigma}} \hat{H} \, d\hat{\sigma} \geq \left(\frac{|\hat{\Sigma}|}{4\pi}\right)^{\frac{1}{2}}.$$

Let $T_0 = \partial_t$. By (23) and (8),

$$-\frac{1}{8\pi} \int_\Sigma \langle \vec{J}_0, T_0\rangle - \langle \overline{\nabla}^0_{T_0^{\|}} n_0, \nu_0\rangle \, d\sigma \geq \left(\frac{|\hat{\Sigma}|}{4\pi}\right)^{\frac{1}{2}} \geq \left(\frac{|\Sigma|}{4\pi}\right)^{\frac{1}{2}}, \tag{24}$$

where $\vec{J}_0$ is the dual of $\vec{H}_0$, and $|\hat{\Sigma}| \geq |\Sigma|$ is due to the fact $d\hat{\sigma} = \sqrt{1 + |\nabla\tau|^2} \, d\sigma$.

For a properly chosen future-directed null vector field l along Σ in $\mathbb{R}^{3,1}$, one knows

$$-\int_{\Sigma} \langle \vec{J}_0, T_0 \rangle \, d\sigma = \int_{\Sigma} \theta_+ \, d\sigma$$

with $\theta_+ = -\langle \vec{H}, l \rangle$ (see Remark 3.1). An inequality involving $-\int_{\Sigma} \langle \vec{J}_0, T_0 \rangle \, d\sigma$ was conjectured by Penrose [28].

Conjecture 3.1 ([28]) For a suitable class of spacelike closed surfaces $\Sigma \subset \mathbb{R}^{3,1}$,

$$-\int_{\Sigma} \langle \vec{J}_0, T_0 \rangle \, d\sigma \geq \left(\frac{|\Sigma|}{4\pi} \right)^{\frac{1}{2}}. \tag{25}$$

Motivation to and results on (25) can be found in [5, 6, 17, 23, 27, 28, 34] and the references therein. In relation to the setting of isometric embeddings into $\mathbb{R}^{3,1}$ from Sect. 2, we want to reflect on the following theorem of Brendle–Hung–Wang [6] (more precisely, Theorem 8.1 in [35]).

Theorem 3.1 ([6, 35]) *Let Σ be a closed surface in the hyperbolic space $\mathbb{H}^3_{-1} \subset \mathbb{R}^{3,1}$. Suppose Σ has positive mean curvature in $\mathbb{H}^3_{-1}$ and is star-shaped with respect to the point* $(0, 0, 0, 1) \in \mathbb{H}^3_{-1}$. *Then,*

$$-\int_{\Sigma} \langle \vec{J}_0, \partial_t \rangle \, d\sigma \geq \left(\frac{|\Sigma|}{4\pi} \right)^{\frac{1}{2}}. \tag{26}$$

Suppose $\Sigma \subset \mathbb{H}^3_{-1} \subset \mathbb{R}^{3,1}$. Let Ω be the domain bounded by Σ in $\mathbb{H}^3_{-1}$. By (8), the surface Hamiltonian of Σ, with respect to ∂_t and n_Ω in $\mathbb{R}^{3,1}$, satisfies

$$\mathcal{H}(\partial_t, n_\Omega) = \frac{1}{8\pi} \int_{\Sigma} \langle \vec{J}_0, \partial_t \rangle \, d\sigma, \tag{27}$$

where $\langle \overline{\nabla}^0_{(\partial_t)^\parallel} n_\Omega, \nu_\Omega \rangle = 0$ as $\mathcal{H}^3_{-1}$ is totally umbilic in $\mathbb{R}^{3,1}$. Thus, by Theorem 3.1, if Σ has positive mean curvature and is star-shaped in $\mathbb{H}^3_{-1}$, then

$$-\mathcal{H}(\partial_t, n_\Omega) \geq \left(\frac{|\Sigma|}{4\pi} \right)^{\frac{1}{2}}. \tag{28}$$

Note that, by (14) and (23),

$$\mathcal{H}(\partial_t, n_\Omega) = -\int_{\hat{\Sigma}} \hat{H} \, d\hat{\sigma} + \int_{\Sigma} (\Delta \tau) \psi \, d\sigma. \tag{29}$$

Here, $\tau = t|_\Sigma$, the time function on Σ, $\hat{\Sigma}$ is the projection of Σ in $\mathbb{R}^3$, and ψ is the function determined by $\cosh\psi = -\langle n_\Omega, n_0\rangle$. As $n_\Omega = X$, the position vector in $\mathbb{R}^{3,1}$, one has $\langle n_\Omega, n_0\rangle = \langle X, N^{-1}\partial_t\rangle = \frac{-\tau}{\sqrt{1+|\nabla\tau|^2}}$. Thus, $\psi = \cosh^{-1}\left(\frac{\tau}{\sqrt{1+|\nabla\tau|^2}}\right)$.

4 Construction of the Wang–Yau Energy

In a Hamiltonian approach toward defining energy for $\Sigma \subset \mathcal{S}$, one considers the difference between two surface Hamiltonians: one is from the physical surface and the other one is from a reference surface. In the construction of the Wang–Yau quasi-local energy [36, 37], the reference is chosen as an isometric embedding of Σ in $\mathbb{R}^{3,1}$, with T_0 and n_0 specified in Sect. 3.2. An intriguing step next is to make a judicious choice of $\{T, n\}$ associated with the surface in the physical spacetime.

4.1 The Choice of $\{T, n\}$

Suppose $\Sigma \subset \mathcal{S}$, and let σ denote the metric on Σ. Given an isometric embedding $\iota : (\Sigma, \sigma) \to \mathbb{R}^{3,1}$ and a constant, future timelike unit vector $T_0 \in \mathbb{R}^{3,1}$, let n_0 be given in (16). Along Σ in $\mathcal{S}$, one wants to construct a pair (T, n) so that it matches (T_0, n_0) along Σ in $\mathbb{R}^{3,1}$ suitably. To do so, one requires

$$T = Nn + T^{\parallel}, \tag{30}$$

a relation satisfied by $T_0 = Nn_0 + T_0^{\parallel}$ in $\mathbb{R}^{3,1}$. Here, $T^{\parallel}$ is the pullback of $T_0^{\parallel}$ to Σ via ι. If τ denotes the time function on Σ with respect to T_0 in $\mathbb{R}^{3,1}$, then

$$T^{\parallel} = -\nabla\tau \quad \text{and} \quad N = \sqrt{1+|\nabla\tau|^2}. \tag{31}$$

Viewing T and T_0 as observers of Σ in $\mathcal{S}$ and $\mathbb{R}^{3,1}$, respectively, one may further require the expansions of Σ, observed in $\mathcal{S}$ and $\mathbb{R}^{3,1}$, to be the same. That is,

$$\langle \vec{H}, T\rangle = \langle \vec{H}_0, T_0\rangle. \tag{32}$$

Without losing generality, taking $T_0 = \partial_t$, and writing $\iota = (x_1, x_2, x_3, \tau)$, then

$$\vec{H}_0 = (\Delta x_1, \Delta x_2, \Delta x_3, \Delta\tau) \quad \text{and} \quad \langle \vec{H}_0, T_0\rangle = -\Delta\tau. \tag{33}$$

In what follows, one assumes the mean curvature vector $\vec{H}$ of Σ in $\mathcal{S}$ is spacelike. Normalizing $\{\vec{H}, \vec{J}\}$ to get a reference frame in $(T\Sigma)^{\perp}$, one can let

$$\nu_{\vec{H}} = -\frac{\vec{H}}{|\vec{H}|} \text{ and } n_{\vec{H}} = \frac{J}{|\vec{H}|}.$$

Writing

$$n = \sinh\phi\, \nu_{\vec{H}} + \cosh\phi\, n_{\vec{H}}, \tag{34}$$

it follows from (30) and (32) that ϕ is determined by

$$\sinh\phi = \frac{\Delta\tau}{|\vec{H}|\sqrt{1+|\nabla\tau|^2}}. \tag{35}$$

Thus, a pair $\{T, n\}$ along $\Sigma \subset \mathcal{S}$, matching $\{T_0, n_0\}$ along Σ in $\mathbb{R}^{3,1}$, is determined by (30), (34), and (35).

4.2 A Variational Property

The choice of T by (32) indeed has a variational characterization. Under the relation $T = Nn + T^{\parallel}$, by (15), one has

$$8\pi\mathcal{H}(T, n) = \int_{\Sigma} \sqrt{1+|\nabla\tau|^2}\langle \nu, \vec{H}\rangle - \langle \overline{\nabla}_{\nabla\tau} n, \nu\rangle\, d\sigma. \tag{36}$$

Consider this as a functional of $\{\nu, n\}$, an orthonormal frame in $(T\Sigma)^{\perp}$ so that n is future-pointing and ν is outward-pointing (meaning that $\langle \nu, \nu_{\vec{H}}\rangle > 0$). Representing ν and n as

$$\begin{cases} \nu = \cosh\phi\, \nu_{\vec{H}} + \sinh\phi\, n_{\vec{H}} \\ n = \sinh\phi\, \nu_{\vec{H}} + \cosh\phi\, n_{\vec{H}}, \end{cases} \tag{37}$$

then

$$\begin{aligned} & -\sqrt{1+|\nabla\tau|^2}\langle \nu, \vec{H}\rangle + \langle \overline{\nabla}_{\nabla\tau} n, \nu\rangle \\ =& \sqrt{1+|\nabla\tau|^2}\cosh\phi |\vec{H}| + \langle \overline{\nabla}_{\nabla\tau} n_{\vec{H}}, \nu_{\vec{H}}\rangle + \langle \nabla\tau, \nabla\phi\rangle. \end{aligned} \tag{38}$$

Integrating on Σ, one has

$$-8\pi\mathcal{H}(T,n)=\int_\Sigma \sqrt{1+|\nabla\tau|^2}|\vec{H}|\cosh\phi-(\Delta\tau)\phi\,d\sigma+\int_\Sigma\langle\overline{\nabla}_{\nabla\tau}n_{\vec{H}},\nu_{\vec{H}}\rangle\,d\sigma. \tag{39}$$

The right side of (39) is easily seen to be a convex functional of ϕ. Thus, one has the following lemma:

Lemma 4.1 ([37]) *Under the relation*

$$T=\sqrt{1+|\nabla\tau|^2}\,n-\nabla\tau,$$

$\mathcal{H}(T,n)$ *is uniquely maximized by the choice* $\{\nu,n\}$ *given by* (37) *with* ϕ *satisfying*

$$\sinh\phi=\frac{\Delta\tau}{|\vec{H}|\sqrt{1+|\nabla\tau|^2}}. \tag{40}$$

4.3 Expression of the Wang–Yau Quasi-local Energy

Given a pair (ι, T_0), let $\{T,n\}$ be chosen above. Then,

$$\begin{aligned}-8\pi\mathcal{H}(T,n)=\int_\Sigma &\sqrt{(1+|\nabla\tau|^2)|\vec{H}|^2+(\Delta\tau)^2}-\phi\,\Delta\tau\\ &+\langle\overline{\nabla}_{\nabla\tau}n_{\vec{H}},\nu_{\vec{H}}\rangle\,d\sigma,\end{aligned} \tag{41}$$

where ϕ is specified by (35) or (40). In terms of the 1-form $\alpha_{\nu_{\vec{H}}}(\cdot)$, the connection term satisfies

$$\int_\Sigma\langle\overline{\nabla}_{\nabla\tau}n_{\vec{H}},\nu_{\vec{H}}\rangle\,d\sigma=\int_\Sigma\alpha_{\nu_{\vec{H}}}(\nabla\tau)\,d\sigma=-\int_\Sigma\tau\,\mathrm{div}[\alpha_{\nu_{\vec{H}}}(\cdot)]\,d\sigma. \tag{42}$$

Here, div$(\cdot)$ denotes the divergence on (Σ,σ).

Thinking Σ as a surface in $\mathbb{R}^{3,1}$, one has an identity

$$\begin{aligned}-8\pi\mathcal{H}(T_0,n_0)=\int_\Sigma &\sqrt{(1+|\nabla\tau|^2)|\vec{H}_0|^2+(\Delta\tau)^2}-\phi_0\,\Delta\tau\\ &+\langle\overline{\nabla}^0_{\nabla\tau}n_{\vec{H}_0},\nu_{\vec{H}_0}\rangle\,d\sigma,\end{aligned} \tag{43}$$

where ϕ_0 is given by

$$\sinh\phi_0=\frac{\Delta\tau}{|\vec{H}_0|\sqrt{1+|\nabla\tau|^2}}. \tag{44}$$

Since the dependence of $\mathcal{H}(T, n)$, $\mathcal{H}(T_0, n_0)$ on (ι, T_0) is only via

$$\tau = -\langle T_0, \iota \rangle, \tag{45}$$

the Wang–Yau quasi-local energy of $\Sigma \subset \mathcal{S}$, defined by

$$\frac{1}{8\pi} \left[\mathcal{H}(T, n) - \mathcal{H}(T_0, n_0) \right],$$

is also a functional of τ. Denote it by $E_{WY}(\Sigma, \tau)$, and it follows from (41)–(43) that

$$\begin{aligned} & 8\pi E_{WY}(\Sigma, \tau) \\ = & \int_\Sigma \sqrt{(1 + |\nabla \tau|^2)|\vec{H}_0|^2 + (\Delta \tau)^2} - \sqrt{(1 + |\nabla \tau|^2)|\vec{H}|^2 + (\Delta \tau)^2} \\ & + (\phi - \phi_0)\Delta\tau + \tau \operatorname{div}[\alpha_{\nu_{\vec{H}}}(\cdot) - \bar{\alpha}_{\nu_{\vec{H}_0}}(\cdot)] \, d\sigma. \end{aligned} \tag{46}$$

Here, $\bar{\alpha}_{\nu_{\vec{H}_0}}(\cdot) = \langle \overline{\nabla}^0_{(\cdot)} n_{\vec{H}_0}, \nu_{\vec{H}_0} \rangle$ is the connection 1-form on Σ in $\mathbb{R}^{3,1}$.

In the case that Σ is embedded in $\mathbb{R}^{3,1}$ as a graph over some $\hat{\Sigma}$ in a hyperplane orthogonal to T_0, by (23), $E_{WY}(\Sigma, \tau)$ also takes the form of

$$E_{WY}(\Sigma, \tau) = \frac{1}{8\pi} \int_{\hat{\Sigma}} \hat{H} \, d\hat{\sigma} + \frac{1}{8\pi} \mathcal{H}(T, n). \tag{47}$$

4.4 *Relation to Other Quasi-local Energy*

Suppose the metric σ has positive Gauss curvature. In this case, (Σ, τ) isometrically embeds into $\mathbb{R}^3 = \{t = 0\}$. Let ι be such an embedding, and let $T_0 = \partial_t$. Then $\tau = 0$ and $E_{WY}(\Sigma, 0)$ becomes

$$E_{WY}(\Sigma, 0) = \int_\Sigma (H_0 - |\vec{H}|) \, d\sigma, \tag{48}$$

where H_0 denotes the mean curvature of the isometric embedding of (Σ, σ) into $\mathbb{R}^3$. $E_{WY}(\Sigma, 0)$ agrees with the quasi-local energy of Σ defined by Liu and Yau [21, 22] (also see the work of Booth–Mann [7], Epp [16], and Kijowski [20]).

Suppose $\mathcal{S}$ satisfies the dominant energy condition. Liu and Yau [21, 22] proved the positivity of the right side of (48), under assumptions that Σ bounds a compact spacelike hypersurface Ω and that $\vec{H}$ is inward-pointing relative to Ω. The latter means $\langle \vec{H}, \nu_\Omega \rangle < 0$, where ν_Ω is the outward unit normal to Σ in Ω.

In the special case that Σ bounds a compact time-symmetric hypersurface Ω and $\vec{H}$ is inward-pointing relative to Ω, $E_{WY}(\Sigma, 0)$ further reduces to

$$E_{WY}(\Sigma, 0) = \int_{\Sigma} (H_0 - H)\, d\sigma, \tag{49}$$

where $H = -\langle \vec{H}, \nu_\Omega \rangle$ is the mean curvature of Σ in Ω. The right side of (49) is known as the Brown–York mass [8, 9] of Σ in Ω. Its positivity was guaranteed by the following theorem of Shi and Tam [33].

Theorem 4.1 ([33]) *Let (Ω, g) be a three-dimensional compact Riemannian manifold with nonnegative scalar curvature, with boundary Σ. Suppose the induced metric σ on Σ has positive Gauss curvature, and the mean curvature H of Σ in Ω is positive. Then,*

$$\int_{\Sigma} H_0\, d\sigma \geq \int_{\Sigma} H\, d\sigma, \tag{50}$$

where H_0 is the mean curvature of the isometric embedding of Σ in $\mathbb{R}^3$. Moreover, equality holds if and only if (Ω, g) is isometric to a domain in $\mathbb{R}^3$.

Shi-Tam's theorem is a fundamental result on compact manifolds with nonnegative scalar curvature, with boundary. Its proof made use of the Riemannian positive mass theorem of Schoen–Yau [31] and Witten [39]. The positive mass theorem itself is an assertion of the positivity of the ADM mass [1, 2] on asymptotically flat manifolds.

The following theorem of Wang–Yau [37] provides a generalization of Theorem 4.1. It plays a key role in the proof of the positivity of the Wang–Yau quasi-local energy.

Theorem 4.2 ([37]) *Let (Ω, g) be a three-dimensional compact Riemannian manifold with boundary Σ. Suppose there exists a vector field Y on Ω such that*

$$R \geq 2|Y|^2 - 2\,\mathrm{div} Y, \tag{51}$$

where R is the scalar curvature of g, $\mathrm{div} Y$ is the divergence of Y, and

$$H > \langle Y, \nu_\Omega \rangle, \tag{52}$$

where H is the mean curvature of Σ in Ω with respect to the outward unit normal ν_Ω. If the induced metric σ on Σ has positive Gauss curvature, then

$$\int_{\Sigma} H_0\, d\sigma \geq \int_{\Sigma} H - \langle Y, \nu_\Omega \rangle\, d\sigma, \tag{53}$$

where H_0 is the mean curvature of the isometric embedding of Σ in $\mathbb{R}^3$.

5 Positivity of $E_{WY}(\Sigma, \tau)$

If the spacetime $\mathcal{S}$ satisfies the dominant energy condition, Wang–Yau [37] proved that $E_{WY}(\Sigma, \tau)$ is always nonnegative for surfaces $\Sigma \subset \mathcal{S}$, under suitable assumptions. The proof in [37] consists of several key ingredients, which includes a boundary value problem of Jang's equation [19, 32], an application of Theorem 4.2 of Wang–Yau, an intriguing physical interpretation of boundary terms from Jang's equation, and the variational characterization of $\{T, n\}$ used in defining $E_{WY}(\Sigma, \tau)$.

5.1 A Motivation to Jang's Equation in $\mathbb{R}^{3,1}$

Suppose $\hat{\Omega} \subset \mathbb{R}^3 = \{t = 0\} \subset \mathbb{R}^{3,1}$. Given a function $f : \hat{\Omega} \to \mathbb{R}$, let Ω be the graph of f in $\mathbb{R}^{3,1}$, i.e.,

$$\Omega = \{(x, f(x)) \mid x \in \hat{\Omega}\} \subset \mathbb{R}^{3,1}.$$

Identifying Ω with $\hat{\Omega}$ via $x \mapsto (x, f(x))$, one can view f as a function on Ω. Suppose Ω is spacelike. Let g denote the induced Riemannian metric on Ω. The following are two basic facts about f on (Ω, g):

(i) $\hat{g} := g + df \otimes df$ is the Euclidean metric on $\hat{\Omega} \subset \mathbb{R}^3$.
(ii) If k_{ij} denotes the second fundamental form of Ω in $\mathbb{R}^{3,1}$ with respect to the future-directed timelike normal, then

$$k = \frac{D^2 f}{\sqrt{1 + |Df|^2}}. \tag{54}$$

Here, D^2 and D denote the Hessian and the gradient on (Ω, g), respectively.

In particular, f on (Ω, g) satisfies an elliptic equation

$$\operatorname{tr}_{\hat{g}} \left(\frac{D^2 f}{\sqrt{1 + |Df|^2}} - k \right) = 0. \tag{55}$$

In local coordinates, (55) takes the form of

$$\left(g^{ij} - \frac{f^i f^j}{1 + |Df|^2} \right) \left(\frac{f_{;ij}}{\sqrt{1 + |Df|^2}} - k_{ij} \right) = 0. \tag{56}$$

Here, $f^i = g^{ij} f_{,i}$ and ";" denotes the covariant differentiation on (Ω, g).

Suppose $\hat{\Omega}$ has boundary $\hat{\Sigma}$. Let $\tau = f|_{\hat{\Sigma}}$ and $\Sigma = \partial\Omega$ be the graph of f over $\hat{\Sigma}$. Then, f on Ω satisfies a boundary condition

$$f = \tau \text{ on } \Sigma. \tag{57}$$

5.2 Jang's Equation on a General (Ω, g, k)

Let Ω be a compact spacelike hypersurface with boundary Σ in a spacetime $\mathcal{S}$. Let g and k denote the metric and the second fundamental form of Ω in $\mathcal{S}$, respectively. Suggested by the preceding fact concerning functions and their graphs in $\mathbb{R}^{3,1}$, one can impose the following equations on (Ω, g):

$$\begin{cases} \left(g^{ij} - \frac{f^i f^j}{1+|Df|^2}\right)\left(\frac{f_{;ij}}{\sqrt{1+|Df|^2}} - k_{ij}\right) = 0 \text{ on } \Omega \\ f = \tau \text{ at } \Sigma. \end{cases} \tag{58}$$

The equation in (58) is known as Jang's equation, first proposed by Jang [19]. It was used by Schoen–Yau in their proof of the spacetime positive mass theorem [32]. The boundary value system (58) was analyzed by Wang–Yau [37].

To focus on the idea behind the positivity of $E_{WY}(\Sigma, \tau)$, we suppose f is a smooth solution to (58). The term $g^{ij} - \frac{f^i f^j}{1+|Df|^2}$, hence the tensor $g + df \otimes df$, suggests that it can be useful to consider the graph of f over Ω in the Riemannian product

$$(\Omega \times \mathbb{R}, g + dt^2).$$

Denote this graph by $\tilde{\Omega}$, and identify Ω and $\tilde{\Omega}$ in the usual way. The following holds:

(i) $\tilde{g} := g + df \otimes df$ is the induced metric on $\tilde{\Omega}$.
(ii) If $\tilde{p}$ denotes the second fundamental form of $\tilde{\Omega}$ in $\Omega \times \mathbb{R}$ with respect to the downward unit normal $\tilde{n}_d$, then

$$\tilde{p} = \frac{D^2 f}{\sqrt{1 + |Df|^2}}.$$

(In what follows, by abuse of notation, we also use $\tilde{n}_d$ to denote the vector field on $\Omega \times \mathbb{R}$, obtained by parallel translating $\tilde{n}_d$ along the $\mathbb{R}$-factor.)

In terms of $\tilde{g}$ and $\tilde{p}$, the PDE in (58) takes the form of

$$\tilde{h} = \mathrm{tr}_{\tilde{g}}\tilde{k}. \tag{59}$$

Here, $\tilde{h} = \mathrm{tr}_{\tilde{g}} \tilde{p}$ is the mean curvature of $\tilde{\Omega}$ in $\Omega \times \mathbb{R}$ with respect to $\tilde{n}_d$, and $\tilde{k} = \pi^*(k)$, where $\pi : \Omega \times \mathbb{R} \to \Omega$ denotes the usual projection map.

As shown by Schoen–Yau [32], a crucial implication of (59) is that the scalar curvature $\tilde{R}$ of $(\tilde{\Omega}, \tilde{g})$ satisfies

$$\tilde{R} - 2|\tilde{Y}|^2 + 2\,\mathrm{div}\tilde{Y} \geq |\tilde{p} - \tilde{k}|^2 + 2(\tilde{\mu} - |\tilde{J}|). \tag{60}$$

Here, $\tilde{Y}$, $\tilde{\mu}$, and $\tilde{J}$ are as follows:

- $\tilde{Y}$ is the vector field on $(\tilde{\Omega}, \tilde{g})$ that is dual to the 1-form

$$\eta(\cdot) = \langle \tilde{D}_{\tilde{n}_d} \tilde{n}_d, \cdot \rangle - \tilde{k}(\tilde{n}_d, \cdot), \tag{61}$$

  where $\tilde{D}$ denotes the connection on $(\Omega \times \mathbb{R}, g + dt^2)$.
- $\tilde{\mu} = \mu \circ \pi$ and $\tilde{J} = \pi^*(J)$ are the lift of μ and J, where

$$\mu = \frac{1}{2}\left(R - |k|^2 + (\mathrm{tr}_g k)^2\right) \text{ and } J = \mathrm{div}_g(k - (\mathrm{tr}_g k) g)$$

  denote the local energy density and local current density on Ω in $\mathcal{S}$.

Remark 5.1 The solvability of Jang's equation with boundary value τ was analyzed in [37, Section 4.3].

5.3 Comparison with a Euclidean Domain

Suppose the spacetime $\mathcal{S}$ satisfies the dominant energy condition. Then, $\mu \geq |J|$ on Ω. As a result, (60) implies

$$\tilde{R} - 2|\tilde{Y}|^2 + 2\,\mathrm{div}\tilde{Y} \geq 0. \tag{62}$$

This indicates that Theorem 4.2 of Wang–Yau is applicable to $(\tilde{\Omega}, \tilde{g})$, provided

(i) $(\Sigma, \sigma + d\tau \otimes d\tau)$ has positive Gauss curvature, here σ is the metric on Σ.
(ii) $\tilde{H} - \langle \tilde{Y}, \tilde{\nu} \rangle > 0$ along $\tilde{\Sigma} = \partial \tilde{\Omega}$. Here, $\tilde{\nu}$ is the outward unit normal to $\tilde{\Sigma}$ in $\tilde{\Omega}$ and $\tilde{H}$ is the mean curvature of $\tilde{\Sigma}$ in $\tilde{\Omega}$ with respect to $\tilde{\nu}$.

Condition (i) also calls for the application of Theorem 2.2 of Wan–Yau, i.e., there exists an isometric embedding

$$\iota : (\Sigma, \sigma) \to \mathbb{R}^{3,1}$$

so that $\iota(\Sigma)$ is the graph of $t = \tau$ over a convex surface $\hat{\Sigma} \subset \mathbb{R}^3 = \{t = 0\}$. Here, the metric on $\hat{\Sigma}$ is $\hat{\sigma} = \sigma + d\tau \otimes d\tau$. Consequently, $\hat{\Sigma}$ is the image of an isometric embedding of $\tilde{\Sigma} = \partial \tilde{\Omega}$ in $\mathbb{R}^3$.

By Theorem 4.2, one therefore has

$$\int_{\hat{\Sigma}} \hat{H}\, d\hat{\sigma} \geq \int_{\tilde{\Sigma}} \tilde{H} - \langle \tilde{Y}, \tilde{\nu} \rangle\, d\hat{\sigma}. \tag{63}$$

5.4 A Physical Interpretation of $\tilde{H} - \langle \tilde{Y}, \tilde{\nu} \rangle$

Given any $\tilde{q} \in \tilde{\Sigma}$, let $q = \pi(\tilde{q}) \in \Sigma$. As $d\hat{\sigma} = \sqrt{1 + |\nabla\tau|^2} d\sigma$, one has

$$-\int_{\tilde{\Sigma}} \left(\tilde{H} - \langle \tilde{Y}, \tilde{\nu} \rangle \right)(\tilde{q})\, d\hat{\sigma}(\tilde{q}) = \int_{\Sigma} \sqrt{1 + |\nabla\tau|^2} \left(-\tilde{H} + \langle \tilde{Y}, \tilde{\nu} \rangle \right)(\tilde{q})\, d\sigma(q).$$

In [37], Wang–Yau made an intriguing discovery that recognizes the term

$$\sqrt{1 + |\nabla\tau|^2} \left(-\tilde{H} + \langle \tilde{Y}, \tilde{\nu} \rangle \right)(\tilde{q}) \tag{64}$$

as a surface Hamiltonian density associated with some frame $\{\nu', n'\}$ along $\Sigma \subset \mathcal{S}$. More precisely, Wang and Yau proved the following:

Theorem 5.1 ([37]) *Along $\Sigma \subset \mathcal{S}$, let $\{\nu', n'\}$ be an orthonormal frame in $(T\Sigma)^\perp$ given by*

$$\begin{cases} \nu' = \cosh\theta\, \nu_\Omega + \sinh\theta\, n_\Omega \\ n' = \sinh\theta\, \nu_\Omega + \cosh\theta\, n_\Omega, \end{cases} \tag{65}$$

where θ is the function on Σ determined by

$$\sinh\theta = -\frac{1}{\sqrt{1 + |\nabla\tau|^2}} \frac{\partial f}{\partial \nu_\Omega}.$$

Then, $\tilde{H} - \langle \tilde{Y}, \tilde{\nu} \rangle$ along $\tilde{\Sigma} = \partial\tilde{\Omega}$ satisfies

$$\begin{aligned} &\sqrt{1 + |\nabla\tau|^2} \left(-\tilde{H} + \langle \tilde{Y}, \tilde{\nu} \rangle \right)(\tilde{q}) \\ &= \sqrt{1 + |\nabla\tau|^2} \langle \vec{H}, \nu' \rangle - \langle \overline{\nabla}_{\nabla\tau} n', \nu' \rangle(q). \end{aligned} \tag{66}$$

As a result,

$$-\int_{\tilde{\Sigma}} \tilde{H} - \langle \tilde{Y}, \tilde{\nu} \rangle\, d\hat{\sigma} = \int_{\Sigma} \sqrt{1 + |\nabla\tau|^2} \langle \vec{H}, \nu' \rangle - \langle \overline{\nabla}_{\nabla\tau} n', \nu' \rangle\, d\sigma. \tag{67}$$

Combined with (63) and the definition of $\mathcal{H}(T, n')$, Theorem 5.1 gives

$$\int_{\hat{\Sigma}} \hat{H} \, d\hat{\sigma} \geq -\mathcal{H}(T, n'). \tag{68}$$

On the other hand, by the maximal property of $\mathcal{H}(T, n)$ (Lemma 4.1),

$$-\mathcal{H}(T, n') \geq -\mathcal{H}(T, n). \tag{69}$$

Recall that, by (47), $8\pi E_{WY}(\Sigma, \tau) = \int_{\hat{\Sigma}} \hat{H} \, d\hat{\sigma} + \mathcal{H}(T, n)$. Therefore, one concludes

$$E_{WY}(\Sigma, \tau) \geq 0, \tag{70}$$

under the assumptions (i) and (ii) in Sect. 5.3.

5.5 Some Comment on $\tilde{Y}$ and ν'

The vector field $\tilde{Y}$, by definition, is a vector field on $\tilde{\Omega}$, which is the graph of the solution to Jang's equation. It may be convenient to compute $\tilde{Y}$ via quantities on Ω directly.

For this purpose, let $\{v_i\}$ be a local frame on Ω. Let $w_i = v_i + f_i \partial_s$, where $f_i = v_i(f)$ and s is the coordinate on $\mathbb{R}$ in $\Omega \times \mathbb{R}$. $\{w_i\}$ forms a local frame on $\tilde{\Omega}$. Using the fact $\tilde{n}_d = \frac{1}{\sqrt{1+|Df|^2}}(Df - \partial_s)$, one has

$$\begin{aligned} \langle \tilde{D}_{\tilde{n}_d} \tilde{n}_d, w_i \rangle &= \frac{1}{1+|Df|^2} \langle \tilde{D}_{(Df-\partial_s)}(Df - \partial_s), w_i \rangle \\ &= \frac{1}{1+|Df|^2} \langle D_{Df} Df, v_i \rangle. \end{aligned} \tag{71}$$

Similarly,

$$\begin{aligned} \tilde{k}(\tilde{n}_d, w_i) &= \frac{1}{\sqrt{1+|Df|^2}} \tilde{k}(Df - \partial_s, v_i + f_i \partial_s) \\ &= \frac{1}{\sqrt{1+|Df|^2}} k(Df, v_i). \end{aligned} \tag{72}$$

Therefore, the 1-form η on $\tilde{\Omega}$, given by (61), is determined by

$$\Phi^*(\eta) = \frac{1}{\sqrt{1+|Df|^2}} \left[\frac{1}{\sqrt{1+|Df|^2}} D^2 f(Df, \cdot) - k(Df, \cdot) \right]. \tag{73}$$

Here, $\Phi : \Omega \to \tilde{\Omega}$ is the map with $\Phi(x) = (x, f(x))$.

As a result, $\tilde{Y} = 0$ on $\tilde{\Omega}$ in the setting of Sect. 5.1, where Ω is the graph of a function f over $\hat{\Omega} \subset \mathbb{R}^3$ in $\mathbb{R}^{3,1}$. In this case, $(\tilde{\Omega}, \tilde{g})$ is isometric to the Euclidean domain $\hat{\Omega}$, and thus $\tilde{H} = \hat{H}$. It would be worthy of computing ν' too. Let $\hat{D}$ denote the gradient on $\mathbb{R}^3$. Using the fact $n_\Omega = \frac{1}{\sqrt{1-|\hat{D}f|^2}}(\hat{D}f + \partial_t)$, it is easily seen that if one writes

$$\nu_0 = \cosh\theta\, \nu_\Omega + \sinh\theta\, n_\Omega, \tag{74}$$

where ν_0 is the vector field given in Sect. 3.2, then

$$\sinh\theta = -\frac{1}{\sqrt{1+|\nabla\tau|^2}}\frac{\partial}{\partial \nu_\Omega} f.$$

In other words, $\nu' = \nu_0$ in this model case. Consequently, Theorem 5.1 may be viewed as a generalization of Lemma 3.1.

We would like to end this note with a brief description of the Wang–Yau quasi-local mass [37]. Assumptions (i) and (ii) in Sect. 5.3, together with the assumption that (58) admits a solution, correspond to the admissibility definition of the time function τ in [37]. By minimizing $E_{WY}(\Sigma, \tau)$ among all admissible time functions τ, the Wang–Yau mass of $\Sigma \subset \mathcal{S}$ is defined to be

$$\mathfrak{m}_{WY}(\Sigma) = \inf\left\{E_{WY}(\Sigma, \tau) \mid \text{admissible } \tau\right\}.$$

The variational feature of $\mathfrak{m}_{WY}(\Sigma)$ leads to many interesting questions. For instance, in the time-symmetric case, whether $E_{WY}(\Sigma, \cdot)$ is minimized by $\tau = 0$, or equivalently whether $\mathfrak{m}_{WY}(\Sigma)$ agrees with the Brown–York mass, seems to be an intriguing question. Other related questions include understanding a relation between critical points of $E_{WY}(\Sigma, \cdot)$ and the concept of time-flat surfaces and the asymptotic behavior of the quasi-local energy. We refer the interested readers to [3, 4, 10–15, 25, 38] for further exploration of this subject.

Acknowledgments I am deeply grateful to the organizers of the second "Domoschool – International Alpine School of Mathematics and Physics" held in Domodossola, on July 15–19, 2019, for the kind invitation and warm hospitality.

The author's research was partially supported by NSF grant DMS-1906423.

References

1. R. Arnowitt, S. Deser, C.W. Misner, Coordinate invariance and energy expressions in general relativity. Phys. Rev. **122**, 997–1006 (1961)
2. R. Arnowitt, S. Deser, C.W. Misner, The dynamics of general relativity, in *Gravitation: An Introduction to Current Research*, ed. by L. Witten, chapter 7 (Wiley, Hoboken, 1962), pp. 227–265

3. H.L. Bray, J.L. Jauregui, Time flat surfaces and the monotonicity of the spacetime Hawking mass. Comm. Math. Phys. **335**(1), 285–307 (2015)
4. H.L. Bray, J.L. Jauregui, M. Mars, Time flat surfaces and the monotonicity of the spacetime Hawking mass II. Ann. Henri Poincaré **17**(6), 1457–1475 (2016)
5. S. Brendle, M.-T. Wang, A Gibbons-Penrose inequality for surfaces in Schwarzschild spacetime. Comm. Math. Phys. **330**(1), 33–43 (2014)
6. S. Brendle, P.-K. Hung, M.-T. Wang, A Minkowski Inequality for Hypersurfaces in the Anti-de Sitter-Schwarzschild Manifold. Commun. Pure Appl. Math. **69**(1), 124–144 (2016)
7. I.S. Booth, R.B. Mann, Moving observers, nonorthogonal boundaries, and quasilocal energy. Phys. Rev. D (3) **59**(6), 064021 (1999)
8. J.D. Brown, J.W. York Jr., Quasilocal energy in general relativity, in *Mathematical Aspects of Classical Field Theory (Seattle, 1991)*, vol. 132. Contemporary Mathematics (American Mathematical Society, Providence, 1992), pp. 129–142
9. J.D. Brown, J.W. York Jr., Quasilocal energy and conserved charges derived from the gravitational action. Phys. Rev. D (3) **47**(4), 1407–1419 (1993)
10. P.-N. Chen, M.-T. Wang, Rigidity and minimizing properties of quasi-local mass, in *Surveys in Differential Geometry 2014*. Regularity and Evolution of Nonlinear Equations, Survey in Differential Geometry, vol. 19 (International Press, Somerville, 2015), pp. 49–61
11. P.-N. Chen, M.-T. Wang, S.-T. Yau, Evaluating quasilocal energy and solving optimal embedding equation at null infinity. Comm. Math. Phys. **308**(3), 845–863 (2011)
12. P.-N. Chen, M.-T. Wang, Y.-K. Wang, Rigidity of time-flat surfaces in the Minkowski spacetime. Math. Res. Lett. **21**(6), 1227–1240 (2014)
13. P.-N. Chen, M.-T. Wang, S.-T. Yau, Minimizing properties of critical points of quasi-local energy. Comm. Math. Phys. **329**(3), 919–935 (2014)
14. P.-N. Chen, M.-T. Wang, S.-T. Yau, Evaluating small sphere limit of the Wang-Yau quasi-local energy. Comm. Math. Phys. **357**(2), 731–774 (2018)
15. P.-N. Chen, M.-T. Wang, Y.-K.Wang, S.-T. Yau, Quasi-local mass on unit spheres at spatial infinity. Comm. Anal. Geom. (2019). arXiv:1901.06954
16. R.J. Epp, Angular momentum and an invariant quasilocal energy in general relativity. Phys. Rev. D **62**(12), 124108 (2000)
17. G.W. Gibbons, Collapsing shells and the isoperimetric inequality for black holes. Class. Quantum Grav. **14**, 2905–2915 (1997)
18. S.W. Hawking, G.T. Horowitz, The gravitational Hamiltonian, action, entropy and surface terms. Class. Quant. Grav. **13**(6), 1487–1498 (1996)
19. P.S. Jang, On the positivity of energy in general relativity. J. Math. Phys. **19**(5), 1152–1155 (1978)
20. J. Kijowski, A simple derivation of canonical structure and quasi-local Hamiltonians in general relativity. Gen. Relativ. Gravit. **29**(3), 307–343 (1997)
21. C.-C.M. Liu, S.-T. Yau, Positivity of quasilocal mass. Phys. Rev. Lett. **90**(23), 231102 (2003)
22. C.-C.M. Liu, S.-T. Yau, Positivity of quasilocal mass II. J. Amer. Math. Soc. **19**, 181–204 (2006)
23. M. Mars, A. Soria, On the Penrose inequality for dust null shells in the Minkowski spacetime of arbitrary dimension. Class. Quant. Grav. **29**(6), 135005 (2012)
24. P. Miao, Quasi-local mass via isometric embeddings: a review from a geometric perspective. Class. Quantum Grav. **32**, 233001 (2015). 18pp
25. P. Miao, L.-F. Tam, N.-Q. Xie, Critical points of Wang-Yau quasi-local energy. Ann. Henri Poincaré **12**(5), 987–1017 (2011)
26. L. Nirenberg, The Weyl and Minkowski problems in differential geometry in the large. Comm. Pure Appl. Math. **6**, 337–394 (1953)
27. M.A. Pelath, K.P. Tod, R.M. Wald, Trapped surfaces in prolate collapse in the Gibbons-Penrose construction. Class. Quantum Grav. **15**, 3917–3934 (1998)
28. R. Penrose, Naked singularities. Ann. New York Acad. Sci. **224**, 125–134 (1973)
29. A.V. Pogorelov, Regularity of a convex surface with given Gaussian curvature. Mat. Sbornik N.S. **31**(73), 88–103 (1952)

30. A.V. Pogorelov, Some results on surface theory in the large. Adv. Math. **1**(fasc. 2), 191–264 (1964)
31. R. Schoen, S.-T. Yau, On the proof of the positive mass conjecture in general relativity. Comm. Math. Phys. **65**(1), 45–76 (1979)
32. R. Schoen, S.-T. Yau, Proof of the positive mass theorem. II. Comm. Math. Phys. **79**(2), 231–260 (1981)
33. Y.-G. Shi, L.-F. Tam, Positive mass theorem and the boundary behaviors of compact manifolds with nonnegative scalar curvature. J. Differ. Geom. **62**, 79–125 (2002)
34. K.P. Tod, Penrose quasi-local mass and the isoperimetric inequality for static black holes. Class. Quantum Grav. **2**, L65–L68 (1985)
35. M.-T. Wang, Quasilocal mass and surface Hamiltonian in spacetime, in *XVIIth International Congress on Mathematical Physics, 2013* (World Scientific Publishing, Hackensack, 2014), pp. 229–238
36. M.-T. Wang, S.-T. Yau, Quasilocal mass in general relativity. Phys. Rev. Lett. **102**(2), 021101 (2009), 4 pp.
37. M.-T. Wang, S.-T. Yau, Isometric embeddings into the Minkowski space and new quasi-local mass. Comm. Math. Phys. **288**(3), 919–942 (2009)
38. M.-T. Wang, S.-T. Yau, Limit of quasilocal mass at spatial infinity. Comm. Math. Phys. **296**(1), 271–283 (2010)
39. E. Witten, A new proof of the positive energy theorem. Comm. Math. Phys. **80**, 381–402 (1981)

Gravitational Self-force in the Schwarzschild Spacetime

Donato Bini and Andrea Geralico

Mathematics Subject Classification (2000) Primary 83C25; Secondary 83C57

Contents

1 Introduction

When a particle or a field evolves on a given gravitational background, its presence induces a perturbation on the background, which in turn acts back on the particle or the field itself, implying modifications to the particle's world line, or the support region of the field. The modified background as well as the modified dynamics can then be fully determined by using the self-force (SF) formalism. To make a long story short, one can say that SF is the actual terminology to mean a complete (either

D. Bini (✉) · A. Geralico
Istituto per le Applicazioni del Calcolo "M. Picone", Rome, Italy

S. L. Cacciatori, A. Kamenshchik (eds.), *Einstein Equations: Local Energy, Self-Force, and Fields in General Relativity*, Tutorials, Schools, and Workshops in the Mathematical Sciences, https://doi.org/10.1007/978-3-031-21845-3_2

analytical or numerical) determination of the gravitational perturbation of a certain background due to a small mass, charge, or field. Whereas a numerical treatment of the problem can always be performed, it is only recently that SF techniques have allowed for analytical results (see, e.g., Ref. [1] for a recent review). Indeed, starting from 2013 [2], novel analytical SF calculations have been carried on successfully in black hole spacetimes (Schwarzschild and Kerr). Several gauge-invariant quantities have been computed within the framework of black hole first-order perturbation theory through a very high post-Newtonian (PN) accuracy, including the redshift function, the periastron advance, and the precession angle of a gyroscope for motion along both circular and eccentric orbits.

The theoretical foundations of the SF program were posed more than a decade before, thanks to the pioneering works of Quinn and Wald, Mino, Sasaki and Tanaka, and Detweiler and Whiting (see, e.g., the Living Review article by Poisson [3] for the fundamental formulation of the SF in curved spacetime, that by Sasaki and Tagoshi [4] for applications to black hole spacetimes, and the review by Barack [5] for a historical presentation of developments in SF research). However, all the various steps which are necessary to accomplish the SF tasks have been elucidated and standardized only very recently. The problem is the old one of solving the differential equations for the backreaction induced on a given gravitational background by a moving body or an evolving field as a consequence of the (linearized) Einstein's equation. Mode-sum decomposition and Fourier analysis are the preliminary tools necessary to obtain a set of (coupled) ordinary differential equations instead of (coupled) partial differential equations. Then, the symmetries of the background allow for separated radial and angular equations and the (simple) motion of the source of the perturbation implies the possibility to fully integrate (in PN sense) the complete set of the perturbation equations. The final goal is that of reconstructing the perturbed metric to compute several orbital gauge-invariants.

2 The Various Steps of Gravitational Self-force Computations

Let us consider a particle with mass μ moving in the background gravitational field of a much larger mass M described by the Schwarzschild spacetime, with metric written in standard spherical-like coordinates

$$g^{(0)}_{\alpha\beta}dx^\alpha dx^\beta = -f(r)dt^2 + \frac{dr^2}{f(r)} + r^2(d\theta^2 + \sin^2\theta d\phi^2)\,, \qquad f(r) = 1 - \frac{2M}{r}\,. \tag{1}$$

We will assume geometrical units throughout all this work, with $G = 1 = c$. The particle moves along a geodesic orbit with parametric equations $x_p^\alpha = x_p^\alpha(\tau)$, τ denoting the proper time parameter, and four velocity $U = \frac{dx_p^\alpha}{d\tau}\partial_\alpha$ $(U\cdot U = -1)$. The associated energy-momentum tensor is then given by

$$
\begin{aligned}
T^{\mu\nu} &= \mu \int_{-\infty}^{\infty} \frac{1}{\sqrt{-g^{(0)}}} \delta^{(4)}(x^\alpha - x_p^\alpha(\tau)) U^\mu U^\nu d\tau \\
&= \mu \frac{U^\mu U^\nu}{r_p^2 \sin\theta_p U^t} \delta(r - r_p(t)) \delta(\theta - \theta_p(t)) \delta(\phi - \phi_p(t)) ,
\end{aligned}
\tag{2}
$$

where $g^{(0)}$ is the metric determinant for the background, and $\delta^{(4)}$ is the four dimensional delta function. According to the Detweiler-Whiting formulation [6], through $O(\frac{\mu}{M})$ the particle effectively moves along a geodesic of a smooth regularly perturbed spacetime with metric

$$
g^{\rm R}_{\alpha\beta} = g^{(0)}_{\alpha\beta} + \frac{\mu}{M} h^{\rm R}_{\alpha\beta} . \tag{3}
$$

The main goal of the SF program is to determine the (regularized) metric perturbation $h^{\rm R}_{\alpha\beta}$. The first-order perturbation can be obtained by using standard tools of perturbation theory following either the Regge–Wheeler–Zerilli (RWZ) [7, 8] or the Teukolsky approach [9–11]. The former is usually adopted in the spherically symmetric case, whereas the latter is most suited for stationary black holes and general type D spacetimes. In both cases, after separation of variables and for simple motion of the source, the problem is reduced to solve a single Schrödinger-like equation for a master function of the radial variable, with a Dirac-delta source term having support at the particle position. We will shortly review below the main steps in the simplest case of circular motion, i.e.,

$$
U = U^t (\partial_t + \Omega \partial_\phi) , \tag{4}
$$

with

$$
U^t = \left(1 - \frac{3M}{r_0}\right)^{-1/2} , \qquad \Omega = \sqrt{\frac{M}{r_0^3}} . \tag{5}
$$

2.1 *Metric Perturbations: The Regge–Wheeler–Zerilli Formalism*

After decomposing both the perturbed metric and the energy-momentum tensor associated with the perturbing mass in tensor harmonics, the perturbation equations can be separated into two different sets, with even parity and odd parity, respectively. All ten metric perturbation components can then be summarized by a single radial function, irrespective of their parity and reflection properties about the equatorial plane. Therefore, the general situation is that of a Regge–Wheeler (–Zerilli, when including a unified treatment of odd and even waves) equation with different source terms involving a Dirac-delta function and its derivatives, that is

$$\mathcal{L}^{(r)}_{(\mathrm{RW})}[R^{(\mathrm{even/odd})}_{lm\omega}] = S^{(\mathrm{even/odd})}_{lm\omega}(r)\,. \tag{6}$$

Here

$$\mathcal{L}^{(r)}_{(\mathrm{RW})} = \left(1-\frac{2M}{r}\right)^2 \frac{d^2}{dr^2} + \frac{2M}{r^2}\left(1-\frac{2M}{r}\right)\frac{d}{dr} + [\omega^2 - V_{(\mathrm{RW})}(r)] \tag{7}$$

denotes the RW operator with the potential

$$V_{(\mathrm{RW})}(r) = \left(1-\frac{2M}{r}\right)\left(\frac{l(l+1)}{r^2} - \frac{6M}{r^3}\right),$$

and the even/odd source terms are of the type

$$S^{(\mathrm{even/odd})}_{lm\omega}(r) = s^{(\mathrm{e/o})}_0(r_0)\delta(r-r_0) + s^{(\mathrm{e/o})}_1(r_0)\delta'(r-r_0) + s^{(\mathrm{e/o})}_2(r_0)\delta''(r-r_0)\,,$$

with $s^{(\mathrm{o})}_2(r_0) = 0$. Recalling the properties of the Green function $G(r,r')$ of $\mathcal{L}^{(r)}_{(\mathrm{RW})}$, i.e.,

$$\mathcal{L}^{(r)}_{(\mathrm{RW})}[G(r,r')] = f(r')\delta(r-r') = \mathcal{L}^{(r')}_{(\mathrm{RW})}[G(r,r')]\,, \tag{8}$$

we have

$$R^{(\mathrm{even/odd})}_{lm\omega}(r) = \int dr' G(r,r') f(r')^{-1} S^{(\mathrm{even/odd})}_{lm\omega}(r')\,. \tag{9}$$

Here

$$G(r,r') = \frac{1}{W}\Big[X_{\mathrm{in}}(r)X_{\mathrm{up}}(r')H(r'-r) + X_{\mathrm{in}}(r')X_{\mathrm{up}}(r)H(r-r')\Big], \tag{10}$$

where W is the (constant) Wronskian

$$W = f(r)\left[X_{\mathrm{in}}(r)\frac{d}{dr}X_{\mathrm{up}}(r) - \frac{d}{dr}X_{\mathrm{in}}(r)X_{\mathrm{up}}(r)\right] = \text{const.} \tag{11}$$

and $H(x)$ is the Heaviside step function. The functions $X_{(\mathrm{in})}(r)$ and $X_{(\mathrm{up})}(r)$ are the solutions of the homogeneous RW equation satisfying the proper regularity conditions at the horizon and at infinity, obtained by using different methods: PN approximation (for generic l, in the weak-field and slow motion regime), Wentzel-Kramers-Brillouin (WKB) approximation (for large l, useful for regularization purposes), Mano, Suzuki, and Takasugi (MST) [12, 13] technique (specified values of l, useful to cure the shortcomings of the PN solutions). Finally, both even-parity and odd-parity solutions are given by

$$R_{lm\omega}^{\text{(even/odd)}}(r) = s_0^{\text{(e/o)}}(r_0)\frac{G(r,r_0)}{f(r_0)} - s_1^{\text{(e/o)}}(r_0)\lim_{r'\to r_0}\left[\frac{d}{dr'}\left(\frac{G(r,r')}{f(r')}\right)\right]$$
$$+s_2^{\text{(e/o)}}(r_0)\lim_{r'\to r_0}\left[\frac{d^2}{dr'^2}\left(\frac{G(r,r')}{f(r')}\right)\right]. \tag{12}$$

For instance, for $r < r_0$ we have

$$R_{lm\omega,-}^{\text{(even/odd)}}(r) = s_0^{\text{(e/o)}}(r_0)\frac{X_{\text{(in)}}(r)X_{\text{(up)}}(r_0)}{Wf(r_0)}$$
$$-s_1^{\text{(e/o)}}(r_0)\lim_{r'\to r_0}\left[\frac{d}{dr'}\left(\frac{X_{\text{(in)}}(r)X_{\text{(up)}}(r')}{Wf(r')}\right)\right]$$
$$+s_2^{\text{(e/o)}}(r_0)\lim_{r'\to r_0}\left[\frac{d^2}{dr'^2}\left(\frac{X_{\text{(in)}}(r)X_{\text{(up)}}(r')}{Wf(r')}\right)\right]$$
$$= C_-^{\text{(e/o)}}(r_0)X_{\text{(in)}}(r), \tag{13}$$

where

$$WC_-^{\text{(e/o)}}(r_0) = s_0^{\text{(e/o)}}(r_0)\frac{X_{\text{(up)}}(r_0)}{f(r_0)} - s_1^{\text{(e/o)}}(r_0)\lim_{r'\to r_0}\left[\frac{d}{dr'}\left(\frac{X_{\text{(up)}}(r')}{f(r')}\right)\right]$$
$$+s_2^{\text{(e/o)}}(r_0)\lim_{r'\to r_0}\left[\frac{d^2}{dr'^2}\left(\frac{X_{\text{(up)}}(r')}{f(r')}\right)\right]. \tag{14}$$

Once the radial function is known for both parities, the perturbed metric components are then computed by Fourier anti-transforming, multiplying by the angular part, and summing over m (between $-l$ and $+l$), and then over l (between 0 and ∞). The final step consists in building up the gauge-invariant quantity one is interested in, and evaluating it at the source location, where the perturbed metric is singular. The sum over m is done by using standard spherical harmonic identities. The sum over l, instead, requires some care. In fact, the series is in general divergent, so that a regularization procedure is needed. Furthermore, the contribution of the non-radiative modes $l = 0, 1$ must be computed separately, corresponding to a shift in the black hole mass and spin due to the energy and angular momentum of the particle.

2.2 *Curvature Perturbations: The Teukolsky Formalism*

The metric perturbation can be also computed in a radiation gauge by using the Teukolsky formalism. Einstein's equations reduce to a single master equation for the perturbed Weyl scalars ψ_0 (spin-weight $s = 2$) or ψ_4 (spin-weight $s = -2$), which can be solved by separation of variables. The radiative part of the metric perturbation ($l \geq 2$) can then be reconstructed from the "Hertz potential," through the Chrzanowski–Cohen–Kegeles (CCK) procedure [14–16] (see also Refs. [17–20]).

Let us start with the Teukolsky master equation governing the perturbations on a Kerr background due to a field of spin-weight s. The radial and angular part can be separated. The eigenfunctions of the angular part are the spin-weighted spherical harmonics for vanishing black hole rotation parameter (spheroidal harmonics in the general Kerr case), so that in the frequency domain

$$\psi_s = \sum_{lm\omega} e^{-i\omega t}\, {}_sR_{lm\omega}(r)\, {}_sY_{lm\omega}(\theta, \phi)\,. \tag{15}$$

The radial equation turns out to be

$$\mathscr{L}_r^s\, {}_sR_{lm\omega}(r) = -8\pi T_{slm\omega}\,, \tag{16}$$

where

$$\mathscr{L}_r^s = \Delta^{-s}\frac{d}{dr}\left(\Delta^{s+1}\frac{d}{dr}\right) + \left[\frac{\omega^2 r^2 - 2is(r-M)\omega r^2}{\Delta} + 4is\omega r - \lambda\right], \tag{17}$$

with $\Delta = r^2 - 2Mr$ and $\lambda = l(l+1) - s(s+1)$, and the source term $T_{slm\omega}$ follows from the harmonic decomposition of the energy-momentum tensor

$$T^{\mu\nu} = \mu \sum_{l,m} \frac{U^\mu U^\nu}{r_0^2 U^t}\delta(r - r_0)\, {}_sY_{lm\omega}(\theta, \phi)\, {}_s\bar{Y}_{lm\omega}\left(\frac{\pi}{2}, \Omega t\right), \tag{18}$$

with $\omega = m\Omega$. In the case $s = +2$ (i.e., for $\psi_{s=2} = \psi_0$), we have

$$T_{s=2} = \mathcal{L}_1\left(\mathcal{L}_2(T_{13}) - \mathcal{L}_3(T_{11})\right) + \mathcal{L}_4\left(\mathcal{L}_5(T_{13}) - \mathcal{L}_6(T_{33})\right), \tag{19}$$

where $T_{11} = T_{ll}$, $T_{13} = T_{lm}$, $T_{33} = T_{mm}$ are the frame components of the stress-energy tensor with respect to the Newman-Penrose principal frame

$$\begin{aligned} l &= \frac{1}{\Delta}(r^2\partial_t + \Delta\partial_r)\,, \\ n &= \frac{1}{2r^2}(r^2\partial_t - \Delta\partial_r)\,, \\ m &= \frac{1}{\sqrt{2}r}\left(\partial_\theta + \frac{i}{\sin\theta}\partial_\phi\right), \end{aligned} \tag{20}$$

with nonvanishing spin coefficients

$$\rho = -\frac{1}{r}\,, \qquad \beta = -\frac{\cos\theta}{2\sqrt{2}r\sin\theta} = -\alpha\,, \qquad \mu = -\frac{\Delta}{2r^3}\,, \qquad \gamma = \frac{M}{2r^2}\,, \tag{21}$$

and associated frame derivatives

$$\mathbf{D} = l^\mu \partial_\mu\,, \qquad \mathbf{\Delta} = n^\mu \partial_\mu\,, \qquad \boldsymbol{\delta} = m^\mu \partial_\mu\,. \tag{22}$$

The differential operators entering Eq. (19) are then given by

$$\begin{aligned} \mathcal{L}_1 &= \boldsymbol{\delta} - 2\beta\,, \quad & \mathcal{L}_2 &= \mathbf{D} - 2\rho\,, \quad & \mathcal{L}_3 &= \boldsymbol{\delta}\,, \\ \mathcal{L}_4 &= \mathbf{D} - 5\rho\,, \quad & \mathcal{L}_5 &= \boldsymbol{\delta} - 2\beta\,, \quad & \mathcal{L}_6 &= \mathbf{D} - \rho\,. \end{aligned}$$

The Teukolsky radial equation (16) can be solved by using the Green's function method. For $s = 2$ the (retarded) Green's function satisfies the equation

$$\mathscr{L}_r^{s=2}(G_{lm}(r, r')) = \frac{1}{\Delta}\delta(r - r')\,, \tag{23}$$

which yields

$$G_{lm}(r, r') = \frac{(\Delta')^2}{W_{lm}}\left[R_{\rm in}(r)R_{\rm up}(r')H(r' - r) + R_{\rm in}(r')R_{\rm up}(r)H(r - r')\right]\,, \tag{24}$$

where $H(x)$ denotes the Heaviside step function, $\Delta' \equiv \Delta(r')$, $R_{\rm in}(r)$ and $R_{\rm up}(r)$ are two independent solutions to the homogeneous radial equation that are ingoing at the outer horizon and outgoing at infinity, respectively, and W_{lm} is the associated (constant) Wronskian. One needs –also in this case– different types of solutions: PN, MST, WKB solutions.

The final solution for the Weyl scalar ψ_0 thus has the form

$$\psi_0 = -8\pi \int (r')^2 T(x', x_0) G(x, x') dr' d(\cos\theta') d\phi'\,, \tag{25}$$

where $T(x', x_0)$ and $G(x, x')$ stand for the full source term and the full Green's function of the Teukolsky radial equation, respectively, x denoting the spatial point with coordinates (r, θ, ϕ). According to the CCK procedure, all the components of the radiative part ($l \geq 2$) of the perturbed metric are then evaluated by applying a suitable differential operator on a scalar quantity Ψ, the Hertz potential, which also satisfies the homogeneous Teukolsky equation but with opposite spin as the Weyl scalar from which it is constructed. The (outgoing radiation gauge) metric perturbation is thus given by

$$h_{\alpha\beta} = -r^4\{n_\alpha n_\beta D_{nn} + \bar{m}_\alpha \bar{m}_\beta D_{\bar{m}\bar{m}} - n_{(\alpha}\bar{m}_{\beta)} D_{n\bar{m}}\}\Psi + \text{c.c.}\,, \tag{26}$$

where

$$\begin{aligned} D_{nn} &= (\bar{\delta} + 2\beta)(\bar{\delta} + 4\beta)\,, \\ D_{\bar{m}\bar{m}} &= (\mathbf{\Delta} + 5\mu - 2\gamma)(\mathbf{\Delta} + \mu - 4\gamma)\,, \\ D_{n\bar{m}} &= (\bar{\delta} + 4\beta)(\mathbf{\Delta} + \mu - 4\gamma) + (\mathbf{\Delta} + 4\mu - 4\gamma)(\bar{\delta} - 4\alpha)\,, \end{aligned} \tag{27}$$

and the Hertz potential Ψ is related to ψ_0 by

$$\psi_0 = \frac{1}{8}\left[\mathcal{L}^4\bar{\Psi} + 12M\partial_t\Psi\right], \tag{28}$$

with

$$\mathcal{L}^4 = \mathcal{L}_1\mathcal{L}_0\mathcal{L}_{-1}\mathcal{L}_{-2}\,, \qquad \mathcal{L}_s = -[\partial_\theta - s\cot\theta + i\csc\theta\partial_\phi] - ia\sin\theta\partial_t\,. \tag{29}$$

Finally, the radial part of the harmonic decomposition of Ψ

$$\Psi = \sum_{lm\omega} e^{-i\omega t}{}_2\mathcal{R}_{lm\omega}(r){}_2Y_{lm\omega}(\theta,\phi) \tag{30}$$

turns out to be

$${}_2\mathcal{R}_{lm\omega} = 8\frac{(-1)^m\mathcal{D}\,{}_2\bar{R}_{l,-m,-\omega} + 12iM\omega\,{}_2R_{lm\omega}}{\mathcal{D}^2 + 144M^2\omega^2}\,, \tag{31}$$

with $\mathcal{D} = l(l-1)(l+2)(l+1)$.

The evaluation of gauge-invariant quantities then proceeds along the lines sketched before for the RWZ approach, requiring the calculation of the contribution due to the non-radiative modes [21] and the regularization procedure after summation over all multipoles.

3 Scalar Self-force: Computational Details

The problem of scalar self-force (SSF) is easier to handle, even if the various steps needed to compute a gauge-invariant quantity are practically the same as in the gravitational case. Let the source of the perturbation be a particle endowed with a scalar charge q, which generates a (massless) scalar field ψ. The latter acts back on the scalar charge world line, determining modifications to the otherwise free evolution.

In order to illustrate how SSF computations are explicitly carried out, let us assume that the scalar charge moves along a timelike circular (equatorial) geodesic orbit, with parametric equations

$$t = \Gamma\tau\,, \quad r = r_0\,, \quad \theta = \frac{\pi}{2}\,, \quad \phi = \Gamma\Omega\tau = \Omega t\,, \tag{32}$$

where τ is a proper time parameter along the orbit and

$$\Gamma = \frac{1}{\sqrt{1-3u}}\,, \qquad M\Omega = u^{3/2}\,, \qquad u = \frac{M}{r_0}\,. \tag{33}$$

The associated 4-velocity of the scalar charge is

$$U = \Gamma k\,, \qquad k = \partial_t + \Omega\partial_\phi \tag{34}$$

and is aligned with the Killing vector k (i.e., a linear combination of the temporal and azimuthal Killing vectors, ∂_t and ∂_ϕ), implying additional simplifications.

3.1 Scalar Wave Equation

The scalar field ψ satisfies a wave equation sourced by the scalar charge density ρ, i.e.,

$$\Box\psi = -4\pi\rho\,, \tag{35}$$

where

$$\rho = q\int \frac{1}{\sqrt{-g}}\delta^{(4)}(x^\alpha - z^\alpha(\tau))d\tau\,. \tag{36}$$

The density ρ determines a perturbation to the scalar field evolution itself. For circular motion it reads

$$\begin{aligned}\rho &= q\int \frac{1}{r^2\sin\theta}\delta(t-\Gamma\tau)\delta(r-r_0)\delta\left(\theta-\frac{\pi}{2}\right)\delta(\phi-\Omega t)d\tau \\ &= \frac{q}{\Gamma r_0^2}\delta(r-r_0)\delta\left(\theta-\frac{\pi}{2}\right)\delta(\phi-\Omega t)\,.\end{aligned} \tag{37}$$

Having this decomposition for ρ, one can show that a corresponding decomposition in (scalar) spherical harmonics of the field ψ exists too. We will show this in detail below.

3.2 Angular Part of the Perturbation

Let us recall the spherical harmonic identity

$$\delta\left(\cos\theta-\cos\theta_0\right)\delta(\phi-\phi_0)=\sum_{l=0}^{\infty}\sum_{m=-l}^{l}Y_{lm}(\theta,\phi)Y^*_{lm}(\theta_0,\phi_0)\,, \tag{38}$$

with

$$\frac{\partial^2 Y_{lm}}{\partial\theta^2}+\cot\theta\frac{\partial Y_{lm}}{\partial\theta}+\left(l(l+1)-\frac{m^2}{\sin^2\theta}\right)Y_{lm}=0\,. \tag{39}$$

Replacing $\theta_0\to\pi/2$, $\phi_0\to 0$, $\phi\to\phi-\Omega t$, one has

$$\delta\left(\cos\theta\right)\delta(\phi-\Omega t)=\sum_{l=0}^{\infty}\sum_{m=-l}^{l}Y_{lm}(\theta,\phi-\Omega t)Y^*_{lm}(\pi/2,0)\,. \tag{40}$$

Therefore, because of the property $\delta\left(\cos\theta\right)=\delta(\theta-\pi/2)$,

$$\delta\left(\theta-\frac{\pi}{2}\right)\delta(\phi-\Omega t)=\sum_{l=0}^{\infty}\sum_{m=-l}^{l}Y_{lm}(\theta,\phi-\Omega t)Y^*_{lm}(\pi/2,0)\,, \tag{41}$$

with

$$\begin{aligned} Y_{lm}(\theta,\phi)&=C_{lm}\,e^{im\phi}\,\mathrm{LegendreP}(l,m,\cos\theta)\,,\\ C_{lm}&=\frac{1}{2}(-1)^m\sqrt{\frac{(2l+1)}{\pi}}\sqrt{\frac{(l-m)!}{(l+m)!}} \end{aligned} \tag{42}$$

and where (see [22] pag. 959, Eq. 8.756_1)

$$\begin{aligned} Y_{lm}(\pi/2,0)&=C_{lm}\;\mathrm{LegendreP}(l,m,0)\\ &=C_{lm}\frac{2^m\sqrt{\pi}}{\Gamma\left(1-\frac{m}{2}+\frac{l}{2}\right)\Gamma\left(1-\frac{m}{2}-\frac{l}{2}\right)}\,, \end{aligned} \tag{43}$$

for a later use. Substituting Eq. (41) in Eq. (37) one gets the following expression for the density ρ

$$\rho=\frac{q}{\Gamma r_0^2}\delta(r-r_0)\sum_{l=0}^{\infty}\sum_{m=-l}^{l}Y_{lm}(\theta,\phi-\Omega t)Y^*_{lm}(\pi/2,0)$$

$$= \frac{q}{\Gamma r_0^2}\delta(r - r_0)\sum_{l,m} e^{-i\omega t} Y_{lm}(\theta, \phi) Y^*_{lm}(\pi/2, 0)$$

$$= \sum_{l,m} \frac{q_{lm}}{4\pi r_0} e^{-i\omega t}\delta(r - r_0) Y_{lm}(\theta, \phi)\,, \tag{44}$$

where we have used the relation

$$Y_{lm}(\theta, \phi - \Omega t) = e^{-im\Omega t} Y_{lm}(\theta, \phi) = e^{-i\omega t} Y_{lm}(\theta, \phi)\,, \tag{45}$$

with $\omega = m\Omega$ and

$$q_{lm} = \frac{4\pi q}{r_0 \Gamma} Y^*_{lm}(\pi/2, 0)\,. \tag{46}$$

Separating then the variables in the scalar field equation, i.e., assuming

$$\psi = \sum_{l,m} \psi_{lm\omega}(r) e^{-i\omega t} Y_{lm}(\theta, \phi)\,, \tag{47}$$

one gets a full decoupling of the temporal and angular variables.

3.3 *Radial Part of the Perturbation*

The remaining problem is to solve the "radial equation"

$$\mathcal{L}(\psi_{lm\omega}) = S^0_{lm}\delta(r - r_0)\,, \tag{48}$$

where $\mathcal{L}$ is the following second-order ordinary differential operator

$$\mathcal{L} = \frac{d^2}{dr^2} + \frac{2(r - M)}{r^2 f}\frac{d}{dr} + \frac{1}{f^2}\left[\omega^2 - f\frac{l(l+1)}{r^2}\right], \tag{49}$$

and

$$S^0_{lm} = -\frac{q_{lm}}{r_0 - 2M} = -\frac{4\pi q}{r_0^2 f(r_0)}\frac{1}{\Gamma} Y^*_{lm}(\pi/2, 0)\,, \tag{50}$$

so that explicitly

$$\frac{d^2}{dr^2}\psi_{lm\omega} + \frac{2(r - M)}{r(r - 2M)}\frac{d}{dr}\psi_{lm\omega} + \left[\frac{\omega^2 r^2}{(r - 2M)^2} - \frac{l(l+1)}{r(r - 2M)}\right]\psi_{lm\omega} = S^0_{lm}\delta(r - r_0)\,. \tag{51}$$

Note that in the radial equation, Eq. (51), there appear three parameters: l, M, $\omega = m\Omega = m\sqrt{\frac{M}{r_0^3}}$. There is also the hidden symmetry

$$l \quad \to \quad -l-1\,, \tag{52}$$

which leaves invariant the equation (and hence act correspondingly on the solutions). With the parameters M and ω two competing (dimensionless) quantities are naturally formed:

$$\frac{M}{r}\,, \qquad \omega r\,, \tag{53}$$

coming but with a different post-Newtonian weight, namely

$$\frac{M}{r} \quad \to \quad \frac{GM}{c^2 r}\,, \qquad \omega r \quad \to \quad \frac{\omega r}{c}\,, \tag{54}$$

characterized by the fact that their product is a constant

$$\frac{GM}{c^2 r}\frac{\omega r}{c} = \frac{GM\omega}{c^3} = \text{constant}\,. \tag{55}$$

As we will see, in the solution for $\psi_{lm\omega}(r)$ these two length-scales "compete among them," and it is possible to identify regimes where one is dominant with respect to the other characterizing so far the corresponding behavior of the solution.

Let us consider two independent solutions of the homogeneous equation, $R^{(l,m,\omega)}_{(\text{in})}(r)$, and $R^{(l,m,\omega)}_{(\text{up})}(r)$, such that

$$\mathcal{L}(R^{(l,m,\omega)}_{(\text{in,up})}(r)) = 0\,. \tag{56}$$

The in-solution is regular at the horizon, while the up-solution is regular at infinity. Such solutions can be of different types, which we will specify later: exact (Heun confluent functions), post-Newtonian (PN, expansion in the inverse of the speed of light), WKB (expansion in large l values), etc. With these two solutions (of whatever type they are) we can form then their (constant) Wronskian

$$\begin{aligned} W &\equiv W^{(l,m,\omega)} \\ &= r^2 f(r)\left[R^{(l,m,\omega)}_{(\text{in})}(r)\frac{d}{dr}R^{(l,m,\omega)}_{(\text{up})}(r) - R^{(l,m,\omega)}_{(\text{up})}(r)\frac{d}{dr}R^{(l,m,\omega)}_{(\text{in})}(r)\right]\,, \end{aligned} \tag{57}$$

and the associated Green's function

$$\begin{aligned} G_{(l,m)}(r,r') = \frac{1}{W}\Big[&R^{(l,m,\omega)}_{(\text{in})}(r)R^{(l,m,\omega)}_{(\text{up})}(r')H(r'-r) \\ &+R^{(l,m,\omega)}_{(\text{in})}(r')R^{(l,m,\omega)}_{(\text{up})}(r)H(r-r')\Big]\,, \end{aligned} \tag{58}$$

satisfying the equation

$$\mathcal{L}(G_{(l,m,\omega)}(r,r')) = \frac{\delta(r-r')}{r^2 f(r)}\,. \tag{59}$$

The solution of the inhomogeneous equation (51) then writes as

$$\begin{aligned}\psi_{lm\omega}(r) &= \int dr' G_{(l,m,\omega)}(r,r')r'^2 f(r') S^0_{lm}\delta(r'-r_0)\\ &= G_{(l,m,\omega)}(r,r_0)r_0^2 f(r_0)S^0_{lm}\,,\end{aligned} \tag{60}$$

so that the full reconstructed scalar field is given by

$$\begin{aligned}\psi(t,r,\theta,\phi) &= \sum_{l,m} G_{(l,m,\omega)}(r,r_0)r_0^2 f(r_0)S^0_{lm}e^{-i\omega t}Y_{lm}(\theta,\phi)|_{\omega=m\Omega} \qquad (61)\\ &= -\frac{4\pi q}{\Gamma}\sum_{l,m} G_{(l,m,m\Omega)}(r,r_0)\,Y^*_{lm}(\pi/2,0)e^{-im\Omega t}Y_{lm}(\theta,\phi)\\ &= -\frac{4\pi q}{\Gamma}\sum_{l,m} G_{(l,m,m\Omega)}(r,r_0)\,Y^*_{lm}(\pi/2,0)e^{im(\phi-\Omega t)}Y_{lm}(\theta,0)\,.\end{aligned}$$

3.4 *Computing the Scalar Field Along the Source World Line*

The dependence on the variables t and ϕ of the scalar field (61) turns out to be of the type $\psi(r,\theta,\phi-\Omega t)$. This implies that the self-force modification ψ_0 to the field *along the source world line* only depends on r_0, i.e., $\psi(r,\theta,\phi-\Omega t)|_{(r=r_0,\theta=\frac{\pi}{2},\phi=\Omega t)} = \psi_0(r_0)$, and is given by

$$\begin{aligned}\psi_0(r_0) &= -\sum_{l,m} G_{(l,m)}(r_0)\frac{4\pi q}{\Gamma}\,Y^*_{lm}(\pi/2,0)e^{-im\Omega t}Y_{lm}(\pi/2,\Omega t)\\ &= -\frac{4\pi q}{\Gamma}\sum_{l,m} G_{(l,m)}(r_0)\,Y^*_{lm}(\pi/2,0)Y_{lm}(\pi/2,0)\\ &= -\frac{4\pi q}{\Gamma}\sum_{l,m} G_{(l,m)}(r_0)\,|Y_{lm}(\pi/2,0)|^2\,,\end{aligned} \tag{62}$$

where

$$G_{(l,m)}(r_0) = G_{(l,m,m\Omega)}(r_0,r_0)\,, \tag{63}$$

and, we recall, $M\Omega = \sqrt{\frac{M^3}{r_0^3}} = u^{3/2}$. As we will see below, the Green's function along the world line can be then factorized as

$$G_{(l,m)}(r_0) = \sum_k G_k(l) m^k \,, \tag{64}$$

so that the summation over m can be carried out separately,

$$\begin{aligned} \psi_0(r_0) &= \sum_{l=0}^{\infty} \psi_0^{(l)}(r_0) = -\frac{4\pi q}{\Gamma} \sum_l \sum_k G_k(l; r_0) C_k(l) \,, \\ C_k(l) &= \sum_{m=-l}^{l} m^k \, |Y_{lm}(\pi/2, 0)|^2 \,. \end{aligned} \tag{65}$$

For $C_k(l)$ we have a closed form relation. In fact, using

$$\begin{aligned} \Gamma_l(z) &= \sum_{m=-l}^{l} e^{mz} |Y_{lm}(\pi/2, 0)|^2 = \frac{2l+1}{4\pi} e^{lz} {}_2F_1\left(\frac{1}{2}, -l; 1; 1 - e^{2z}\right) \\ &= \frac{2l+1}{4\pi} \left[1 + \frac{l(l+1)}{2} \frac{z^2}{2} + \frac{l(l+1)(3l^2+3l-2)}{8} \frac{z^4}{4!} \right. \\ &\quad \left. + \frac{l(l+1)(5l^4+10l^3-5l^2-10l+8)}{16} \frac{z^6}{6!} + O(z^8) \right] , \end{aligned} \tag{66}$$

and expanding the left-hand-side in series of z formally yields

$$\begin{aligned} \Gamma_l(z) = \sum_{m=-l}^{l} \sum_{k=0}^{\infty} \frac{z^k}{k!} m^k |Y_{lm}(\pi/2, 0)|^2 &= C_0(l) + C_1(l) z + C_2(l) \frac{z^2}{2} + C_3(l) \frac{z^3}{3!} \\ &\quad + C_4(l) \frac{z^4}{4!} + C_5(l) \frac{z^5}{5!} + O(z^6) \,. \end{aligned} \tag{67}$$

All odd powers give a vanishing contribution $C_{2k+1}(l) = 0$,

$$\Gamma_l(z) = C_0(l) + C_2(l) \frac{z^2}{2} + C_4(l) \frac{z^4}{4!} + O(z^6) \,, \tag{68}$$

while the even powers give

$$\begin{aligned} C_2(l) &= \frac{L'L}{8\pi} \,, \\ C_4(l) &= \frac{L'L}{4\pi} \left(-\frac{1}{4} + \frac{3}{8} L \right) , \end{aligned}$$

$$
\begin{aligned}
C_6(l) &= \frac{L'L}{4\pi}\left(\frac{1}{2}-\frac{5}{8}L+\frac{5}{16}L^2\right), \\
C_8(l) &= \frac{L'L}{4\pi}\left(-\frac{17}{8}+\frac{77}{32}L-\frac{35}{32}L^2+\frac{35}{128}L^3\right),
\end{aligned} \tag{69}
$$

where we have introduced the notation

$$
L = l(l+1), \qquad L' = 1+2l, \tag{70}
$$

allowing for more compact expressions. After summing over m, the scalar field (62) thus becomes

$$
\begin{aligned}
\psi_0(r_0) &= -\frac{4\pi q}{\Gamma}\sum_{l=0}^{\infty}[G_0(l;r_0)C_0(l)+G_2(l;r_0)C_2(l) \\
&\quad +G_4(l;r_0)C_4(l)+G_6(l;r_0)C_6(l)+\ldots] \\
&= \sum_{l=0}^{\infty}\psi_0^{(l)}(r_0),
\end{aligned} \tag{71}
$$

with

$$
\psi_0^{(l)}(r_0) = -\frac{4\pi q}{\Gamma}\sum_{k=0}^{\infty}G_{2k}(l;r_0)C_{2k}(l). \tag{72}
$$

An additional complication of the present computation concerns the summation over l. In fact, the latter diverges, and –in order to be carried out– one needs a regularizing procedure so to subtract the corresponding singular part. More in detail, one would have in general

$$
\psi_0^{(l)}(u) = \underbrace{\ldots + a(u)l^2 + b(u)l + c(u) + \frac{d(u)}{l}}_{\text{divergent}} + \underbrace{\frac{e(u)}{l^2} + O\left(\frac{1}{l^3}\right)}_{\text{convergent}}, \tag{73}
$$

where we have replaced r_0 in terms of $u = M/r_0$. The divergent terms, representing the singular part of the (self) field,

$$
\sum_{l=0}^{\infty} D_l(u) = \sum_{l=0}^{\infty}\left(\ldots + a(u)l^2 + b(u)l + c(u) + \frac{d(u)}{l}\right) \tag{74}
$$

should be removed, leading to the regularized field $\psi_0^{(\text{reg})}(u)$, namely

$$
\psi_0^{(\text{reg})}(u) = \sum_l [\psi_0^{(l)}(u) - D_l(u)]. \tag{75}
$$

From Ref. [23] we know (a priori) the singular part of the $l^{\rm th}$ multipole piece in $\psi_0^{(l)}(u)$ given by $c(u)$ only

$$D_l(u) = c(u) = qu\sqrt{\frac{1-3u}{1-2u}}\frac{2}{\pi}\,\text{EllipticK}(k^2)\,, \qquad k^2 = \frac{u}{1-2u}\,, \tag{76}$$

whose expansion in power series of u reads

$$D_l(u) = q\left(u - \frac{1}{4}u^2 - \frac{39}{64}u^3 - \frac{385}{256}u^4 - \frac{61559}{16384}u^5 - \frac{622545}{65536}u^6 - \frac{25472511}{1048576}u^7 \right.$$
$$\left. - \frac{263402721}{4194304}u^8 - \frac{176103411255}{1073741824}u^9\right) + O(u^{10})\,. \tag{77}$$

Unfortunately, additional difficulties arise. In fact, even if one has subtracted the singular field, the l summation should be performed carefully, since in the terms

$$G_0(l;u)C_0(l) + G_2(l;u)C_2(l) + G_4(l;u)C_4(l) + \ldots = \sum_k c_k(l)u^k \tag{78}$$

there can be coefficients diverging for some value of l, i.e., terms of the type

$$c_2(l) = \frac{\bar{c}_2(l)}{l-2}\,, \qquad c_3(l) = \frac{\bar{c}_3(l)}{l-3}\,, \qquad c_4(l) = \frac{\bar{c}_4(l)}{l-4}\,, \ldots \tag{79}$$

The presence of diverging coefficients is associated with the solution of the homogeneous radial equation considered, i.e., with the PN-type solutions mainly. To overcome this problem, a convenient use of exact (hypergeometric-like) and PN (polynomial) solutions should be performed. We will discuss this problem in detail in next sections.

3.5 Solutions of the Radial Homogeneous Equation

In order apply the procedure outlined above one needs to compute explicitly the (exact or approximated) solutions of the homogeneous radial equation (51), i.e.,

$$\frac{d^2}{dr^2}\psi_{lm\omega} + \frac{2(r-M)}{r(r-2M)}\frac{d}{dr}\psi_{lm\omega} + \left[\frac{\omega^2 r^2}{(r-2M)^2} - \frac{l(l+1)}{r(r-2M)}\right]\psi_{lm\omega} = 0\,. \tag{80}$$

As stated above, these solutions can be of different types.

3.5.1 Heun Confluent (Exact) Solutions

Equation (80) belongs to the confluent Heun class of equations and its solution (formally) writes as

$$\psi_{lm\omega}(r) = e^{i\omega r}\left[C_1(-x)^{i\epsilon}H_c^{(1)}(\epsilon, L; x) + C_2(-x)^{-i\epsilon}H_c^{(2)}(\epsilon, L; x)\right], \quad (81)$$

where

$$\begin{aligned} H_c^{(1)}(\epsilon, L; x) &= \text{HeunC}\left(-2i\epsilon, 2i\epsilon, 0, -2\epsilon^2, -L + 2\epsilon^2, x\right), \\ H_c^{(2)}(\epsilon, L; x) &= \text{HeunC}\left(-2i\epsilon, -2i\epsilon, 0, -2\epsilon^2, -L + 2\epsilon^2, x\right), \end{aligned} \quad (82)$$

with C_1 and C_2 integration constants and

$$L = l(l+1), \qquad \epsilon = 2M\omega, \qquad x = 1 - \frac{r}{2M}. \quad (83)$$

The confluent Heun functions can be Taylor-expanded around $r = 2M$ (only). For example,

$$\begin{aligned} H_c^{(1)}(\epsilon, L; x) &\approx 1 + \frac{(L - 2i\epsilon)}{(1 + 2i\epsilon)}\left(\frac{r}{2M} - 1\right) \\ &+ \frac{(2i\epsilon - 2L - 8\epsilon^2 - 8i\epsilon L + L^2)}{4(1 + 2i\epsilon)(1 + i\epsilon)}\left(\frac{r}{2M} - 1\right)^2 \\ &+ O\left(\frac{r}{2M} - 1\right)^3. \end{aligned} \quad (84)$$

3.5.2 PN Solutions

PN solutions follow from a PN expansion of the radial equation. This is accomplished by restoring physical units, i.e., introducing the "weight" $\eta = 1/c$ (with proper powers) in M and ω, that is

$$M \to \eta^2 M, \qquad \omega \to \eta\omega, \quad (85)$$

and expanding in η all quantities. The homogeneous radial equation (80) then reads

$$\frac{d^2}{dr^2}\psi_{lm\omega} + \frac{2(r - \eta^2 M)}{r(r - 2M\eta^2)}\frac{d}{dr}\psi_{lm\omega} + + \left[\frac{\eta^2\omega^2 r^2}{(r - 2M\eta^2)^2} - \frac{l(l+1)}{r(r - 2M\eta^2)}\right]\psi_{lm\omega} = 0, \quad (86)$$

and we will look for solutions up to a fixed order in the η-expansion, e.g., η^8.

Equation (86) admits solutions which begin as polynomial in r,

$$R^{(l,m,\omega)}_{\rm (in)}(r) \approx r^l\,, \qquad R^{(l,m,\omega)}_{\rm (up)}(r) \approx r^{-(l+1)}\,, \tag{87}$$

but then must include some logarithmic term, even if l is not considered as an integer. These solutions are regular for $r \to 0$ or $r \to \infty$.[1] Moreover, replacing $l \to -l-1$ one passes from in to up solutions. In the general case polynomial (fundamental) solutions of Eq. (86) are thus given by

$$\begin{aligned} R^{(l,m,\omega)}_{\rm (in)}(r) &= r^l[1+\eta^2 A_1+\eta^4 A_2+\eta^6 A_3+\eta^8 A_4+\eta^{10} A_5+\eta^{12} A_6+\ldots] \\ R^{(l,m,\omega)}_{\rm (up)}(r) &= R^{(-l-1,m)}_{\rm (in)}(r)\,, \end{aligned} \tag{88}$$

where $A_k = A_k(M/r, \omega^2 r^2; l)$, $k = 1\ldots 5$. The logarithms entering for "generic" (non-integer) l appear at fractional order η^6 ($\eta^6 A_3$), both in $R_{\rm (in)}$ and $R_{\rm (up)}$. In addition, $R^{(l,m,\omega)}_{\rm (up)}(r)$ contains for generic l "dangerous denominators": $1/l$ in $\eta^4 A_2$, $1/(l(l+1))$ in $\eta^6 A_3$, $1/(l(l+1)(l+2))$ in $\eta^8 A_4$, etc. When l is integer these denominators yield additional logarithms.

We find it convenient to introduce the notation

$$X_1 = \frac{M}{r}\,, \qquad X_2 = \omega^2 r^2\,, \qquad X_1^2 X_2 = M^2\omega^2 = \text{const}\,. \tag{89}$$

Each $A_k = A_k(X_1, X_2; l)$ is a k-order polynomial in the variables (X_1, X_2), with coefficients depending on l (modulo log terms), as indicated below:

$$\begin{aligned} R^{(l,m,\omega)}_{\rm (in)}(r) &= r^l[1+\eta^2\, {}^{(1)}C^{(lm)}_A X_A+\eta^4\, {}^{(2)}C^{(lm)}_{AB} X_A X_B \\ &\quad +\eta^6({}^{(3)}C^{(lm)}_{ABC}+{}^{(3)\ln}C^{(lm)}_{ABC}\ln(r/R))X_A X_B X_C \\ &\quad +\eta^8({}^{(4)}C^{(lm)}_{ABCD}+{}^{(4)\ln}C^{(lm)}_{ABCD}\ln(r/R))X_A X_B X_C X_D \\ &\quad +\eta^{10}({}^{(5)}C^{(lm)}_{ABCDE}+{}^{(5)\ln}C^{(lm)}_{ABCDE}\ln(r/R))X_A X_B X_C X_D X_E \\ &\quad +\ldots]\,, \end{aligned} \tag{90}$$

with R a constant (irrelevant for the present computation) with the dimensions of a length, later to be identified as $M\eta^2$ (when $G = 1$). The PN-order of $(X_1^2 X_2)$ is η^6. Actually, starting from η^6 (and successively, after every each η^6 orders) our polynomial solution would include a constant term (like $X_1^2 X_2 = M^2\omega^2$ =const.): the latter always appears with an accompanying logarithm. Therefore, the PN solution shows

[1] Note that PN solutions "do not know" the horizon at $r = 2M$, but they are sensitive to the origin $r = 0$ only. The horizon, in fact, will be a regular singular point of the radial equation only when *all* PN terms will be summed. Practically, we may say that high-PN orders already start "seeing" the horizon, marked by the presence of successive $l - n$ diverging coefficients at various PN n orders.

- $\ln(r/R_1))$ at $O(\eta^6)$;
- $\ln^2(r/R_2))$ at $O(\eta^{12})$;
- $\ln^3(r/R_3))$ at $O(\eta^{18})$, etc.

Each log can come with a scale; different scales are allowed but inessential. To be more explicit, the first coefficients of the expansion are listed below.

$$
\begin{aligned}
A_1 &= -\frac{1}{2}\frac{[\omega^2 r^3 + 2Ml(3+2l)]}{r(3+2l)}\,,\\
A_2 &= \frac{r^4\omega^4}{8(5+2l)(3+2l)} + \frac{M\omega^2 r(-10-5l+l^2)}{2(l+1)(3+2l)} + \frac{M^2 l(l-1)^2}{r^2(-1+2l)}\,,\\
A_3 &= -\frac{\omega^6 r^6}{48(5+2l)(3+2l)(2l+7)} - \frac{\omega^4 r^3 M(3l^3-27l^2-142l-136)}{24(5+2l)(3+2l)(l+2)(l+1)}\\
&\quad -\frac{M^3 l(l-1)(l-2)^2}{3r^3(-1+2l)} - \frac{2\omega^2 M^2(15l^2+15l-11)}{(3+2l)(2l+1)(-1+2l)}\ln(r/R)\,,\\
A_4 &= \frac{r^8\omega^8}{384(5+2l)(3+2l)(2l+9)(2l+7)}\\
&\quad + \frac{\omega^6 r^5 M(-60l^3-1548l-1108+5l^4-625l^2)}{240(2l+7)(5+2l)(3+2l)(l+3)(l+2)(l+1)}\\
&\quad - \frac{r^2 M^2\omega^4}{24(5+2l)(3+2l)^3(l+2)(l+1)(-1+2l)(2l+1)}(24l^8+156l^7\\
&\quad -1802l^6-14843l^5-37099l^4-33535l^3+381l^2+11838l+1392)\\
&\quad - \frac{M^3\omega^2(2l^5-21l^4-13l^3+24-186l^2+44l)}{6r(-1+2l)(2l+1)}\\
&\quad + \frac{M^4 l(l-1)(l-2)^2(-3+l)^2}{6r^4(-1+2l)(2l-3)}\\
&\quad + \frac{\omega^2 M^2(15l^2+15l-11)(4l^2M+6lM+\omega^2 r^3)}{r(3+2l)^2(-1+2l)(2l+1)}\ln(r/R)\,,\\
A_5 &= A_5^{\ln}\ln(r/R)\\
&\quad - \frac{\omega^{10} r^{10}}{3840(2l+3)(2l+7)(5+2l)(2l+9)(2l+11)}\\
&\quad - \frac{M\omega^8 r^7(35l^5-490l^4-8855l^3-40754l^2-73032l-43968)}{13440(l+1)(2l+3)(2l+7)(2+l)(5+2l)(l+3)(2l+9)(l+4)}\\
&\quad + \frac{M^2 r^4\omega^6\bar{A}_5^{(1)}}{120(l+1)(2l+3)^3(2l+7)(2+l)(5+2l)^2(l+3)(1+2l)(-1+2l)}\\
&\quad + \frac{M^3 r\omega^4\bar{A}_5^{(2)}}{24(l+1)^2(2l+3)^3(5+2l)l(1+2l)(-1+2l)}
\end{aligned}
$$

$$
\begin{aligned}
&+\frac{M^4\omega^2\bar{A}_5^{(3)}}{12r^2(2l+3)(2l-3)l(-1+2l)^3(1+2l)}\\
&-\frac{(l-1)(l-2)(l-3)^2(-4+l)^2M^5l}{30r^5(2l-3)(-1+2l)}\,,
\end{aligned}
\tag{91}
$$

where

$$
\begin{aligned}
\bar{A}_5^{(1)} &= 80l^{10}+1040l^9-5280l^8-128132l^7-738413l^6-2020400l^5\\
&\quad -2775100l^4-1506838l^3+363673l^2+586770l+75240\,,\\
\bar{A}_5^{(2)} &= 40l^{10}-364l^9-2030l^8+1007l^7+541l^6-77809l^5-226083l^4\\
&\quad -205386l^3-2940l^2+61728l+6480\,,\\
\bar{A}_5^{(3)} &= 40l^{10}-556l^9+850l^8-5057l^7+11821l^6+5167l^5-30399l^4\\
&\quad +21098l^3-2316l^2-1800l+432\,,
\end{aligned}
\tag{92}
$$

and

$$
\begin{aligned}
A_5^{\ln} = &-\frac{\omega^6r^4M^2(15l^2+15l-11)}{4(2l+3)^2(5+2l)(-1+2l)(1+2l)}\\
&-\frac{\omega^4M^3r(-5l-10+l^2)(15l^2+15l-11)}{(l+1)(2l+3)^2(-1+2l)(1+2l)}\\
&-\frac{2lM^4\omega^2(15l^2+15l-11)(l-1)^2}{r^2(2l+3)(-1+2l)^2(1+2l)}\,.
\end{aligned}
\tag{93}
$$

Note that A_4 contains terms like $1/[(l+1)(l+2)(l+3)]$: when $l\to -l-1$ they become $1/[l(l-1)(l-2)]$ and hence the up-solution is singular starting from this order and it can be used only up to η^6 included (i.e., the 3PN order). Moreover, using the notation $X_1=\frac{M}{r}$ and $X_2=\omega^2r^2$ introduced above, we have, for example,

$$
A_1=-\frac{1}{2}\frac{X_2+2X_1l(3+2l)}{(3+2l)}\,,
\tag{94}
$$

which is a first order polynomial in the variables (X_1,X_2), with coefficients depending on l.

3.5.3 MST or Hypergeometric-Like Solutions

The PN solutions do not correctly include the boundary conditions, since they provide solutions which are regular at the origin ($r=0$) rather than at the horizon ($r=2M$). Solutions which incorporate the correct boundary conditions are obtained following Mano, Suzuki, and Takasugi [12, 13] (see also Refs. [24–26]). They are called MST or hypergeometric-like solutions. Let us introduce the notation

$$\epsilon = 2M\omega\eta^3 \equiv \epsilon_0\eta^3\,, \tag{95}$$

and the following three "radial" variables x, y, and z

$$x = 1 - \frac{r}{2M} = 1 - \frac{1}{y} = 1 - \frac{z}{\epsilon}\,, \qquad z = \omega r\,, \qquad y = \frac{2M}{r}\,. \tag{96}$$

Restoring physical units, $y = (2M/r)\eta^2$ and $z = \omega r\eta$ are "small quantities" useful for series expansions, differently from $x = 1 - r/(2M\eta^2)$, which is "large" (and negative, so that it will appear as $-x$). One should discuss separately in and up solutions.

The MST in-solutions are written in the form

$$\psi_{lm}(x) = C_{(\mathrm{in})}(x) R_{(\mathrm{in})}^{(l,m,\omega)}(x)\,, \tag{97}$$

where

$$\begin{aligned}
C_{(\mathrm{in})} &= e^{i\epsilon x}(-x)^{-i\epsilon} = e^{i\epsilon[x-\ln(-x)]} \\
&= e^{i\left[-\eta(\omega r)+\eta^3 2M\omega\left(1-\ln\left(\frac{r}{2M\eta^2}-1\right)\right)\right]} \\
&= 1 - i\omega r\eta - \frac{1}{2}\omega^2 r^2\eta^2 + \left[2M\omega\left(1-\ln\left(\frac{r}{2M\eta^2}\right)\right) + \frac{1}{6}\omega^3 r^3\right] i\eta^3 \\
&\quad + \left[2\omega^2 Mr\left(1-\ln\left(\frac{r}{2M\eta^2}\right)\right) + \frac{\omega^4 r^4}{24}\right]\eta^4 \\
&\quad + \left[\frac{4M^2\omega}{r} - M\omega^3 r^2\left(1-\ln\left(\frac{r}{2M\eta^2}\right)\right) - \frac{1}{120}\omega^5 r^5\right] i\eta^5 \\
&\quad + \left[4M^2\omega^2 - 2M^2\omega^2\left(1-\ln\left(\frac{r}{2M\eta^2}\right)\right)^2\right. \\
&\quad \left. -\frac{1}{3}\omega^4 Mr^3\left(1-\ln\left(\frac{r}{2M\eta^2}\right)\right) - \frac{\omega^6 r^6}{720}\right]\eta^6 + \ldots\,,
\end{aligned} \tag{98}$$

and

$$R_{(\mathrm{in})}^{(l,m,\omega)}(x) = \sum_{n=-\infty}^{\infty} a_n \phi_{n+\nu}^{(\mathrm{in})}(x)\,, \tag{99}$$

where $\phi_{n+\nu}^{(\mathrm{in})}(x) = \mathrm{hypergeom}([a,b],[c],x) \equiv F(a,b;c;x)$, with

$$\begin{aligned}
a &= n + \nu + 1 - i\epsilon\,, \\
b &= -n - \nu - i\epsilon\,, \\
c &= 1 - 2i\epsilon = a + b\,.
\end{aligned} \tag{100}$$

The hypergeometric function is conveniently rewritten using the identity 15.3.8 of Ref. [22] (see pag. 559) so that its argument becomes $y = 1/(1-x)$, namely

$$
\begin{aligned}
F(a, b; c; x) &= y^a \frac{\Gamma(c)\Gamma(b-a)}{\Gamma(b)\Gamma(c-a)} F(a, c-b; a-b+1; y) \\
&+ y^b \frac{\Gamma(c)\Gamma(a-b)}{\Gamma(a)\Gamma(c-b)} F(b, c-a; b-a+1; y)
\end{aligned} \tag{101}
$$

with $x = 1 - 1/y$ or $y = 1/(1-x)$ [Note: Passing from the first to the second expression in Eq. (101) is obtained by the map $a \to b$, $b \to a$; $c = a + b$ is instead unchanged]. Explicitly

$$
\begin{aligned}
\phi^{(\mathrm{in})}_{n+\nu}(y) &= y^{n+\nu+1-i\epsilon} \frac{\Gamma(1-2i\epsilon)\Gamma(-2n-2\nu-1)}{\Gamma^2(-n-\nu-i\epsilon)} H_1(y) \\
&+ y^{-n-\nu-i\epsilon} \frac{\Gamma(1-2i\epsilon)\Gamma(2n+2\nu+1)}{\Gamma^2(n+\nu+1-i\epsilon)} H_2(y)\,,
\end{aligned} \tag{102}
$$

where

$$
\begin{aligned}
H_1(y) &= \mathrm{hypergeom}([n+\nu+1-i\epsilon, n+\nu+1-i\epsilon], [2n+2\nu+2], y) \\
H_2(y) &= \mathrm{hypergeom}([-n-\nu-i\epsilon, -n-\nu-i\epsilon], [-2n-2\nu+2], y)\,.
\end{aligned} \tag{103}
$$

Differently from x, which is small as approaching the horizon, the variable $y = (2M/r)\eta^2$ is a "PN-small quantity," and one consider the series representation of the hypergeometric functions around $y = 0$. Therefore, the hypergeometric functions in these expressions are replaced by their associated series expansion around $y = 0$, up the order N. In the final expression $\psi_{lm\omega}$ is expanded up to η^N. The coefficients a_n are determined by solving the following three-point recurrence relation:

$$
\alpha^\nu_n a_{n+1} + \beta^\nu_n a_n + \gamma^\nu_n a_{n-1} = 0\,, \tag{104}
$$

where

$$
\begin{aligned}
\alpha^\nu_n &= \frac{i\epsilon(n+\nu+1+i\epsilon)^2(n+\nu+1-i\epsilon)}{(n+\nu+1)(2n+2\nu+3)}\,, \\
\beta^\nu_n &= -l(l+1) + (n+\nu)(n+\nu+1) + 2\epsilon^2 + \frac{\epsilon^4}{(n+\nu)(n+\nu+1)}\,, \\
\gamma^\nu_n &= \frac{-i\epsilon(n+\nu+i\epsilon)(n+\nu-i\epsilon)^2}{(n+\nu)(2n+2\nu-1)}\,.
\end{aligned} \tag{105}
$$

In these coefficients n always enters as $n + \nu$ and ϵ as $i\epsilon$.

An approximated solution of the recursive relation (104) can be found for any given l. Actually, one fixes "a priori" the largest value of $n = n_{\max}$ to be used (corresponding the maximum PN-order to be reached with the solutions of the radial equation) and puts

$$a_{n_{\max}} = 0 = a_{-n_{\max}}\,, \qquad a_0 = 1\,. \tag{106}$$

Then solves the recursion equations for $n_{\max}, n_{\max} - 1, n_{\max} - 2, \ldots 3, 2, 1$ and for $-n_{\max}, -n_{\max}+1, -n_{\max}+2, \ldots -3, -2, -1$. Both series of equations include a relation for a_0 (for $n = 1$ and $n = -1$) which is then taken as a compatibility condition between ν and l. The latter, when l is not explicitly chosen, implies

$$\begin{aligned}
\nu &= l + \frac{1}{2l+1}\left[-2 + \frac{(l+1)^4}{(2l+1)(2l+2)(2l+3)} - \frac{l^4}{(2l-1)2l(2l+1)}\right]\epsilon^2 + O(\epsilon^4)\\
&= l + \frac{1}{2l+1}\left[-2 + H(l) - H(l-1)\right]\epsilon^2 + O(\epsilon^4)\,,
\end{aligned} \tag{107}$$

where

$$H(l) = \frac{(l+1)^4}{(2l+1)(2l+2)(2l+3)}\,. \tag{108}$$

As an example, we list below the relations ν vs l for specific values of $l = 0, 1, 2, 3, 4$ up to $O(\epsilon^{12})$. The structure of the coefficients a_n is shown in Fig. 1 in the case $l = 4$ for $n_{\max} = 17$. As soon as the values of l increases the corresponding expressions (both for ν and a_n) become more and more involved (i.e., written in terms of "large fractions"), saturating soon the computational facilities of any computer available today.

$$\begin{aligned}
\nu_0 &= -\frac{7}{6}\epsilon^2 - \frac{9449}{7560}\epsilon^4 - \frac{102270817}{33339600}\epsilon^6 - \frac{4988909608861}{588110544000}\epsilon^8\\
&\quad - \frac{72237319625071987}{2593567499040000}\epsilon^{10} - \frac{2008359560158182591511}{21153904986614400000}\epsilon^{12}\,,\\
\nu_1 &= 1 - \frac{19}{30}\epsilon^2 - \frac{1325203}{3591000}\epsilon^4 - \frac{1876733084209}{4083505650000}\epsilon^6 - \frac{29333897359675088897}{40863314359098000000}\epsilon^8\\
&\quad - \frac{8456229829644845965886767 09}{678049353372765874500000000}\epsilon^{10}\\
&\quad - \frac{447936092572020422484529225371046957}{192953716502507138725465668750000000}\epsilon^{12}\,,\\
\nu_2 &= 2 - \frac{79}{210}\epsilon^2 - \frac{708247}{9261000}\epsilon^4 - \frac{423940501889}{11090886778125}\epsilon^6\\
&\quad - \frac{59271881276715766819}{221018171352891412500 0}\epsilon^8\\
&\quad - \frac{564858584456642264465384132611}{25168170200278032013086000000000}\epsilon^{10}
\end{aligned}$$

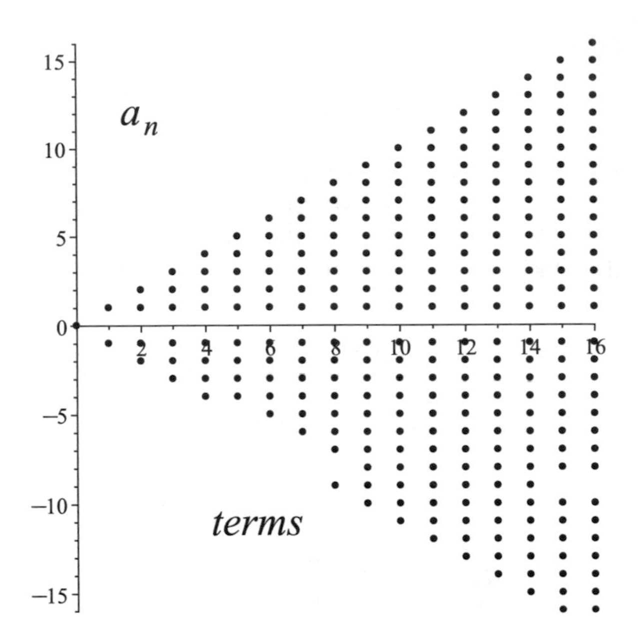

Fig. 1 Assuming $a_{-17} = 0 = a_{17}$ (as well as $a_0 = 1$), the various a_n obtained by solving the three-point recurrence relation are expressed as a sum of terms corresponding to different powers of ϵ. For example: in a_{-16} (horizontal line to be drawn corresponding to the value -16 on the ordinate axis) only two terms enter corresponding to the powers ϵ^{15} and ϵ^{16}; in a_{-15} only three terms enter corresponding to the powers ϵ^{14}, ϵ^{15} and ϵ^{16}; etc. Some irregularities (absence/presence of expected terms) in the behavior a_n vs ϵ^n are shown in the plot. In general, to reach a specified PN-precision the necessary $a_n$terms are not necessarily symmetric with respect to a_0

$$
\begin{aligned}
&-\frac{26450326014126037391692269487786665402461}{1278949899712679600763473140424400000000000}\epsilon^{12},\\
\nu_3 &= 3-\frac{169}{630}\epsilon^2-\frac{74380421}{2750517000}\epsilon^4-\frac{1008725842043489}{156110268213900000}\epsilon^6\\
&-\frac{7518839302383545659213}{3544121463974343252000000}\epsilon^8\\
&-\frac{53477783012565497991161254235239}{63506761644931323272606052000000000}\epsilon^{10}\\
&-\frac{3880391139017304179697545363306389370120999}{1009001285608227759721797153906115440000000000000}\epsilon^{12},\\
\nu_4 &= 4-\frac{289}{1386}\epsilon^2-\frac{435454879}{34612505928}\epsilon^4-\frac{38349366083366489}{21609407767240893600}\epsilon^6\\
&-\frac{31009517594899209604558009}{91740604801957499688876576000}\epsilon^8\\
&-\frac{80442119512143591442375139998688729}{1057148247265900978506752691728190720000}\epsilon^{10}
\end{aligned}
$$

$$-\frac{928921644507684167764716631472578770486566725 7}{48320998055359439271968601711020086727557432320000 0}\epsilon^{12}.$$

Similarly, the MST up-solution is written in the form

$$\psi_{lm\omega}(z) = C_{(\mathrm{up})}(z) R_{(\mathrm{up})}^{(l,m,\omega)}(z)\,, \tag{109}$$

where

$$\begin{aligned} C_{(\mathrm{up})} &\equiv \eta \tilde{C}_{(\mathrm{up})} \\ &= 2^{\nu} e^{-\pi\epsilon} e^{-i\pi(\nu+1)} e^{iz} z^{\nu+i\epsilon} (z-\epsilon)^{-i\epsilon} \\ &= (2z)^{\nu} e^{-\pi\epsilon} e^{-i\pi(\nu+1)} e^{iz} \left(1 - \frac{\epsilon}{z}\right)^{-i\epsilon}, \end{aligned} \tag{110}$$

and is further rescaled (and denoted by a tilde) as

$$\begin{aligned} \tilde{C}_{(\mathrm{up})} &= 2\omega r + 2i\omega^2 r^2 \eta - \omega^3 r^3 \eta^2 - \left(4\pi M\omega + \frac{1}{3} i\omega^3 r^3\right) \omega r \eta^3 \\ &+ \left(\frac{1}{12}\omega^4 r^4 - 4i\pi M\omega^2 r\right) \omega r \eta^4 + \left[2\pi M\omega^4 r^3 + \frac{1}{60} i\omega^6 r^6 + 8i\omega^2 M^2\right] \eta^5 \\ &+ \left[-\frac{76}{15}\omega^3 M^2 r(\ln(2\omega r\eta) - i\pi) - \frac{\omega^7 r^7}{360} + \frac{2}{3} i\pi\omega^5 M r^4 \right. \\ &\left. +4(\pi^2 - 2)\omega^3 M^2 r\right] \eta^6 + O(\eta^7)\,, \end{aligned} \tag{111}$$

in order to start from η^0. The up-solutions are first written in terms of a Kummer function U

$$\mathrm{KummerU}(\nu + 1 - i\epsilon, 2\nu + 2, -2iz) = \Psi(\nu + 1 - i\epsilon, 2\nu + 2, -2iz)\,. \tag{112}$$

The KummerU function (also denoted by Ψ and known as "decaying confluent hypergeometric function") is then related to hypergeometric functions by the Eq. 13.1.3, pag. 504 of Ref. [22], that is

$$\begin{aligned} \Psi(a, b, x) = \frac{\pi}{\sin \pi b} &\left(\frac{\mathrm{hypergeom}([a], [b], x)}{\Gamma(a - b + 1)\Gamma(b)} \right. \\ &\left. -x^{1-b} \frac{\mathrm{hypergeom}([a - b + 1], [2 - b], x)}{\Gamma(a)\Gamma(2 - b)} \right), \end{aligned} \tag{113}$$

conveniently rewritten as

$$\Psi(a, b, x) = \frac{\Gamma(1 - b)}{\Gamma(a - b + 1)} \mathrm{hypergeom}([a], [b], x)$$

$$+x^{1-b}\frac{\Gamma(b-1)}{\Gamma(a)}\text{hypergeom}([a-b+1],[2-b],x))\,. \tag{114}$$

In our case

$$a=n+\nu+1-i\epsilon\,,\qquad b=2(n+\nu+1)\,, \tag{115}$$

so that

$$a-b+1=-n-\nu-i\epsilon\,,\qquad 2-b=-2n-2\nu\,. \tag{116}$$

Up-solutions are finally given by

$$R^{(l,m,\omega)}_{(\text{up})}(z)=\sum_{n=-\infty}^{\infty}\frac{(\nu+1-i\epsilon)_n}{(\nu+1+i\epsilon)_n}(2iz)^n a_n\Psi(a,b,-2iz)\equiv\sum_{n=-\infty}^{\infty}a_n\phi^{(\text{up})}_{n+\nu}\,, \tag{117}$$

with a_n the same quantities as in the in-case [whereas a and b in the Ψ function here are different from the similar quantities used in the in-case] and $(Q)_n$ is the Pochhammer symbol

$$(Q)_n=\frac{\Gamma(Q+n)}{\Gamma(Q)}\,. \tag{118}$$

Note that, recalling that

$$\nu=l+\delta\nu\,, \tag{119}$$

we can replace, for example, the term x^{1-b} in the Ψ function by

$$x^{1-b}\approx x^{-1-2n-2l}\left(1-2\delta\nu\ln x+2(\delta\nu)^2\ln^2 x\right) \tag{120}$$

and the hypergeometric functions by their series representation up to a fixed number $n_{\max}$ of terms around $x=0$ since $x\to -2iz=-2i\omega_0 r\eta$ is a "small quantity."

3.6 *Analytical Versus Numerical Results*

Let us turn to the calculation of the scalar field (62) at the source position. We have already discussed how to remove the singular part of the field, leading to the regularized value (75), i.e.,

$$\psi_0^{(\text{reg})}(u)=\sum_l[\psi_0^{(l)}(u)-D_l(u)]=\sum_l\psi^{(l)}_{\text{reg}}(u)\,, \tag{121}$$

which has a gauge-invariant character and can be associated with physical, measurable quantities. The subtraction of the Detweiler singular field $D_l(r_0)$ (see Eq. (76)) is enough to ensure that the series converges. We want now to compute the final sum of this series over l.

Let us consider the various regularized PN contributions (i.e., obtained by using as fundamental solutions of the homogeneous radial equation the PN solutions)

$$S_N = \sum_{l=N}^{\infty} \psi_{\rm reg}^{(l)}(u)\,, \tag{122}$$

with

$$\begin{aligned}
S_0 &= 0\,,\\
S_1 &= \left(\frac{1}{4}u^2 - \frac{61}{192}u^3\right) q\,,\\
S_2 &= \left[\frac{1}{10}u^2 + \frac{181}{672}u^3 + \left(\frac{126253}{86400} - \frac{7}{32}\pi^2\right) u^4\right] q\,,\\
S_3 &= \left[\frac{9}{140}u^2 + \frac{127}{1344}u^3 + \left(\frac{9741157}{3725568} - \frac{7}{32}\pi^2\right) u^4\right.\\
&\quad \left. + \left(-\frac{4938262487443}{2712213504000} + \frac{29}{512}\pi^2\right) u^5\right] q\,,\\
S_4 &= \left[\frac{1}{21}u^2 + \frac{661}{11088}u^3 + \left(-\frac{7}{32}\pi^2 + \frac{709490963}{302702400}\right) u^4\right.\\
&\quad + \left(\frac{27506934151}{93884313600} + \frac{29}{512}\pi^2\right) u^5\\
&\quad \left. + \left(-\frac{279}{1024}\pi^2 + \frac{4383505633491 83}{912931065446400}\right) u^6\right] q\,,\\
S_5 &= \left[\frac{5}{132}u^2 + \frac{25525}{576576}u^3 + \left(-\frac{7}{32}\pi^2 + \frac{13045986313}{5708102400}\right) u^4\right.\\
&\quad + \left(-\frac{232049019479863}{1434605959987200} + \frac{29}{512}\pi^2\right) u^5\\
&\quad + \left(-\frac{279}{1024}\pi^2 + \frac{6200419652418108697}{1417390688467353600}\right) u^6\\
&\quad \left. + \left(\frac{652137534418305593073 89}{440048782704328468070 4} + \frac{76585}{262144}\pi^4 - \frac{42084587}{8847360}\pi^2\right) u^7\right] q\,.
\end{aligned} \tag{123}$$

The way to read the above results is the following. If we want the sum of the series representing the regular field we cannot use only PN solutions: in that case, for example, one can sum from 1 to infinity but the accuracy reached is only up to u^3 terms included, and, in any case, the contribution corresponding to $l = 0$ should be added separately (taken from MST solutions). The sum of PN solutions from $l = 2$ to infinity will allow to include terms $O(u^4)$ but the terms corresponding to $l = 0, 1$ should be added separately (taken from MST solutions). Similarly, the sum of PN solutions from $l = 3$ to infinity will allow to include terms $O(u^5)$ but the terms corresponding to $l = 0, 1, 2$ should be added separately (taken from MST solutions), etc. The MST contributions for specific values of $l = 0, 1, 2\ldots$ are listed below.

$$\begin{aligned}\psi^{(0)}_{\rm reg}(u) = \Bigg(&-\frac{1}{4}u^2 - \frac{131}{192}u^3 - \frac{335}{256}u^4 - \frac{196141}{81920}u^5 - \frac{4203457}{983040}u^6 - \frac{269861411}{36700160}u^7\\ &-\frac{1716337981}{146800640}u^8 - \frac{1036093931639}{67645734912}u^9\Bigg) q\,,\end{aligned}$$

$$\begin{aligned}\psi^{(1)}_{\rm reg}(u) = \Bigg[&\frac{3}{20}u^2 - \frac{263}{448}u^3 + \left(\frac{309619}{172800} - \frac{2}{3}\ln(u) - \frac{4}{3}\ln(2) - \frac{4}{3}\gamma\right)u^4\\ &+\left(\frac{74}{15}\gamma - \frac{5810627131}{4257792000} + \frac{37}{15}\ln(u) + \frac{74}{15}\ln(2)\right)u^5\\ &-\frac{38}{45}\pi u^{11/2}\\ &+\left(-\frac{2789}{1050}\ln(2) - \frac{2789}{1050}\gamma - \frac{2789}{2100}\ln(u) - \frac{21001766837951}{1549836288000}\right)u^6\\ &+\frac{703}{225}\pi u^{13/2}\\ &+\Bigg(\frac{304}{45}\gamma\ln(2) + \frac{152}{45}\ln(2)\ln(u) + \frac{157603582067170153}{3347646382080000} + \frac{152}{45}\ln(2)^2\\ &-\frac{16}{3}\zeta(3) - \frac{1792079}{56700}\gamma - \frac{1943279}{56700}\ln(2)\\ &-\frac{38}{27}\pi^2 + \frac{38}{45}\ln(u)^2 + \frac{152}{45}\gamma^2 - \frac{2094479}{113400}\ln(u) + \frac{152}{45}\gamma\ln(u)\Bigg)u^7\\ &-\frac{52991}{31500}\pi u^{15/2}\\ &+\Bigg(-\frac{5624}{225}\gamma\ln(2) - \frac{2812}{225}\ln(2)\ln(u) - \frac{683655801703030048136\,9}{876413822828544000\,00}\\ &-\frac{2812}{225}\ln(2)^2 + \frac{296}{15}\zeta(3) + \frac{14433220177}{174636000}\gamma\\ &+\frac{12803284177}{174636000}\ln(2) + \frac{703}{135}\pi^2 - \frac{703}{225}\ln(u)^2 - \frac{2812}{225}\gamma^2\end{aligned}$$

$$
\begin{aligned}
&+\frac{11173348177}{349272000}\ln(u)-\frac{2812}{225}\gamma\ln(u)\Big)u^8\\
&-\frac{18829}{21000}\pi u^{17/2}\Big]q\,,
\end{aligned}
$$

$$
\begin{aligned}
\psi^{(2)}_{\rm reg}(u)=&\Big[\frac{1}{28}u^2+\frac{235}{1344}u^3-\frac{35809397}{31046400}u^4\\
&+\left(\frac{4745938586173}{542442700800}-\frac{32}{15}\ln(u)-\frac{128}{15}\ln(2)-\frac{64}{15}\gamma\right)u^5\\
&+\left(\frac{272}{21}\ln(u)+\frac{1088}{21}\ln(2)-\frac{149810782311919}{6509312409600}+\frac{544}{21}\gamma\right)u^6\\
&-\frac{5056}{1575}\pi u^{13/2}\\
&+\left(-\frac{131324}{6615}\ln(u)-\frac{636640539318575085631}{10224827932999680000}-\frac{262648}{6615}\gamma-\frac{525296}{6615}\ln(2)\right)u^7\\
&+\frac{42976}{2205}\pi u^{15/2}\\
&+\Big(\frac{161792}{1575}\gamma\ln(2)+\frac{80896}{1575}\ln(2)\ln(u)\\
&+\frac{3375415163210325785333497}{42428945990775472128000}+\frac{161792}{1575}\ln(2)^2\\
&-\frac{1024}{15}\zeta(3)-\frac{1692896596}{5457375}\gamma-\frac{3385793192}{5457375}\ln(2)-\frac{10112}{945}\pi^2+\frac{10112}{1575}\ln(u)^2\\
&+\frac{40448}{1575}\gamma^2-\frac{846448298}{5457375}\ln(u)+\frac{40448}{1575}\gamma\ln(u)\Big)u^8\\
&-\frac{67624}{3675}\pi u^{17/2}\Big]q\,,
\end{aligned}
\tag{124}
$$

$$
\begin{aligned}
\psi^{(3)}_{\rm reg}(u)=&\Big[\frac{1}{60}u^2+\frac{221}{6336}u^3+\frac{327912173}{1210809600}u^4-\frac{7370880752321}{3487131648000}u^5\\
&+\left(\frac{68417846729249363}{2608374472704000}-\frac{2734}{525}\ln(u)-\frac{729}{70}\ln(3)-\frac{5468}{525}\ln(2)-\frac{5468}{525}\gamma\right)u^6\\
&+\Big(\frac{209149}{4725}\ln(u)-\frac{674626673763142103301119}{5828151921809817600000}+\frac{418298}{4725}\gamma\\
&+\frac{418298}{4725}\ln(2)+\frac{12393}{140}\ln(3)\Big)u^7
\end{aligned}
$$

$$
\begin{aligned}
&-\frac{2772107}{330750}\pi u^{15/2}\\
&+\left(-\frac{92736697810888171768000037}{104906734592576716800000}-\frac{7258653}{30800}\ln(3)\right.\\
&\left.-\frac{1102504759}{9355500}\ln(u)-\frac{1102504759}{4677750}\gamma-\frac{1102504759}{4677750}\ln(2)\right)u^8\\
&+\frac{2772107}{220500}\pi u^{17/2}\Bigg]q\,,
\end{aligned}
\tag{125}
$$

$$
\begin{aligned}
\psi^{(4)}_{\rm reg}(u) = &\left[\frac{3}{308}u^2+\frac{983}{64064}u^3+\frac{7172009}{122943744}u^4+\frac{304439656059107}{669482781327360}u^5\right.\\
&-\frac{2575929576296522669}{661448987951431680}u^6\\
&+\left(-\frac{459008}{19845}\gamma+\frac{7315317259754538077097871}{1100121956760821170176 00}\right.\\
&\left.-\frac{229504}{19845}\ln(u)-\frac{1376768}{19845}\ln(2)\right)u^7\\
&+\left(\frac{7901056}{31185}\gamma+\frac{3950528}{31185}\ln(u)\right.\\
&\left.-\frac{6387125503292297173925911657 7}{155865278833873143390535680}+\frac{677120}{891}\ln(2)\right)u^8\\
&]q\,.
\end{aligned}
\tag{126}
$$

The final result is obtained as follows

$$
\begin{aligned}
\psi_{\rm reg}(u) &= \psi^{(0)}_{\rm reg}(u)+S_1+O(u^4)\\
\psi_{\rm reg}(u) &= \psi^{(0)}_{\rm reg}(u)+\psi^{(1)}_{\rm reg}(u)+S_2+O(u^5)\\
\psi_{\rm reg}(u) &= \psi^{(0)}_{\rm reg}(u)+\psi^{(1)}_{\rm reg}(u)+\psi^{(2)}_{\rm reg}(u)+S_3+O(u^6)\\
\psi_{\rm reg}(u) &= \psi^{(0)}_{\rm reg}(u)+\psi^{(1)}_{\rm reg}(u)+\psi^{(2)}_{\rm reg}(u)+\psi^{(3)}_{\rm reg}(u)+S_4+O(u^7)\\
\psi_{\rm reg}(u) &= \psi^{(0)}_{\rm reg}(u)+\psi^{(1)}_{\rm reg}(u)+\psi^{(2)}_{\rm reg}(u)+\psi^{(3)}_{\rm reg}(u)+\psi^{(4)}_{\rm reg}(u)+S_5+O(u^8)
\end{aligned}
\tag{127}
$$

and includes more and more MST solutions if one wants to obtain a high-PN order final result. Again the way to read the above results is the following. Summing from 1 to infinity the accuracy reached is only up to u^3 terms included; summing from $l=2$ to infinity will allow to include terms $O(u^4)$; summing from $l=3$ to infinity will allow to include terms $O(u^5)$, etc.

The final result corresponding to contributions up to $O(u^8)$ (i.e., $u^{7.5}$ terms included) is then the following [27]

$$
\begin{aligned}
\psi_0^{(\mathrm{reg})}(u) = \Bigg[& -u^3 + \left(\frac{35}{18} - \frac{7}{32}\pi^2 - \frac{4}{3}\gamma - \frac{4}{3}\ln(2) - \frac{2}{3}\ln(u)\right) u^4 \\
& + \left(\frac{1141}{360} + \frac{29}{512}\pi^2 + \frac{2}{3}\gamma - \frac{18}{5}\ln(2) + \frac{1}{3}\ln(u)\right) u^5 \\
& - \frac{38}{45}\pi u^{11/2} \\
& + \left(-\frac{23741}{1680} - \frac{279}{1024}\pi^2 + \frac{77}{6}\gamma + \frac{1627}{42}\ln(2) - \frac{729}{70}\ln(3)\right. \\
& \left. + \frac{77}{12}\ln(u)\right) u^6 \\
& - \frac{3}{35}\pi u^{13/2} \\
& + \left[-\frac{1515589307}{27216000} - \frac{6059603}{983040}\pi^2 + \frac{76585}{262144}\pi^4\right. \\
& + \left(-\frac{5321}{900} + \frac{152}{45}\gamma + \frac{304}{45}\ln(2) + \frac{152}{45}\ln(u)\right)\gamma \\
& + \left(-\frac{1786621}{18900} + \frac{152}{45}\ln(2) + \frac{152}{45}\ln(u)\right)\ln(2) + \frac{12393}{140}\ln(3) \\
& \left. - \frac{16}{3}\zeta(3) + \left(-\frac{10121}{1800} + \frac{38}{45}\ln(u)\right)\ln(u)\right] u^7 \\
& + \frac{35633}{3780}\pi u^{15/2} + O(u^8)\Bigg] q \,.
\end{aligned}
\tag{128}
$$

Note that here the inverse radial variable u is also a gauge-invariant variable, since it is related to the orbital frequency $u = (M\Omega)^{2/3}$ which is a standardly used gauge-invariant variable. Replacing ordinary logarithms by "eulerlogs," i.e.,

$$
\mathrm{eulerlog}_m(x) = \gamma + \ln(2) + \frac{1}{2}\ln(x) + \ln(m)\,, \qquad m = 1, 2, 3, \ldots, \tag{129}
$$

as was first introduced in Ref. [28], allows to absorb also the Euler γ constant and helps also to highlight the study of the transcendental structure of the various PN orders. For example, the lowest order ($O(u^4)$) contains only $\mathrm{eulerlog}_1(u)$, at $O(u^5)$ a combination of $\mathrm{eulerlog}_1(u)$ and $\mathrm{eulerlog}_2(u)$ appears, etc. Unfortunately, starting from $O(u^7)$ this replacement is not enough to completely remove the Euler γ terms, meaning that the transcendental structure is more involved.

Scalar self-force effects on a Schwarzschild background have been numerically studied in Ref. [29]. The comparison between the above analytical results and the numerical results of Ref. [29] shows a reasonable agreement (see Table 1 and Fig. 2). It is also interesting to study the behavior of this scalar field at the light-ring $u = 1/3$

Table 1 Comparison between the analytical prediction (128) for the regularized scalar field and the numerical values taken from Table I of Ref. [29]. The difference $\Delta\psi_0 = \psi_0^{\rm num} - \psi_0^{\rm analytic}$ and the relative error $\Delta\psi_0/\psi_0^{\rm analytic}$ are shown in the 3rd and 4th column, respectively. The superscript "reg" has been suppressed here to ease notation and we have set $q = 1$ for simplicity

u	$\psi_0^{\rm analytic}$	$\Delta\psi_0$	$\Delta\psi_0/\psi_0^{\rm analytic}$
1/4	-0.02304519610	-9.43×10^{-4}	0.0409
1/5	-0.01022371010	-1.05×10^{-5}	0.00102
1/6	-0.005468782560	1.40×10^{-5}	-0.00255
1/7	-0.003282635718	7.29×10^{-6}	-0.00222
1/8	-0.002130877461	3.37×10^{-6}	-0.00158
1/10	-0.001050586634	7.94×10^{-7}	-7.55×10^{-4}
1/14	$-3.701411742 \times 10^{-4}$	7.66×10^{-8}	-2.07×10^{-4}
1/20	$-1.246786056 \times 10^{-4}$	5.81×10^{-9}	-4.66×10^{-5}
1/30	$-3.661740186 \times 10^{-5}$	3.02×10^{-10}	-8.24×10^{-6}
1/50	$-7.889525256 \times 10^{-6}$	7.26×10^{-12}	-9.20×10^{-7}
1/70	$-2.877222881 \times 10^{-6}$	8.81×10^{-13}	-3.06×10^{-7}
1/100	$-9.884245218 \times 10^{-7}$	2.18×10^{-14}	-2.21×10^{-8}
1/200	$-1.239865750 \times 10^{-7}$	-2.50×10^{-14}	2.02×10^{-7}

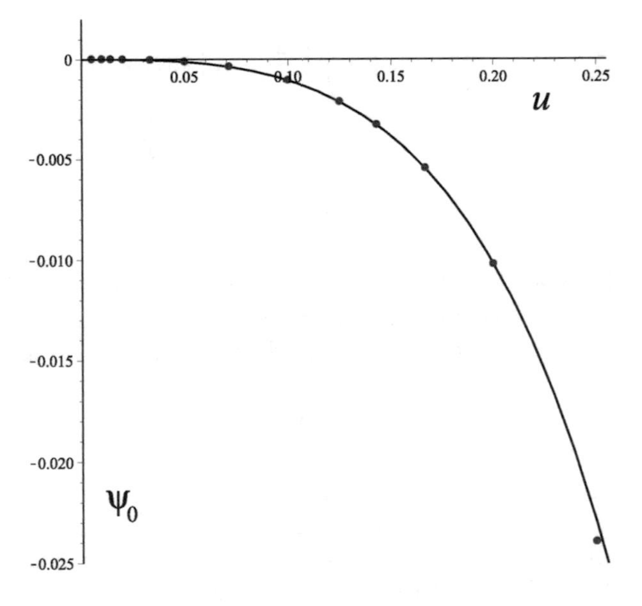

Fig. 2 The superposition of the numerical data and our analytical results (see Table 1)

(see Ref. [30] where a corresponding study has been done for the case of a massive particle orbiting a Schwarzschild black hole which is quite different from that of the scalar field). We provide below a simple numerical fit of the data of Table 1

$$\psi_0^{\text{reg, fit}} = -\frac{u^3}{(1-3u)^2}(1 - 7.84u + 47.36u^2 - 8.65u^3 + 81.77u^3 \ln(u)) , \tag{130}$$

(with a maximal residual of about 2.4×10^{-4}), suggesting a blow-up of the form $(1-3u)^{-2}$. However, this should be considered as an indication only, whereas a more conclusive statement would require strong field numerical data still unavailable.

4 So Far So Good: A List of SF Accomplishments

Gravitational self-force accomplishments concern the redshift variable (z_1), the gyroscope precession ($\Delta\psi$), the periastron advance (Δk) and the tidal invariants ($\Delta\lambda_i^{E^2,B^2,E^3}$). Any of these quantities, generically denoted as X, has an expression of the type

$$\begin{aligned} X = & X^{a^0,e^0} + X^{a^0,e^2} + X^{a^0,e^4} + \dots \\ & + X^{a^2,e^0} + X^{a^2,e^2} + X^{a^2,e^4} + \dots \\ & + X^{a^4,e^0} + X^{a^4,e^2} + X^{a^4,e^4} + \dots \\ & + \dots \end{aligned} \tag{131}$$

where the term X^{a^0,e^0} refers to circular orbits in the Schwarzschild spacetime, X^{a^0,e^2} to slightly eccentric orbits (including second order corrections in eccentricity) in the Schwarzschild spacetime, X^{a^0,e^4} to slightly eccentric orbits (including fourth order corrections in eccentricity) in the Schwarzschild spacetime, X^{a^2,e^0} to circular orbits in the Kerr spacetime including corrections of order a^2 in the Kerr rotational parameter, etc. We summarize in Table 2 the current knowledge of the gauge-invariant quantities z_1, $\Delta\psi$, ΔK and $\Delta\lambda_i^{E^2,B^2,E^3}$.

Table 2 List of gravitational self-force calculations of orbital invariants for both circular and eccentric orbits in Schwarzschild and Kerr spacetimes and a sample of references

Schwarzschild	$a^0\,e^0$	$a^0\,e^2$	$a^0\,e^4$	$a^0\,e^6$	$a^0\,e^8$
z_1	[2, 30–35]	[36]	[37]	[38]	
$\Delta\psi$	[39, 40]	[41, 42]			
Δk	[37, 43]				
$\Delta\lambda_i^{E^2,B^2,E^3}$	[35, 44–47]	[48]			
Kerr	$a^2\,e^0$	$a^2\,e^2$	$a^2\,e^4$	$a^4\,e^0$	$a^4\,e^2$
z_1	[19, 49–51]	[20, 52]	[53]		
$\Delta\psi$	[54]	[55]		[54]	
Δk	[56, 57]				
$\Delta\lambda_i^{E^2,B^2,E^3}$	[58]			[58]	

5 Concluding Remarks

The self-force formalism has been developed to deal with the effects of the perturbation induced by particles or fields on a given gravitational background as well as on their own motion. The novelty of the last few years is that the metric perturbation can be fully reconstructed and several orbital invariants can be analytically computed (i.e., the multipolar decomposition infinite summations can be explicitly performed). This information can be then translated into other approaches (e.g., Hamiltonian formalisms, like the effective one-body [59, 60]), which are currently used to model the two-body gravitational interaction. The main applications of these big computational efforts concern the modeling of gravitational wave signals which are being detected by Earth-based interferometers, like those of the LIGO/Virgo collaboration.

Acknowledgments We thank Thibault Damour for helpful discussions on analytical self-force computations.

References

1. L. Barack, A. Pound, Self-force and radiation reaction in general relativity. Rept. Prog. Phys. **82**(1), 016904 (2019). https://doi.org/10.1088/1361-6633/aae552, arXiv:1805.10385 [gr-qc]
2. D. Bini, T. Damour, Analytical determination of the two-body gravitational interaction potential at the fourth post-Newtonian approximation. Phys. Rev. D **87**(12), 121501 (2013). https://doi.org/10.1103/PhysRevD.87.121501, arXiv:1305.4884 [gr-qc]
3. E. Poisson, The Motion of point particles in curved space-time. Living Rev. Rel. **7**, 6 (2004). https://doi.org/10.12942/lrr-2004-6, gr-qc/0306052
4. M. Sasaki, H. Tagoshi, Analytic black hole perturbation approach to gravitational radiation. Living Rev. Rel. **6**, 6 (2003). https://doi.org/10.12942/lrr-2003-6, gr-qc/0306120
5. L. Barack, Gravitational self force in extreme mass-ratio inspirals. Class. Quant. Grav. **26**, 213001 (2009). https://doi.org/10.1088/0264-9381/26/21/213001, arXiv:0908.1664 [gr-qc]
6. S.L. Detweiler, B.F. Whiting, Self-force via a Green's function decomposition. Phys. Rev. D **67**, 024025 (2003). https://doi.org/10.1103/PhysRevD.67.024025, gr-qc/0202086
7. T. Regge, J.A. Wheeler, Stability of a Schwarzschild singularity. Phys. Rev. **108**, 1063 (1957). https://doi.org/10.1103/PhysRev.108.1063
8. F.J. Zerilli, Gravitational field of a particle falling in a Schwarzschild geometry analyzed in tensor harmonics. Phys. Rev. D **2**, 2141 (1970). https://doi.org/10.1103/PhysRevD.2.2141
9. S.A. Teukolsky, Perturbations of a rotating black hole. 1. Fundamental equations for gravitational electromagnetic and neutrino field perturbations. Astrophys. J. **185**, 635 (1973). https://doi.org/10.1086/152444
10. W.H. Press, S.A. Teukolsky, Perturbations of a Rotating Black Hole. II. Dynamical Stability of the Kerr Metric. Astrophys. J. **185**, 649 (1973). https://doi.org/10.1086/152445
11. S.A. Teukolsky, W.H. Press, Perturbations of a rotating black hole. III - Interaction of the hole with gravitational and electromagnetic radiation. Astrophys. J. **193**, 443 (1974). https://doi.org/10.1086/153180
12. S. Mano, H. Suzuki, E. Takasugi, Analytic solutions of the Regge-Wheeler equation and the post-Minkowskian expansion. Prog. Theor. Phys. **96**, 549 (1996). gr-qc/9605057

13. S. Mano, H. Suzuki, E. Takasugi, Analytic solutions of the Teukolsky equation and their low frequency expansions. Prog. Theor. Phys. **95**, 1079 (1996). gr-qc/9603020
14. J.M. Cohen, L.S. Kegeles, Electromagnetic fields in curved spaces - a constructive procedure. Phys. Rev. D **10**, 1070 (1974). https://doi.org/10.1103/PhysRevD.10.1070
15. P.L. Chrzanowski, Vector Potential and metric perturbations of a rotating black hole. Phys. Rev. D **11**, 2042 (1975). https://doi.org/10.1103/PhysRevD.11.2042
16. L.S. Kegeles, J.M. Cohen, Constructive procedure for perturbations of space-times. Phys. Rev. D **19**, 1641 (1979). https://doi.org/10.1103/PhysRevD.19.1641
17. T.S. Keidl, A.G. Shah, J.L. Friedman, D.H. Kim, L.R. Price, Gravitational self-force in a radiation gauge. Phys. Rev. D **82**(12), 124012 (2010). Erratum: Phys. Rev. D **90**(10), 109902 (2014). https://doi.org/10.1103/PhysRevD.82.124012, https://doi.org/10.1103/PhysRevD.90.109902, arXiv:1004.2276 [gr-qc]
18. A.G. Shah, T.S. Keidl, J.L. Friedman, D.H. Kim, L.R. Price, Conservative, gravitational self-force for a particle in circular orbit around a Schwarzschild black hole in a Radiation Gauge. Phys. Rev. D **83**, 064018 (2011). https://doi.org/10.1103/PhysRevD.83.064018, arXiv:1009.4876 [gr-qc]
19. A.G. Shah, J.L. Friedman, T.S. Keidl, EMRI corrections to the angular velocity and redshift factor of a mass in circular orbit about a Kerr black hole. Phys. Rev. D **86**, 084059 (2012). https://doi.org/10.1103/PhysRevD.86.084059, arXiv:1207.5595 [gr-qc]
20. M. van de Meent, A.G. Shah, Metric perturbations produced by eccentric equatorial orbits around a Kerr black hole. Phys. Rev. D **92**(6), 064025 (2015). https://doi.org/10.1103/PhysRevD.92.064025, arXiv:1506.04755 [gr-qc]
21. D. Bini, A. Geralico, Gauge-fixing for the completion problem of reconstructed metric perturbations of a Kerr spacetime. arXiv:1908.03191 [gr-qc]
22. M. Abramowitz, I. Stegun, *Handbook of Mathematical Functions* (Dover Publications Inc., New York, 1970)
23. S.L. Detweiler, E. Messaritaki, B.F. Whiting, Self-force of a scalar field for circular orbits about a Schwarzschild black hole. Phys. Rev. D **67**, 104016 (2003). https://doi.org/10.1103/PhysRevD.67.104016, gr-qc/0205079
24. N. Sago, H. Nakano, M. Sasaki, Gauge problem in the gravitational self-force. 1. Harmonic gauge approach in the Schwarzschild background. Phys. Rev. D **67**, 104017 (2003). https://doi.org/10.1103/PhysRevD.67.104017, gr-qc/0208060
25. H. Nakano, N. Sago, M. Sasaki, Gauge problem in the gravitational self-force. 2. First post-Newtonian force under Regge-Wheeler gauge. Phys. Rev. D **68**, 124003 (2003). https://doi.org/10.1103/PhysRevD.68.124003, gr-qc/0308027
26. W. Hikida, H. Nakano, M. Sasaki, Self-force regularization in the Schwarzschild spacetime. Class. Quant. Grav. **22**(15), S753 (2005). https://doi.org/10.1088/0264-9381/22/15/009, gr-qc/0411150
27. D. Bini, G. Carvalho, A. Geralico, Scalar field self-force effects on a particle orbiting a Reissner-Nordström black hole. Phys. Rev. D **94**(12), 124028 (2016). https://doi.org/10.1103/PhysRevD.94.124028, arXiv:1610.02235 [gr-qc]
28. T. Damour, B.R. Iyer, A. Nagar, Improved resummation of post-Newtonian multipolar waveforms from circularized compact binaries. Phys. Rev. D **79**, 064004 (2009), arXiv:0811.2069 [gr-qc]
29. L.M. Diaz-Rivera, E. Messaritaki, B.F. Whiting, S.L. Detweiler, Scalar field self-force effects on orbits about a Schwarzschild black hole. Phys. Rev. D **70**, 124018 (2004), gr-qc/0410011
30. S. Akcay, L. Barack, T. Damour, N. Sago, Gravitational self-force and the effective-one-body formalism between the innermost stable circular orbit and the light ring. Phys. Rev. D **86**, 104041 (2012). arXiv:1209.0964 [gr-qc]
31. S.L. Detweiler, A Consequence of the gravitational self-force for circular orbits of the Schwarzschild geometry. Phys. Rev. D **77**, 124026 (2008). https://doi.org/10.1103/PhysRevD.77.124026, arXiv:0804.3529 [gr-qc]

32. D. Bini, T. Damour, High-order post-Newtonian contributions to the two-body gravitational interaction potential from analytical gravitational self-force calculations. Phys. Rev. D **89**(6), 064063 (2014). https://doi.org/10.1103/PhysRevD.89.064063, arXiv:1312.2503 [gr-qc]
33. D. Bini, T. Damour, Analytic determination of the eight-and-a-half post-Newtonian self-force contributions to the two-body gravitational interaction potential. Phys. Rev. D **89**(10), 104047 (2014). https://doi.org/10.1103/PhysRevD.89.104047, arXiv:1403.2366 [gr-qc]
34. D. Bini, T. Damour, Detweiler's gauge-invariant redshift variable: Analytic determination of the nine and nine-and-a-half post-Newtonian self-force contributions. Phys. Rev. D **91**, 064050 (2015). https://doi.org/10.1103/PhysRevD.91.064050, arXiv:1502.02450 [gr-qc]
35. C. Kavanagh, A.C. Ottewill, B. Wardell, Analytical high-order post-Newtonian expansions for extreme mass ratio binaries. Phys. Rev. D **92**(8), 084025 (2015). https://doi.org/10.1103/PhysRevD.92.084025, arXiv:1503.02334 [gr-qc]
36. D. Bini, T. Damour, A. Geralico, Confirming and improving post-Newtonian and effective-one-body results from self-force computations along eccentric orbits around a Schwarzschild black hole. Phys. Rev. D **93**(6), 064023 (2016). https://doi.org/10.1103/PhysRevD.93.064023, arXiv:1511.04533 [gr-qc]
37. D. Bini, T. Damour, A. Geralico, New gravitational self-force analytical results for eccentric orbits around a Schwarzschild black hole. Phys. Rev. D **93**(10), 104017 (2016). https://doi.org/10.1103/PhysRevD.93.104017 arXiv:1601.02988 [gr-qc]
38. D. Bini, T. Damour, A. Geralico, Novel approach to binary dynamics: application to the fifth post-Newtonian level. Phys. Rev. Lett. **123**(23), 231104 (2019). https://doi.org/10.1103/PhysRevLett.123.231104, arXiv:1909.02375 [gr-qc]
39. D. Bini, T. Damour, Two-body gravitational spin-orbit interaction at linear order in the mass ratio. Phys. Rev. D **90**(2), 024039 (2014). https://doi.org/10.1103/PhysRevD.90.024039, arXiv:1404.2747 [gr-qc]
40. D. Bini, T. Damour, Analytic determination of high-order post-Newtonian self-force contributions to gravitational spin precession. Phys. Rev. D **91**(6), 064064 (2015). https://doi.org/10.1103/PhysRevD.91.064064, arXiv:1503.01272 [gr-qc]
41. C. Kavanagh, D. Bini, T. Damour, S. Hopper, A.C. Ottewill, B. Wardell, Spin-orbit precession along eccentric orbits for extreme mass ratio black hole binaries and its effective-one-body transcription. Phys. Rev. D **96**(6), 064012 (2017). https://doi.org/10.1103/PhysRevD.96.064012, arXiv:1706.00459 [gr-qc]
42. D. Bini, T. Damour, A. Geralico, Spin-orbit precession along eccentric orbits: improving the knowledge of self-force corrections and of their effective-one-body counterparts. Phys. Rev. D **97**(10), 104046 (2018). https://doi.org/10.1103/PhysRevD.97.104046, arXiv:1801.03704 [gr-qc]
43. T. Damour, Gravitational self-force in a Schwarzschild background and the effective one body formalism. Phys. Rev. D **81**, 024017 (2010). https://doi.org/10.1103/PhysRevD.81.024017, arXiv:0910.5533 [gr-qc]
44. D. Bini, T. Damour, Gravitational self-force corrections to two-body tidal interactions and the effective one-body formalism. Phys. Rev. D **90**, 124037 (2014). https://doi.org/10.1103/PhysRevD.90.124037, arXiv:1409.6933 [gr-qc]
45. S.R. Dolan, P. Nolan, A.C. Ottewill, N. Warburton, B. Wardell, Tidal invariants for compact binaries on quasicircular orbits. Phys. Rev. D **91**, 023009 (2015). https://doi.org/10.1103/PhysRevD.91.023009, arXiv:1406.4890 [gr-qc]
46. P. Nolan, C. Kavanagh, S.R. Dolan, A.C. Ottewill, N. Warburton, B. Wardell, Octupolar invariants for compact binaries on quasicircular orbits. Phys. Rev. D **92**, 123008 (2015). https://doi.org/10.1103/PhysRevD.92.123008, arXiv:1505.04447 [gr-qc]
47. A.G. Shah, A. Pound, Linear-in-mass-ratio contribution to spin precession and tidal invariants in Schwarzschild spacetime at very high post-Newtonian order. Phys. Rev. D **91**(12), 124022 (2015). https://doi.org/10.1103/PhysRevD.91.124022, arXiv:1503.02414 [gr-qc]
48. D. Bini, A. Geralico, Gravitational self-force corrections to tidal invariants for particles on eccentric orbits in a Schwarzschild spacetime. Phys. Rev. D **98**(6), 064026 (2018). https://doi.org/10.1103/PhysRevD.98.064026, arXiv:1806.06635 [gr-qc]

49. A.G. Shah, Talk delivered at the XIV Marcel Grossmann meeting, Rome (2015)
50. D. Bini, T. Damour, A. Geralico, Spin-dependent two-body interactions from gravitational self-force computations. Phys. Rev. D **92**(12), 124058 (2015). Erratum: Phys. Rev. D **93**(10), 109902 (2016). https://doi.org/10.1103/PhysRevD.93.109902, https://doi.org/10.1103/PhysRevD.92.124058, arXiv:1510.06230 [gr-qc]
51. C. Kavanagh, A.C. Ottewill, B. Wardell, Analytical high-order post-Newtonian expansions for spinning extreme mass ratio binaries. Phys. Rev. D **93**(12), 124038 (2016). https://doi.org/10.1103/PhysRevD.93.124038, arXiv:1601.03394 [gr-qc]
52. D. Bini, T. Damour, A. Geralico, High post-Newtonian order gravitational self-force analytical results for eccentric equatorial orbits around a Kerr black hole. Phys. Rev. D **93**(12), 124058 (2016). https://doi.org/10.1103/PhysRevD.93.124058, arXiv:1602.08282 [gr-qc]
53. D. Bini, A. Geralico, New gravitational self-force analytical results for eccentric equatorial orbits around a Kerr black hole: redshift invariant. Phys. Rev. D **100**(10), 104002 (2019). https://doi.org/10.1103/PhysRevD.100.104002, arXiv:1907.11080 [gr-qc]
54. D. Bini, T. Damour, A. Geralico, C. Kavanagh, M. van de Meent, Gravitational self-force corrections to gyroscope precession along circular orbits in the Kerr spacetime. Phys. Rev. D **98**(10), 104062 (2018). https://doi.org/10.1103/PhysRevD.98.104062, arXiv:1809.02516 [gr-qc]
55. D. Bini, A. Geralico, New gravitational self-force analytical results for eccentric equatorial orbits around a Kerr black hole: gyroscope precession. Phys. Rev. D **100**(10), 104003 (2019). https://doi.org/10.1103/PhysRevD.100.104003, arXiv:1907.11082 [gr-qc]
56. M. van de Meent, Self-force corrections to the periapsis advance around a spinning black hole. Phys. Rev. Lett. **118**(1), 011101 (2017). https://doi.org/10.1103/PhysRevLett.118.011101, arXiv:1610.03497 [gr-qc]
57. D. Bini, A. Geralico, Analytical determination of the periastron advance in spinning binaries from self-force computations. Phys. Rev. D **100**(12), 121502 (2019). https://doi.org/10.1103/PhysRevD.100.121502, arXiv:1907.11083 [gr-qc]
58. D. Bini, A. Geralico, Gravitational self-force corrections to tidal invariants for particles on circular orbits in a Kerr spacetime. Phys. Rev. D **98**(6), 064040 (2018). https://doi.org/10.1103/PhysRevD.98.064040, arXiv:1806.08765 [gr-qc]
59. A. Buonanno, T. Damour, Effective one-body approach to general relativistic two-body dynamics. Phys. Rev. D **59**, 084006 (1999). https://doi.org/10.1103/PhysRevD.59.084006, gr-qc/9811091
60. A. Buonanno, T. Damour, Transition from inspiral to plunge in binary black hole coalescences. Phys. Rev. D **62**, 064015 (2000). https://doi.org/10.1103/PhysRevD.62.064015, gr-qc/0001013

Geometry and Analysis in Black Hole Spacetimes

Lars Andersson

Contents

Based on lectures given at the 2019 Summer School on Einstein Equations at Domodossola, Italy

L. Andersson (✉)
Albert Einstein Institute, Potsdam, Germany
e-mail: laan@aei.mpg.de

S. L. Cacciatori, A. Kamenshchik (eds.), *Einstein Equations: Local Energy, Self-Force, and Fields in General Relativity*, Tutorials, Schools, and Workshops in the Mathematical Sciences, https://doi.org/10.1007/978-3-031-21845-3_3

1 Introduction

A solution to the Einstein vacuum equations is a Lorentzian spacetime $(\mathcal{M}, g_{ab})$, satisfying $R_{ab} = 0$, where R_{ab} is the Ricci tensor of g_{ab}. The Einstein equation is the Euler-Lagrange equation of the diffeomorphism invariant Einstein–Hilbert action functional, given by the integral of the scalar curvature of $(\mathcal{M}, g_{ab})$,

$$\int_{\mathcal{M}} R d\mu_g .$$

The diffeomorphism invariance, or general covariance, of the action has the consequence that Cauchy data for the Einstein equation must satisfy a set of constraint equations. After introducing suitable gauge conditions, the Einstein equations can be reduced to a hyperbolic system of evolution equations. For sufficiently regular Cauchy data satisfying the constraints, the Cauchy problem for the Einstein equation has a unique solution which is maximal among all regular, vacuum Cauchy developments. This general result, however, does not give any detailed information about the properties of the maximal development.

There are two main conjectures about the maximal development. The strong cosmic censorship conjecture states that a generic maximal development is inextendible, as a regular vacuum spacetime. There are examples where the maximal development is extendible and has non-unique extensions, which furthermore may contain closed timelike curves. In these cases, predictability fails for the Einstein equations, but if strong cosmic censorship holds, they are non-generic. At present, this is only known to hold in the context of families of spacetimes with symmetry restrictions, see [3, 62] and the references therein. The weak cosmic censorship conjecture, on the other hand, states that a generic maximal development of asymptotically flat data has no singularities in the domain of outer communication, i.e., the region of spacetime visible to observers at null infinity, cf. [69]. Results on black hole stability can thus be viewed as supporting weak cosmic censorship.

The Schwarzschild solution is static, spherically symmetric, asymptotically flat and has a single free parameter M which represents the mass of the black hole.

By Birkhoff's theorem it is the unique solution of the vacuum Einstein equations with these properties. In 1963 Roy Kerr [42] discovered a new, explicit family of asymptotically flat solutions of the vacuum Einstein equations which are stationary, axisymmetric, and rotating. The Kerr family of solutions is parametrized by the mass M, and the azimuthal angular momentum per unit mass a. In the limit $a = 0$, the Kerr solution reduces to the spherically symmetric Schwarzschild solution.

If $|a| \leq M$, the Kerr spacetime contains a black hole, while if $|a| > M$, there is a ringlike singularity which is naked, in the sense that it fails to be hidden from observers at infinity. This situation would violate the weak cosmic censorship conjecture, and one, therefore, expects that an overextreme Kerr spacetime is unstable, and in particular, that it cannot arise through a dynamical process from regular Cauchy data.

For a geodesic $\gamma^a(\lambda)$ with velocity $\dot\gamma^a = d\gamma^a/d\lambda$, in a stationary axisymmetric spacetime,[1] there are three conserved quantities, the mass $\boldsymbol{\mu}^2 = \dot\gamma^a\dot\gamma_b$, energy $\boldsymbol{e} = \dot\gamma^a(\partial_t)_a$, and angular momentum $\boldsymbol{\ell}_z = \dot\gamma^a(\partial_\phi)_a$. In a general axisymmetric spacetime, geodesic motion is chaotic. However, as was discovered by Brandon Carter in 1968, there is a fourth conserved quantity for geodesics in the Kerr spacetime, the Carter constant $\boldsymbol{k}$, see Sect. 5 for details. By Liouville's theorem, this allows one to integrate the geodesic equations by quadrature, and thus geodesics in the Kerr spacetime do not exhibit a chaotic behavior.

2 Background

2.1 Minkowski Space

Minkowski space $\mathbb{M}$ is $\mathbb{R}^4$ with metric which in a Cartesian coordinate system $(x^a) = (t, x^i)$ takes the form[2]

$$d\tau_{\mathbb{M}}^2 = dt^2 - (dx^1)^2 - (dx^2)^2 - (dx^3)^2.$$

A tangent vector ν^a is timelike, null, or spacelike when $g_{ab}\nu^a\nu^b > 0, = 0$, or < 0, respectively. Vectors with $g_{ab}\nu^a\nu^b \geq 0$ are called causal.

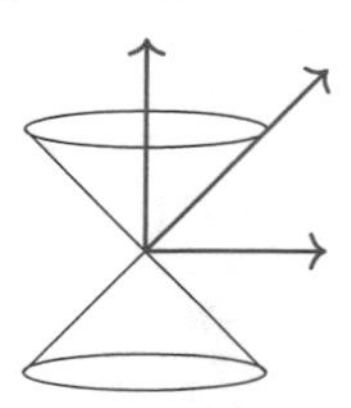

[1] We use signature $+ - - -$, in particular timelike vectors have positive norm.

[2] Here and below we shall use line elements, e.g., $d\tau_{\mathbb{M}}^2 = (g_{\mathbb{M}})_{ab}dx^a dx^b$ and metrics, e.g., $(g_{\mathbb{M}})_{ab}$ interchangeably.

Let u, v be given by

$$u = t - r, \quad v = t + r,$$

with $r^2 = (x^1)^2 + (x^2)^2 + (x^3)^2$. In terms of these coordinates the line element takes the form

$$d\tau_{\mathbb{M}}^2 = dudv - r^2 d\Omega_{S^2}^2. \tag{1}$$

A complex null tetrad is given by

$$l^a = \sqrt{2}(\partial_u)^a = \frac{1}{\sqrt{2}}\left((\partial_t)^a + (\partial_r)^a\right), \tag{2a}$$

$$n^a = \sqrt{2}(\partial_v)^a = \frac{1}{\sqrt{2}}\left(((\partial_t)^a - (\partial_r)^a\right), \tag{2b}$$

$$m^a = \frac{1}{\sqrt{2}r}\left((\partial_\theta)^a + \frac{i}{\sin\theta}(\partial_\phi)^a\right) \tag{2c}$$

normalized so that $n^a l_a = 1 = -m^a \bar{m}_a$, with all other inner products of tetrad legs zero. Complex null tetrads with this normalization play a central role in the Newman-Penrose and Geroch-Held-Penrose formalisms, see Sect. 4. In these notes we will use such tetrads unless otherwise stated.

In terms of a null tetrad, we have

$$g_{ab} = 2(l_{(a}n_{b)} - m_{(a}\bar{m}_{b)}). \tag{3}$$

Introduce compactified null coordinates $\mathcal{U}, \mathcal{V}$, given by

$$\mathcal{U} = \arctan u, \quad \mathcal{V} = \arctan v.$$

These take values in $\{(-\pi/2, \pi/2) \times (-\pi/2, \pi/2)\} \cap \{\mathcal{V} \geq \mathcal{U}\}$, and we can thus present Minkowski space in a *causal diagram*, see Fig. 1. Here each point represents an S^2 and we have drawn null vectors at 45° angles. A compactification of Minkowski space is now given by adding the null boundaries[3] $\mathscr{I}^\pm$, spatial infinity i_0 and timelike infinity $i^\pm$ as indicated in the figure. Explicitly,

$$\mathscr{I}^+ = \{\mathcal{V} = \pi/2\}$$

$$\mathscr{I}^- = \{\mathcal{U} = -\pi/2\}$$

$$i_0 = \{\mathcal{V} = \pi/2, \mathcal{U} = -\pi/2\}$$

[3] Here $\mathscr{I}$ is pronounced "Scri" for "script I."

Fig. 1 Causal diagram of Minkowski space

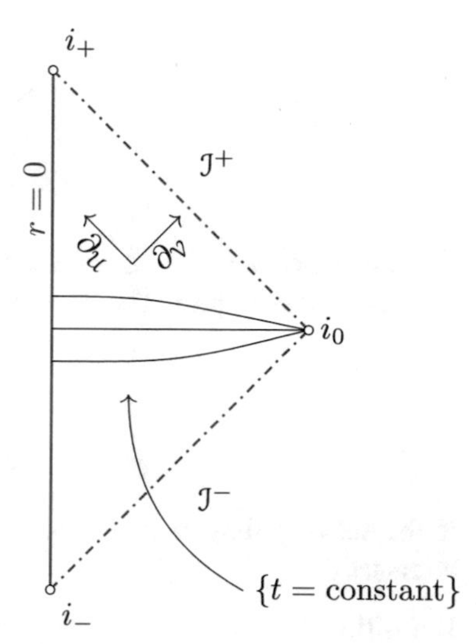

$$i_\pm = \{(\mathcal{V}, \mathcal{U}) = \pm(\pi/2, \pi/2)\}.$$

In Fig. 1, we have also indicated schematically the t-level sets which approach spatial infinity i_0. Causal diagrams are a useful tool which, if applied with proper care, can be used to understand the structure of quite general spacetimes. Such diagrams are often referred to as Penrose, or Carter-Penrose diagrams.

2.2 *Lorentzian Geometry and Causality*

We now consider a smooth Lorentzian 4-manifold $(\mathcal{M}, g_{ab})$ with signature $+---$. Each tangent space in a 4-dimensional spacetime is isometric to Minkowski space $\mathbb{M}$, and we can carry intuitive notions of causality over from $\mathbb{M}$ to $\mathcal{M}$. We say that a smooth curve $\gamma^a(\lambda)$ is causal if the velocity vector $\dot{\gamma}^a = d\gamma^a/d\lambda$ is causal. Two points in $\mathcal{M}$ are causally related if they can be connected by a piecewise smooth causal curve. The concept of causal curves is most naturally defined for C^0 curves. A C^0 curve γ^a is said to be causal if each pair of points on γ^a are causally related. We may define timelike curve and timelike related points in the analogous manner.

We now assume that $\mathcal{M}$ is time oriented, i.e., that there is a globally defined timelike vector field on $\mathcal{M}$. This allows us to distinguish between future and past directed causal curves, and to introduce a notion of the causal and timelike future of a spacetime point. The corresponding past notions are defined analogously. If q is in the causal future of p, we write $p \preccurlyeq q$. This introduces a partial order on $\mathcal{M}$. The causal future $J^+(p)$ of p is defined as $J^+(p) = \{q : p \preccurlyeq q\}$ while the timelike future $I^+(p)$ is defined in the analogous manner, with timelike replacing causal. A subset $\Sigma \subset \mathcal{M}$ is achronal

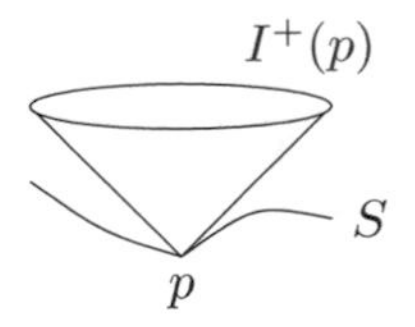

if there is no pair $p, q \in \mathcal{M}$ such that $q \in I^+(p)$, i.e., Σ does not intersect its timelike future or past. The domain of dependence $D(S)$ of $S \subset \mathcal{M}$

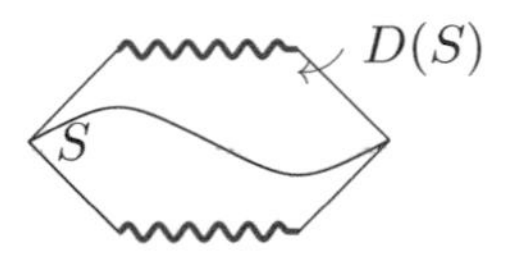

is the set of points p such that any inextendible causal curve starting at p must intersect S.

Definition 2.1 A spacetime $\mathcal{M}$ is globally hyperbolic if there is a closed, achronal $\Sigma \subset \mathcal{M}$ such that $\mathcal{M} = D(\Sigma)$. In this case, Σ is called a Cauchy surface.

Due to results of Bernal and Sanchez [19], global hyperbolicity is characterized by the existence of a smooth, Cauchy time function $\tau : \mathcal{M} \to \mathbb{R}$. A function τ on $\mathcal{M}$ is a time function if $\nabla^a \tau$ is timelike everywhere, and it is Cauchy if the level sets $\Sigma_t = \tau^{-1}(t)$ are Cauchy surfaces. If τ is smooth, its level sets are then smooth and spacelike. It follows that a globally hyperbolic spacetime $\mathcal{M}$ is globally foliated by Cauchy surfaces, and in particular is diffeomorphic to a product $\Sigma \times \mathbb{R}$. In the following, unless otherwise stated, we shall consider only globally hyperbolic spacetimes.

If a globally hyperbolic spacetime $\mathcal{M}$ is a subset of a spacetime $\mathcal{M}'$, then the boundary $\partial\mathcal{M}$ in $\mathcal{M}'$ is called the Cauchy horizon.

Example 2.2 Let O be the origin in Minkowski space, and let $\mathcal{M} = I^+(O) = \{t > r\}$ be its timelike future. Then $\mathcal{M}$ is globally hyperbolic with Cauchy time function $\tau = \sqrt{t^2 - r^2}$. Further, $\mathcal{M}$ is a subset of Minkowski space $\mathbb{M}$, which is a globally hyperbolic space with Cauchy time function t. Minkowski space is geodesically complete and hence inextendible. The boundary $\{t = r\}$ is the Cauchy horizon $\partial\mathcal{M}$ of $\mathcal{M}$. Past inextendible causal geodesics (i.e., past causal rays) in $\mathcal{M}$ end on $\partial\mathcal{M}$. In particular, $\mathcal{M}$ is incomplete. However, $\mathcal{M}$ is extendible, as a smooth flat spacetime, with many inequivalent extensions.

We remark that for a globally hyperbolic spacetime, which is extendible, the extension is in general non-unique. In the particular case considered in Example 2.2, $\mathbb{M}$ is an extension of $\mathcal{M}$, which is also happens to be maximal and globally hyperbolic. In the vacuum case, there is a unique maximal globally hyperbolic extension, cf. Sect. 2.5 below. However, a maximal extension is in general non-unique, and may fail to be globally hyperbolic.

2.3 Conventions and Notation

We shall use mostly abstract indices, but will sometimes work with coordinate indices, and unless confusion arises will not be too specific about this. We raise and lower indices with g_{ab}, for example ,$\xi^a = g^{ab}\xi_b$, with $g^{ab}g_{bc} = \delta^a{}_c$, where $\delta^a{}_c$ is the Kronecker delta, i.e., the tensor with the property that $\delta^a{}_c\xi^c = \xi^a$ for any ξ^a.

Let $\epsilon_{a\cdots d}$ be the Levi-Civita symbol, i.e., the skew symmetric expression which in any coordinate system has the property that $\epsilon_{1\cdots n} = 1$. The volume form of g_{ab} is $(\mu_g)_{abcd} = \sqrt{|g|}\epsilon_{abcd}$. Given $(\mathcal{M}, g_{ab})$ we have the canonically defined Levi-Civita covariant derivative ∇_a. For a vector v^a, this is of the form

$$\nabla_a v^b = \partial_a v^b + \Gamma^b_{ac} v^c,$$

where $\Gamma^b_{ac} = \frac{1}{2}g^{bd}(\partial_a g_{dc} + \partial_c g_{db} - \partial_d g_{ac})$ is the Christoffel symbol. In order to fix the conventions used here, we recall that the Riemann curvature tensor is defined by

$$(\nabla_a\nabla_b - \nabla_b\nabla_a)\xi_c = R_{abc}{}^d\xi_d.$$

The Riemann tensor R_{abcd} is skew symmetric in the pairs of indices ab, cd, $R_{abcd} = R_{[ab]cd} = R_{ab[cd]}$, is pairwise symmetric $R_{abcd} = R_{cdab}$, and satisfies the first Bianchi identity $R_{[abc]d} = 0$. Here square brackets $[\cdots]$ denote antisymmetrization. We shall similarly use round brackets $(\cdots)$ to denote symmetrization. Further, we have $\nabla_{[a}R_{bc]de} = 0$, the second Bianchi identity. A contraction gives $\nabla^a R_{abcd} = 0$. The Ricci tensor is $R_{ab} = R^c{}_{acb}$ and the scalar curvature $R = R^a{}_a$. We further let $S_{ab} = R_{ab} - \frac{1}{4}Rg_{ab}$ denote the trace free part of the Ricci tensor. The Riemann tensor can be decomposed as follows,

$$\begin{aligned} R_{abcd} = & -\tfrac{1}{12}g_{ad}g_{bc}R + \tfrac{1}{12}g_{ac}g_{bd}R + \tfrac{1}{2}g_{bd}S_{ac} - \tfrac{1}{2}g_{bc}S_{ad} \\ & -\tfrac{1}{2}g_{ad}S_{bc} + \tfrac{1}{2}g_{ac}S_{bd} + C_{abcd}. \end{aligned} \tag{4}$$

This defines the Weyl tensor C_{abcd} which is a tensor with the symmetries of the Riemann tensor, and vanishing traces, $C^c{}_{acb} = 0$. Recall that $(\mathcal{M}, g_{ab})$ is locally conformally flat if and only if $C_{abcd} = 0$. It follows from the contracted second Bianchi identity that the Einstein tensor $G_{ab} = R_{ab} - \frac{1}{2}Rg_{ab}$ is conserved, $\nabla^a G_{ab} = 0$.

2.4 Einstein Equation

The Einstein equation in geometrized units with $G = c = 1$, where G, c denote Newtons constant and the speed of light, respectively, cf. [68, Appendix F], is the system

$$G_{ab} = 8\pi T_{ab}. \tag{5}$$

This equation relates geometry, expressed in the Einstein tensor G_{ab} on the left-hand side, to matter, expressed via the energy-momentum tensor T_{ab} on the right-hand side. For example, for a self-gravitating Maxwell field F_{ab}, $F_{ab} = F_{[ab]}$, we have

$$T_{ab} = \frac{1}{4\pi}(F_{ac}F_{bc} - \frac{1}{4}F_{cd}F^{cd}g_{ab}).$$

The source-free Maxwell field equations

$$\nabla^a F_{ab} = 0, \quad \nabla_{[a}F_{bc]} = 0$$

imply that T_{ab} is conserved, $\nabla^a T_{ab} = 0$. The contracted second Bianchi identity implies that $\nabla^a G_{ab} = 0$, and hence the conservation property of T_{ab} is implied by the coupling of the Maxwell field to gravity. These facts can be seen to follow from the variational formulation of Einstein gravity, given by the action

$$I = \int_{\mathcal{M}} \frac{R}{16\pi} d\mu_g - \int_{\mathcal{M}} L_{\text{matter}} d\mu_g,$$

where L_{matter} is the Lagrangian describing the matter content in the spacetime. In the case of Maxwell theory, this is given by

$$L_{\text{Maxwell}} = \frac{1}{4\pi} F_{cd}F^{cd}.$$

Recall that in order to derive the Maxwell field equation, as an Euler-Lagrange equation, from this action, it is necessary to introduce a vector potential for F_{ab}, by setting $F_{ab} = 2\nabla_{[a}A_{b]}$, and carrying out the variation with respect to A_a. It is a general fact that for generally covariant (i.e., diffeomorphism invariant) Lagrangian field theories which depend on the spacetime location only via the metric and its derivatives, the symmetric energy-momentum tensor

$$T_{ab} = \frac{1}{\sqrt{g}} \frac{\partial L_{\text{matter}}}{\partial g^{ab}}$$

is conserved when evaluated on solutions of the Euler-Lagrange equations.

As a further example of a matter field, we consider the scalar field, with action

$$L_{\text{scalar}} = \tfrac{1}{2}\nabla^c\psi\nabla_c\psi,$$

where ψ is a function on $\mathcal{M}$. The corresponding energy-momentum tensor is

$$T_{ab} = \nabla_a\psi\nabla_b\psi - \tfrac{1}{2}\nabla^c\psi\nabla_c\psi g_{ab}$$

and the Euler-Lagrange equation is the free scalar wave equation

$$\nabla^a \nabla_a \psi = 0. \tag{6}$$

As (6) is another example of a field equation derived from a covariant action which depends on the spacetime location only via the metric g_{ab} or its derivatives, the symmetric energy-momentum tensor is conserved for solutions of the field equation.

In both of the just mentioned cases, the energy-momentum tensor satisfies the dominant energy condition, $T_{ab}\nu^a\zeta^b \geq 0$ for future directed causal vectors ν^a, ζ^a. This implies the null energy condition

$$R_{ab}\nu^a\nu^b \geq 0 \quad \text{if } \nu_a\nu^a = 0. \tag{7}$$

These energy conditions hold for most classical matter.

There are many interesting matter systems which are worthy of consideration, such as fluids, elasticity, kinetic matter models including Vlasov, as well as fundamental fields such as Yang-Mills, to name just a few. We consider only spacetimes which satisfy the null energy condition, and for the most part we shall in these notes be concerned with the vacuum Einstein equations,

$$R_{ab} = 0. \tag{8}$$

2.5 The Cauchy Problem

Given a spacelike hypersurface[4] Σ in $\mathcal{M}$ with timelike normal T^a, induced metric h_{ab} and second fundamental form k_{ab}, defined by $k_{ab}X^aY^b = \nabla_a T_b X^a Y^b$ for X^a, Y^b tangent to Σ, the Gauss, and Gauss-Codazzi equations imply the constraint equations

$$R[h] + (k_{ab}h^{ab})^2 - k_{ab}k^{ab} = 16\pi T_{ab}T^aT^b \tag{9a}$$

$$\nabla[h]_a(k_{bc}h^{bc}) - \nabla[h]^b k_{ab} = T_{ab}T^b. \tag{9b}$$

A 3-manifold Σ together with tensor fields h_{ab}, k_{ab} on Σ solving the constraint equations is called a Cauchy data set. The constraint equations for general relativity are analogs of the constraint equations in Maxwell and Yang-Mills theory, in that they lead to Hamiltonians which generate gauge transformations.

Consider a 3 + 1 split of $\mathcal{M}$, i.e., a 1-parameter family of Cauchy surfaces Σ_t, with a coordinate system $(x^a) = (t, x^i)$, and let

[4] If there is no room for confusion, we shall denote abstract indices for objects on Σ by $a, b, c, \ldots$.

$$(\partial_t)^a = NT^a + X^a$$

be the split of $(\partial_t)^a$ into a normal and tangential piece. The fields (N, X^a) are called lapse and shift. The definition of the second fundamental form implies the equation

$$\mathcal{L}_{\partial_t} h_{ab} = -2Nk_{ab} + \mathcal{L}_X h_{ab}.$$

In the vacuum case, the Hamiltonian for gravity can be written in the form

$$\int N\mathcal{H} + X^a \mathcal{J}_a + \text{ boundary terms},$$

where $\mathcal{H}$ and $\mathcal{J}$ are the densitized left-hand sides of (9). If we consider only compactly supported perturbations in deriving the Hamiltonian evolution equation, the boundary terms mentioned above can be ignored. However, for (N, X^a) not tending to zero at infinity, and considering perturbations compatible with asymptotic flatness, the boundary term becomes significant, cf. Sect. 2.6.

The resulting Hamiltonian evolution equations, written in terms of h_{ab} and its canonical conjugate $\pi^{ab} = \sqrt{h}(k^{ab} - (h^{cd}k_{cd}h^{ab}))$ are usually called the ADM evolution equations.

Let $\Sigma \subset \mathcal{M}$ be a Cauchy surface. Given functions ϕ_0, ϕ_1 on Σ and F on $\mathcal{M}$, the Cauchy problem is the problem of finding solutions to the wave equation

$$\nabla^a \nabla_a \psi = F, \quad \psi\big|_\Sigma = \phi_0, \quad \mathcal{L}_{\partial_t}\psi\big|_\Sigma = \phi_1.$$

Assuming suitable regularity conditions, the solution is unique and stable with respect to initial data. This fact extends to a wide class of non-linear hyperbolic PDE's including quasilinear wave equations, i.e., equations of the form

$$A^{ab}[\psi]\partial_a\partial_b\psi + B[\psi, \partial\psi] = 0$$

with A^{ab} a Lorentzian metric depending on the field ψ.

Given a vacuum Cauchy data set, (Σ, h_{ab}, k_{ab}), a solution of the Cauchy problem for the Einstein vacuum equations is a spacetime metric g_{ab} with $R_{ab} = 0$, such that (h_{ab}, k_{ab}) coincides with the metric and second fundamental form induced on Σ from g_{ab}. Such a solution is called a vacuum extension of (Σ, h_{ab}, k_{ab}).

Due to the fact that R_{ab} is covariant, the symbol of R_{ab} is degenerate. In order to get a well-posed Cauchy problem, it is necessary to either impose gauge conditions or introduce new variables. A standard choice of gauge condition is the harmonic coordinate condition. Let $\widehat{g}_{ab}$ be a given metric on $\mathcal{M}$. The identity map $\mathbf{i} : \mathcal{M} \to \mathcal{M}$ is harmonic if and only if the vector field

$$V^a = g^{bc}(\Gamma^a_{bc} - \widehat{\Gamma}^a_{bc})$$

vanishes. Here $\Gamma^a_{bc}, \widehat{\Gamma}^a_{bc}$ are the Christoffel symbols of the metrics $g_{ab}, \widehat{g}_{ab}$. Then V^a is the tension field of the identity map $\mathbf{i} : (\mathcal{M}, g_{ab}) \to (\mathcal{M}, \widehat{g}_{ab})$. This is harmonic if and only if

$$V^a = 0. \tag{10}$$

Since harmonic maps with a Lorentzian domain are often called wave maps, the gauge condition (10) is sometimes called wave map gauge. A particular case of this construction, which can be carried out if $\mathcal{M}$ admits a global coordinate system (x^a), is given by letting $\widehat{g}_{ab}$ be the Minkowski metric defined with respect to (x^a). Then $\widehat{\Gamma}^a_{bc} = 0$ and (10) is simply

$$\nabla^b \nabla_b x^a = 0, \tag{11}$$

which is usually called the wave coordinate gauge condition.

Going back to the general case, let $\widehat{\nabla}$ be the Levi-Civita covariant derivative defined with respect to $\widehat{g}_{ab}$. We have the identity

$$R_{ab} = -\tfrac{1}{2}\frac{1}{\sqrt{g}}\widehat{\nabla}_a \sqrt{g} g^{ab} \widehat{\nabla}_b g_{ab} + S_{ab}[g, \widehat{\nabla} g] + \nabla_{(a} V_{b)}, \tag{12}$$

where S_{ab} is an expression which is quadratic in first derivatives $\widehat{\nabla}_a g_{cd}$. Setting $V^a = 0$ in (12) yields R^{harm}_{ab}, and (8) becomes a quasilinear wave equation

$$R^{\text{harm}}_{ab} = 0. \tag{13}$$

By standard results, the Eq. (13) has a locally well-posed Cauchy problem in Sobolev spaces H^s for $s > 5/2$. Using more sophisticated techniques, well-posedness can be shown to hold for any $s > 2$ [44]. Recently a local existence has been proved under the assumption of curvature bounded in L^2 [46]. Given a Cauchy data set (Σ, h_{ab}, k_{ab}), together with initial values for lapse and shift N, X^a on Σ, it is possible to find $\mathcal{L}_t N, \mathcal{L}_t X^a$ on Σ such that the V^a are zero on Σ. A calculation now shows that due to the constraint equations, $\mathcal{L}_{\partial_t} V^a$ is zero on Σ. Given a solution to the reduced Einstein vacuum equation (13), one finds that V^a solves a wave equation. This follows from $\nabla^a G_{ab} = 0$, due to the Bianchi identity. Hence, due to the fact that the Cauchy data for V^a is trivial, it holds that $V^a = 0$ on the domain of the solution. Thus, in fact the solution to (13) is a solution to the full vacuum Einstein equation (8). This proves local well-posedness for the Cauchy problem for the Einstein vacuum equation. This fact was first proved by Yvonne Choquet-Bruhat [32], see [63] for background and history.

Global uniqueness for the Einstein vacuum equations was proved by Choquet-Bruhat and Geroch [22]. The proof relies on the local existence theorem sketched above, patching together local solutions. A partial order is defined on the collection of vacuum extensions, making use of the notion of common domain. The common

Fig. 2 Partial Cauchy surface touching ∂U

domain U of two extensions $\mathcal{M}$, $\mathcal{M}'$ is the maximal subset in $\mathcal{M}$ which is isometric to a subset in $\mathcal{M}'$. We can then define a partial order by saying that $\mathcal{M} \leq \mathcal{M}'$ if the maximal common domain is $\mathcal{M}$. One sees from the construction that each totally ordered subset has an upper bound. Thus, a maximal element exists by Zorn's lemma. This is proven to be unique by an application of the local well-posedness theorem for the Cauchy problem sketched above. For a contradiction, let $\mathcal{M}$, $\mathcal{M}'$ be two inequivalent extensions, and let U be the maximal common domain. Due to the Hausdorff property of spacetimes, this leads to a contradiction. By finding a partial Cauchy surface which touches the boundary of U, see Fig. 2 and making use of local uniqueness, one finds a contradiction to the maximality of U. It should be noted that here, uniqueness holds up to isometry, in keeping with the general covariance of the Einstein vacuum equations. These facts extend to the Einstein equations coupled to hyperbolic matter equations. See [64] for a construction of the maximal globally hyperbolic extension which does not rely on Zorn's lemma, see also [73]. The global uniqueness result can be generalized to Einstein-matter systems, provided the matter field equation is hyperbolic and that its solutions do not break down. General results on this topic are lacking, see, however, [58] and the references therein. The minimal regularity needed for global uniqueness is a subtle issue, which has not been fully addressed. In particular, results on local well-posedness are known, see, e.g., [45] and the references therein, which require less regularity than the best results on global uniqueness.

2.6 Asymptotically Flat Data

The Kerr black hole represents an isolated system, and the appropriate data for the black hole stability problem should, therefore, be asymptotically flat. To make this precise we suppose there is a compact set K in $\mathcal{M}$ and a map $\Phi : \mathcal{M} \setminus K \to \mathbb{R}^3 \setminus B(R, 0)$, where $B(R, 0)$ is a Euclidean ball. This defines a Cartesian coordinate system on the end $\mathcal{M} \setminus K$ so that $h_{ab} - \delta_{ab}$ falls off to zero at infinity, at a suitable rate. Here δ_{ab} is the Euclidean metric in the Cartesian coordinate system constructed above. Similarly, we require that k_{ab} falls off to zero.

Let x^a be the chosen Euclidean coordinate system and let r be the Euclidean radius $r = (\delta_{ab} x^a x^b)^{1/2}$. Following Regge and Teitelboim [61], see also [18], we assume that $g_{ab} = \delta_{ab} + h_{ab}$ with

$$h_{ab} = O(1/r), \quad \partial_a h_{bc} = O(1/r^2),$$

$$k_{ab} = O(1/r^2).$$

Further, we impose the parity conditions

$$h_{ab}(x) = h_{ab}(-x), \quad k_{ab}(x) = -k_{ab}(-x). \tag{14}$$

These falloff and parity conditions guarantee that the ADM 4-momentum and angular momentum are well defined. It was shown in [39] that data satisfying the parity condition conditions (14) are dense among data which satisfy an asymptotic flatness condition in terms of weighted Sobolev spaces.

Let ξ^a be an element of the Poincare Lie algebra and assume that $NT^a + X^a$ tends in a suitable sense to ξ^a at infinity. Then the action for Einstein gravity can be written in the form

$$\int_{\mathcal{M}} R d\mu_g = P_a \xi^a + \int \pi^{ij} \dot{h}_{ij} - \int N\mathcal{H} + X^i \mathcal{J}_i.$$

Here we may view P_a as a map to the dual of the Poincare Lie algebra, i.e., a momentum map. Evaluating $P_a\xi^a$ on a particular element of the Poincare Lie algebra gives the corresponding momentum. These can also be viewed as charges at infinity. We have

$$P^0 = \frac{1}{16\pi} \lim_{r\to\infty} \int_{S_r} (\partial_i g_{ji} - \partial_j g_{ii}) d\sigma^i \tag{15a}$$

$$P^i = \frac{1}{8\pi} \lim_{r\to\infty} \int_{S_r} \pi_{ij} d\sigma^j, \tag{15b}$$

where $d\sigma^i$ denotes the hypersurface area element of a family of spheres (which can be taken to be coordinate spheres) S_r foliating a neighborhood of infinity. See [54] and the references therein for a recent discussion of the conditions under which these expressions are well defined.

The energy and linear momentum (P^0, P^i) provide the components of a 4-vector P^a, the ADM 4-momentum. Assuming the dominant energy condition, then under the above asymptotic conditions, P^a is future causal, and timelike unless the maximal development $(\mathcal{M}, g_{ab})$ is isometric to Minkowski space. Further, P^a transforms as a Minkowski 4-vector, and the ADM mass is given by $M = \sqrt{P^a P_a}$. The boost theorem [23] implies, given an asymptotically flat Cauchy data set, that one may find in a boosted slice Σ' in its development such that the data is in the rest frame, i.e., $P^a = M(\partial_t)^a$.

Since the constraint quantities $\mathcal{H}, \mathcal{J}_i$ vanish for solutions of the Einstein equations, the gravitational Hamiltonian takes the value $P_a\xi^a$, and hence the ADM mass and momenta defined by (15) are conserved for an evolution with lapse and shift $(N, X^i) \to (1, 0)$ at infinity.

2.7 Komar Integrals

Assume that v^a is a Killing vector field. Then we have $\nabla_a v_b = \nabla_{[a} v_{b]}$. A calculation shows

$$\nabla^a(\nabla_a \xi_b - \nabla_b \xi_a) = -2R_{bc}\xi^c.$$

Hence, in vacuum,

$$\int_S e_{abcd} \nabla^c \xi^d$$

depends only on the homology class of the two-surface S. The analogous fact for the source-free Maxwell equation, were we have $\nabla^a F_{ab} = 0$, $\nabla_{[a} F_{bc]} = 0$, is the conservation of the charge integrals $\int_S F_{ab}$, $\int_S \epsilon_{abcd} F^{cd}$, which again depend only on the homology class of S. These statements are immediate consequences of Stokes theorem.

If we consider asymptotically flat spacetimes, we have in the stationary case, with $\xi^a = (\partial_t)^a$,

$$P^a \xi_a = -\frac{1}{8\pi} \int_S \epsilon_{abcd} \nabla^c \xi^d,$$

where on the left-hand side we have the ADM 4-momentum evaluated at infinity. Similarly, in the axially symmetric case, with $\eta^a = (\partial_\phi)^a$,

$$J = -\frac{1}{16\pi} \int_S \epsilon_{abcd} \nabla^c \eta^d.$$

These integrals again depend only on the homology class of S. See [41, §6] for background to these facts. For a non-symmetric, but asymptotically flat spacetime, letting S tend to infinity through a sequence of suitably round spheres yields the linkage integrals, which again reproduce the ADM momenta [72].

3 Black Holes

3.1 The Schwarzschild Solution

In Schwarzschild coordinates (t, r, θ, ϕ), the Schwarzschild metric takes the form

$$g_{ab} dx^a dx^b = f dt^2 - f^{-1} dr^2 - r^2 d\Omega^2_{S^2} \tag{16}$$

with $f = 1 - 2M/r$. Here $d\Omega^2_{S^2} = d\theta^2 + \sin^2\theta d\phi^2$ is the line element on the unit 2-sphere. The coordinate r is the area radius, defined by $4\pi r^2 = A(S(r,t))$, where $S(r,t)$ is the 2-sphere with constant t, r. The Schwarzschild metric is asymptotically flat and the parameter M coincides with the ADM mass.

In order to get a better understanding of the Schwarzschild spacetime, it is instructive to consider its maximal extension. In order to do this, we first introduce the tortoise coordinate r_*,

$$r_* = r + 2M\log(\frac{r}{2M} - 1). \tag{17}$$

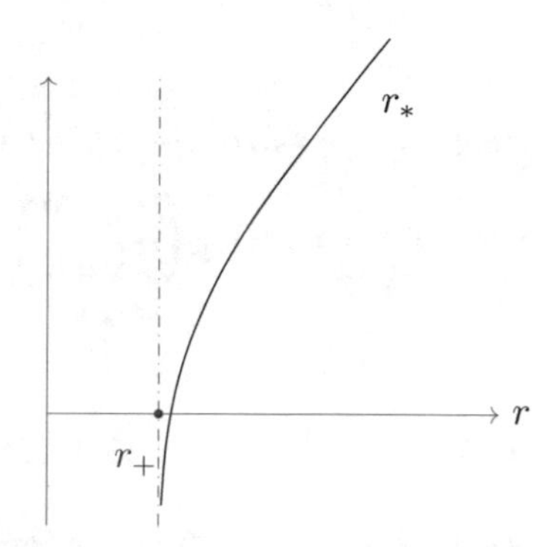

This solves $dr_* = f^{-1}dr$, $r_*(4M) = 4M$. As $r \searrow 2M$, r_* diverges logarithmically to $-\infty$, and for large r, $r_* \sim r$. Inverting (17) yields

$$r = 2M\mathrm{W}\left(e^{\frac{r_*}{2M}-1}\right) + 2M, \tag{18}$$

where W is the principal branch of the Lambert W function.[5] We can now introduce null coordinates

$$u = t - r_*, \quad v = t + r_*.$$

A null tetrad is given by

$$l^a = \sqrt{\frac{2}{f}}\partial^a_v,$$

$$n^a = \sqrt{\frac{2}{f}}\partial^a_u,$$

[5] The Lambert W function, or product logarithm, is defined as the solution of $W(x)e^{W(x)} = x$ for $x > 0$. It satisfies $W'(x) = W(x)/((W(x)+1)x)$. The principal branch is analytic at $x = 0$ and is real valued in the range $(-e^{-1}, \infty)$ with values in $(-1, \infty)$. In particular, $W(0) = 0$. See [25].

$$m^a = \frac{1}{\sqrt{2}r}(\partial_\theta^a + \frac{i}{\sin\theta}\partial_\phi^a).$$

On the exterior region in Schwarzschild, (u, v) take values in the range $(-\infty, \infty) \times (-\infty, \infty)$. Let $\mathcal{U}, \mathcal{V}$ be a pair of coordinates taking values in $(-\pi/2, \pi/2)$, and related to u, v by

$$\begin{aligned} u &= -4M\log(-\tan\mathcal{U}), \quad \mathcal{U} \in (-\pi/2, 0) \\ v &= 4M\log(\tan\mathcal{V}), \quad \mathcal{V} \in (0, \pi/2). \end{aligned}$$

We have

$$\begin{aligned} t &= \tfrac{1}{2}(v+u) &&= 4M\log\left(-\tan\mathcal{V}\tan\mathcal{U}\right) \\ r_* &= \tfrac{1}{2}(v-u) &&= 4M\log\left(-\frac{\tan\mathcal{V}}{\tan\mathcal{U}}\right). \end{aligned}$$

In terms of $\mathcal{U}, \mathcal{V}$ we have

$$r = 2M\mathrm{W}(-e^{-1}\tan\mathcal{U}\tan\mathcal{V}) + 2M \tag{19}$$

and $r > 0$ thus corresponds to $\tan\mathcal{U}\tan\mathcal{V} < 1$. The line element now takes the form

$$g_{ab}dx^a dx^b = \frac{d\mathcal{U}d\mathcal{V}}{\cos^2\mathcal{U}\cos^2\mathcal{V}}\frac{32M^3}{r}e^{-\frac{r}{2M}} - r^2 d\Omega^2_{S^2}. \tag{20}$$

The form (20) of the Schwarzschild line element is non-degenerate in the range

$$(\mathcal{U}, \mathcal{V}) \in (-\pi/2, \pi/2) \times (-\pi/2, \pi/2) \cap \{-\pi/2 < \mathcal{U} + \mathcal{V} < \pi/2\}. \tag{21}$$

In particular, the location $r = 2M$ of the coordinate singularity in the line element (16) corresponds to $\mathcal{U}\mathcal{V} = 0$. The line element (20) has a coordinate singularity, which is also a curvature singularity, at $r = 0$ (corresponding to $\tan\mathcal{U}\tan\mathcal{V} = 1$), and at $\mathcal{U} = \pm\pi/2$, $\mathcal{V} = \pm\pi/2$ (corresponding to u, v taking unbounded values). Figure 3 shows the region given in (21), with lines of constant t, r indicated. Using the causal diagram for the extended Schwarzschild solution, one can easily find the null infinities $\mathcal{I}^\pm$, spatial infinity i_0, timelike infinities $i_\pm$, the horizons $\mathcal{H}^\pm$ at $r = 2M$, which are indicated. Region I is the domain of outer communication, i.e., $I^-(\mathcal{I}^+) \cap I^+(\mathcal{I}^-)$, while region II is the future trapped (or black hole) region, $\mathcal{M}^{\mathrm{Schw}} \setminus I^-(\mathcal{I}^+)$.

The level sets of t hit the bifurcation sphere $\mathcal{B}$ located at $\mathcal{U} = \mathcal{V} = 0$, where $\partial_t = 0$. In particular, we see that the Schwarzschild coordinates are degenerate, since the level sets of t do not foliate the extended Schwarzschild spacetime. On the other hand, a global Cauchy foliation of the maximally extended Schwarzschild spacetime is given by the level sets of the Kruskal time function $\mathcal{T} = \frac{1}{2}(\mathcal{V} + \mathcal{U})$.

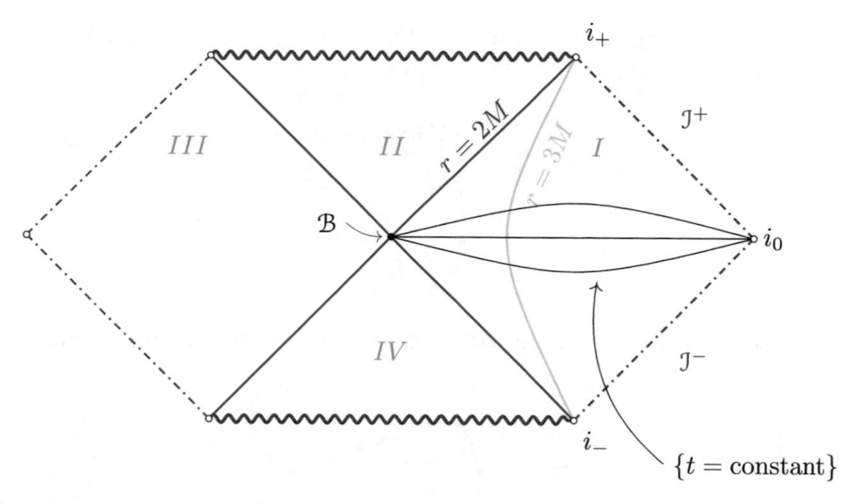

Fig. 3 Causal diagram of the maximally extended Schwarzschild solution

Although the region covered by the null coordinates $\mathcal{U}, \mathcal{V}$ is compact, the line element (20) is of course isometric to the form given in (16). A conformal factor $\Phi = \cos \mathcal{U} \cos \mathcal{V}$ may now be introduced, which brings $\mathcal{J}^{\pm}$ to a finite distance. Letting $\tilde{g}_{ab} = \Phi^2 g_{ab}$, and adding these boundary pieces to $(\mathcal{M}, \tilde{g}_{ab})$ provides a *conformal compactification*[6] of the maximally extended Schwarzschild spacetime.

3.1.1 Orbiting Null Geodesics

Consider a null geodesic γ^a in the Schwarzschild spacetime. Due to the spherical symmetry of the Schwarzschild spacetime, we may assume without loss of generality that $\dot{\theta} = 0$ and set $\theta = \pi/2$, so that γ^a moves in the equatorial plane. We have that the geodesic energy and azimuthal angular momentum $\boldsymbol{e} = -\xi^a \dot{\gamma}_a$ and $\boldsymbol{\ell}_z = \eta^a \dot{\gamma}_a$ are conserved. We have

$$\boldsymbol{\ell}_z = \eta^a \dot{\gamma}^b g_{ab} = r^2 \dot{\phi}.$$

In fact the same is true for the momenta corresponding to each of the three rotational Killing fields. Thus, we may consider the total squared angular momentum $\boldsymbol{L}^2$ given by

$$\boldsymbol{L}^2 = 2r^2 m_{(a} \bar{m}_{b)} \dot{\gamma}^a \dot{\gamma}^b = r^4 (g_{S^2})_{ab} \dot{\gamma}^a \dot{\gamma}^b. \tag{22}$$

[6] There are subtleties concerning the regularity of the conformal boundary of Schwarzschild, and the naive choice of conformal factor mentioned above does not lead to an analytic compactification. See [38] for recent developments.

For geodesics moving in the equatorial plane, we have $\boldsymbol{L}^2 = \boldsymbol{\ell}_z{}^2$. Rewriting $g_{ab}\dot{\gamma}^a\dot{\gamma}^b = 0$ using (3) and these definitions gives

$$\dot{r}^2 + V = \boldsymbol{e}^2, \tag{23}$$

where

$$V = \frac{f}{r^2}\boldsymbol{L}^2.$$

Equation (23) can be viewed as the equation for a particle moving in a potential V.

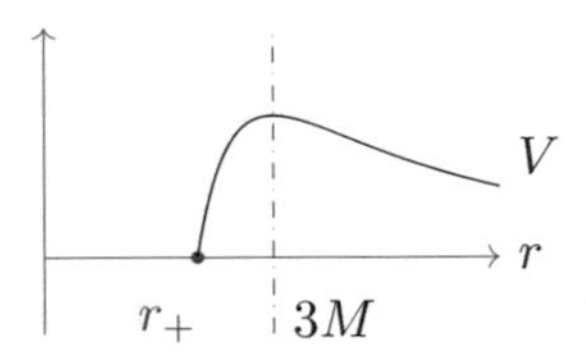

An analysis shows that V has a unique critical point at $r = 3M$, and hence a null geodesic with $\dot{r} = 0$ in the Schwarzschild spacetime must orbit at $r = 3M$. We call such null geodesics trapped. The critical point $r = 3M$ is a local maximum for V and hence the orbiting null geodesics are unstable. The sphere $r = 3M$ is called the *photon sphere*. A similar analysis can be performed for massive particles orbiting the Schwarzschild black hole, see [68, Chapter 6] for further details.

The geometric optics correspondence between waves packets and null geodesics indicates that the phenomenon of trapped null geodesics is an obstacle to dispersion, i.e., the tendency for waves to leave every stationary region. For waves of finite energy, the fact that the trapped orbits are unstable can be used to show that such waves in fact disperse. This is a manifestation of the uncertainty principle.

3.2 Black Hole Stability

The *black hole stability conjecture* states that Cauchy data sufficiently close, in a suitable sense, to Kerr Cauchy data[7] have a maximal development which is future asymptotic to a Kerr spacetime, see Fig. 4. It is important to note that the parameters of the "limiting" Kerr spacetime cannot be determined in any effective manner from the initial data.

As discussed above, cf. Sect. 2.7, if we restrict to axial symmetry, then angular momentum is quasi-locally conserved. This means that if we further restrict to zero angular momentum, the end state of the evolution must be a Schwarzschild

[7] See [5], see also, e.g., [14, 53] for discussions of the problem of characterizing Cauchy data as Kerr data.

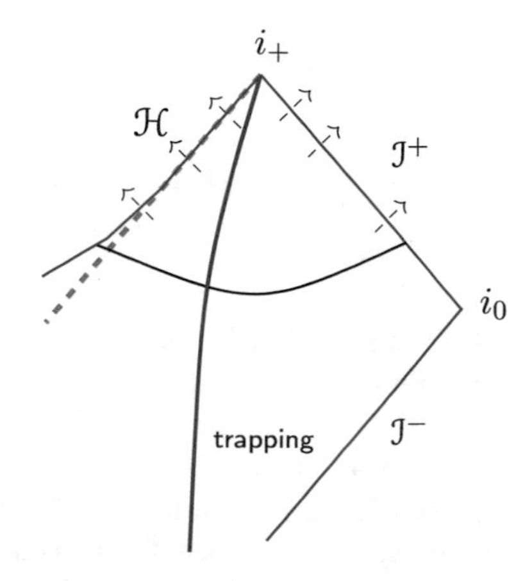

Fig. 4 Causal diagram of a spacetime that is future asymptotic to the Kerr solution

black hole. Thus, the *black hole stability conjecture for the axially symmetric case* is that the maximal developments of sufficiently small (in a suitable sense), axially symmetric, deformations of Schwarzschild Cauchy data with zero angular momentum, are asymptotic to the future to a Schwarzschild spacetime. In this case, due to the loss of energy through $\mathcal{I}^+$, the mass of the "limiting" Schwarzschild black hole cannot be determined directly from the Cauchy data. See Sect. 8.3 below.

3.3 The Kerr Metric

In this section we shall discuss the Kerr metric, which is the main object of our considerations. Although many features of the geometry and analysis on black hole spacetimes are seen in the Schwarzschild case, there are many new and fundamental phenomena present in the Kerr case. Among those are complicated trapping, i.e., the fact that trapped null geodesics fill up an open spacetime region, the fact that the Kerr metric admits only two Killing fields, but a hidden symmetry manifested in the Carter constant, and the fact that the stationary Killing vector field ξ^a fails to be timelike in the whole domain of outer communications, which leads to a lack of a positive conserved energy for waves in the Kerr spacetimes. This fact is the origin of superradiance and the Penrose process. See [67] for a recent survey.

The Kerr metric describes a family of stationary, axisymmetric, asymptotically flat vacuum spacetimes, parametrized by ADM mass M and angular momentum per unit mass a. The expressions for mass and angular momentum introduced in Sect. 2.6 when applied in Kerr geometry yield M and $J = aM$. In Boyer-Lindquist coordinates (t, r, θ, ϕ), the Kerr metric takes the form

$$g_{ab} = \frac{(\Delta - a^2 \sin^2\theta) dt_a dt_b}{\Sigma} + \frac{2a \sin^2\theta (a^2 + r^2 - \Delta) dt_{(a} d\phi_{b)}}{\Sigma} \tag{24}$$

$$- \frac{\Sigma dr_a dr_b}{\Delta} - \Sigma d\theta_a d\theta_b - \frac{\sin^2\theta\big((a^2+r^2)^2 - a^2\sin^2\theta\Delta\big)d\phi_a d\phi_b}{\Sigma},$$

where $\Delta = a^2 - 2Mr + r^2$ and $\Sigma = a^2\cos^2\theta + r^2$. The volume element is

$$\sqrt{|\det g_{ab}|} = \Sigma \sin\theta. \tag{25}$$

There is a ring-shaped singularity at $r = 0$, $\theta = \pi/2$. For $|a| \leq M$, the Kerr spacetime contains a black hole, with event horizon at $r = r_+ \equiv M + \sqrt{M^2 - a^2}$, while for $|a| > M$, the singularity is naked in the sense that it is causally connected to observers at infinity. The area of the horizon is $A_{\mathrm{Hor}} = 4\pi(r_+^2 + a^2)$. This achieves its maximum of $16\pi M^2$ when $a = 0$, providing one of the ingredients in the heuristic argument for the Penrose inequality, see Sect. 3.2. The case $|a| = M$ is called extreme. We shall here be interested only in the subextreme case, $|a| < M$, as this is the only case where we expect black hole stability to hold.

The Boyer-Lindquist coordinates are analogous to the Schwarzschild coordinates Sect. 3.1 and upon setting $a = 0$, (24) reduces to (16). The line element takes a simple form in Boyer-Lindquist coordinates, but similarly for the Schwarzschild coordinates, the Boyer-Lindquist coordinates have the drawback that they are not regular at the horizon.

The Kerr metric admits two Killing vector fields $\xi^a = (\partial_t)^a$ (stationary) and $\eta^a = (\partial_\phi)^a$ (axial). Although the stationary Killing field ξ^a is timelike near infinity, since $g_{ab}\xi^a\xi^b \to 1$ as $r \to \infty$, ξ^a becomes spacelike for r sufficiently small, when $1 - 2M/\Sigma < 0$. In the Schwarzschild case $a = 0$, this occurs at the event horizon $r = 2M$. However, for a rotating Kerr black hole with $0 < |a| \leq M$, there is a region, called the ergoregion, outside the event horizon where ξ^a is spacelike. The ergoregion is bounded by the surface $M + \sqrt{M^2 - a^2\cos^2\theta}$ which touches the horizon at the poles $\theta = 0, \pi$, see Fig. 5. In the ergoregion, null and timelike geodesics can have negative energy with respect to ξ^a. The fact that there is no globally timelike vector field in the Kerr exterior is the origin of superradiance, i.e., the fact that waves which scatter off the black hole can leave the ergoregion with larger energy (as measured by a stationary observer at infinity) than was sent in. This effect was originally found by an analysis based on separation of variables, but can be demonstrated rigorously, see [31]. However, it is a subtle effect and not easy to demonstrate numerically, see [49].

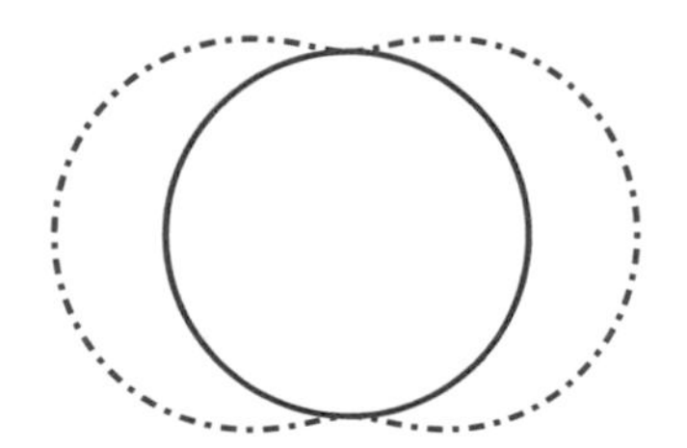

Fig. 5 The ergoregion

Let $\omega_H = a/(r_+^2 + a^2)$ be the rotation speed of the black hole. The Killing field $\chi^a = \xi^a + \omega_H \eta^a$ is null on the event horizon in Kerr, which is, therefore, a Killing horizon. For $|a| < M$, there is a neighborhood of the horizon in the black hole exterior where χ^a is timelike. The surface gravity κ, defined by $\kappa^2 = -\frac{1}{2}(\nabla^a \chi^b)(\nabla_a \chi_b)$ takes the value $\kappa = (r_+ - M)/(r_+^2 + a^2)$, and is in the subextreme case $|a| < M$ nonzero. By general results, a Killing horizon with non-vanishing surface gravity is bifurcate, i.e., there is a cross-section where the null generator vanishes. In the Schwarzschild case, this is the 2-sphere $\mathcal{U} = \mathcal{V} = 0$. See [57, 60] for background on the geometry of the Kerr spacetime, see also [56].

4 Spin Geometry

The 2-spinor formalism, and the closely related GHP formalism, are important tools in Lorentzian geometry and the analysis of black hole spacetimes, and we will introduce them here. A detailed of this material is given by Penrose and Rindler [59]. Following the conventions there, we use the abstract index notation with lower case Latin letters $a, b, c, \dots$ for tensor indices, and unprimed and primed upper-case Latin letters $A, B, C, \dots, A', B', C', \dots$ for spinor indices. Tetrad and dyad indices are boldface Latin letters following the same scheme, $\mathbf{a}, \mathbf{b}, \mathbf{c}, \dots, \mathbf{A}, \mathbf{B}, \mathbf{C}, \dots, \mathbf{A}', \mathbf{B}', \mathbf{C}', \dots$. For coordinate indices we use Greek letters $\alpha, \beta, \gamma, \dots$.

4.1 Spinors on Minkowski Space

Consider Minkowski space $\mathbb{M}$, i.e., $\mathbb{R}^4$ with coordinates $(x^\alpha) = (t, x, y, z)$ and metric

$$g_{\alpha\beta} dx^\alpha dx^\beta = dt^2 - dx^2 - dy^2 - dz^2.$$

Define a complex null tetrad (i.e., frame) $(g_{\mathbf{a}}{}^a)_{\mathbf{a}=0,\cdots,3} = (l^a, n^a, m^a, \bar{m}^a)$, as in (2) above, normalized so that $l^a n_a = 1$, $m^a \bar{m}_a = -1$, so that

$$g_{ab} = 2(l_{(a} n_{b)} - m_{(a} \bar{m}_{b)}). \tag{26}$$

Similarly, let $\epsilon_{\mathbf{A}}{}^A$ be a dyad (i.e., frame) in $\mathbb{C}^2$, with dual frame $\epsilon_A{}^{\mathbf{A}}$. The complex conjugates will be denoted $\bar{\epsilon}_{\mathbf{A}'}{}^{A'}, \bar{\epsilon}_{A'}{}^{\mathbf{A}'}$ and again form a basis in another 2-dimensional complex space denoted $\bar{\mathbb{C}}^2$, and its dual. We can identify the space of complex 2×2 matrices with $\mathbb{C}^2 \otimes \bar{\mathbb{C}}^2$. By construction, the tensor products $\epsilon_{\mathbf{A}}{}^A \bar{\epsilon}_{\mathbf{A}'}{}^{A'}$ and $\epsilon_A{}^{\mathbf{A}} \bar{\epsilon}_{A'}{}^{\mathbf{A}'}$ forms a basis in $\mathbb{C}^2 \otimes \bar{\mathbb{C}}^2$ and its dual.

Now, with $x^{\mathbf{a}} = x^a g_a{}^{\mathbf{a}}$, writing

$$x^{\mathbf{a}} g_{\mathbf{a}}{}^{\mathbf{AA}'} \equiv \begin{pmatrix} x^0 & x^2 \\ x^3 & x^1 \end{pmatrix} \tag{27}$$

defines the soldering forms, also known as Infeld-van der Waerden symbols $g_a{}^{AA'}$, (and analogously $g_{AA'}{}^a$). By a slight abuse of notation we may write $x^{AA'} = x^a$ instead of $x^{\mathbf{AA}'} = x^{\mathbf{a}} g_{\mathbf{a}}{}^{\mathbf{AA}'}$ or, dropping reference to the tetrad, $x^{AA'} = x^a g_a{}^{AA'}$. In particular, we have that $x^a \in \mathbb{M}$ corresponds to a 2×2 complex Hermitian matrix $x^{\mathbf{AA}'} \in \mathbb{C}^2 \otimes \bar{\mathbb{C}}^2$. Taking the complex conjugate of both sides of (27) gives

$$\bar{x}^a = \bar{x}^{A'A} = (x^{AA'})^*,$$

where $*$ denotes Hermitian conjugation. This extends to a correspondence $\mathbb{C}^4 \leftrightarrow \mathbb{C}^2 \otimes \bar{\mathbb{C}}^2$ with complex conjugation corresponding to Hermitian conjugation.

Note that

$$\det(x^{\mathbf{AA}'}) = x^0 x^1 - x^2 x^3 = x^a x_a / 2. \tag{28}$$

We see from the above that the group

$$\mathrm{SL}(2, \mathbb{C}) = \left\{ A = \begin{pmatrix} a & b \\ c & d \end{pmatrix}, \quad a, b, c, d \in \mathbb{C}, \quad ad - bc = 1 \right\}$$

acts on $X \in \mathbb{C}^2 \otimes \bar{\mathbb{C}}^2$ by

$$X \mapsto AXA^*.$$

In view of (28) this exhibits $\mathrm{SL}(2, \mathbb{C})$ as a double cover of the identity component of the Lorentz group $\mathrm{SO}_0(1, 3)$, the group of linear isometries of $\mathbb{M}$. In particular, $\mathrm{SL}(2, \mathbb{C})$ is the spin group of $\mathbb{M}$. The canonical action

$$(A, v) \in \mathrm{SL}(2, \mathbb{C}) \times \mathbb{C}^2 \mapsto Av \in \mathbb{C}^2$$

of $\mathrm{SL}(2, \mathbb{C})$ on $\mathbb{C}^2$ is the spinor representation. Elements of $\mathbb{C}^2$ are called (Weyl) spinors. The conjugate representation given by

$$(A, v) \in \mathrm{SL}(2, \mathbb{C}) \times \mathbb{C}^2 \mapsto \bar{A}v \in \mathbb{C}^2$$

is denoted $\bar{\mathbb{C}}^2$.

Spinors[8] of the form $x^{AA'} = \alpha^A \beta^{A'}$ correspond to matrices of rank one, and hence to complex null vectors. Denoting $o^A = \epsilon_{\mathbf{0}}{}^A$, $\iota^A = \epsilon_{\mathbf{1}}{}^A$, we have from the above that

$$l^a = o^A o^{A'}, \quad n^a = \iota^A \iota^{A'}, \quad m^a = o^A \iota^{A'}, \quad \bar{m}^a = \iota^A o^{A'}. \tag{29}$$

This gives a correspondence between a null frame in $\mathbb{M}$ and a dyad in $\mathbb{C}^2$.

The action of SL(2, $\mathbb{C}$) on $\mathbb{C}^2$ leaves invariant a complex area element, a skew symmetric bispinor. A unique such spinor ϵ_{AB} is determined by the normalization

$$g_{ab} = \epsilon_{AB} \bar{\epsilon}_{A'B'}.$$

The inverse ϵ^{AB} of ϵ_{AB} is defined by $\epsilon_{AB}\epsilon^{CB} = \delta_A{}^C$, $\epsilon^{AB}\epsilon_{AC} = \delta_C{}^B$. As with g_{ab} and its inverse g^{ab}, the spin-metric ϵ_{AB} and its inverse ϵ^{AB} is used to lower and raise spinor indices,

$$\lambda_B = \lambda^A \epsilon_{AB}, \quad \lambda^A = \epsilon^{AB} \lambda_B.$$

We have

$$\epsilon_{AB} = o_A \iota_B - \iota_A o_B.$$

In particular,

$$o_A \iota^A = 1. \tag{30}$$

An element $\phi_{A\cdots DA'\cdots D'}$ of $\bigotimes^k \mathbb{C}^2 \bigotimes^l \bar{\mathbb{C}}^2$ is called a spinor of valence (k, l). The space of totally symmetric[9] spinors $\phi_{A\cdots DA'\cdots D'} = \phi_{(A\cdots D)(A'\cdots D')}$ is denoted $S_{k,l}$. The spaces $S_{k,l}$ for k, l non-negative integers yield all irreducible representations of SL(2, $\mathbb{C}$). In fact, one can decompose any spinor into "irreducible pieces," i.e., as a linear combination of totally symmetric spinors in $S_{k,l}$ with factors of ϵ_{AB}. The above mentioned correspondence between vectors and spinors extends to tensors of any type, and hence the just mentioned decomposition of spinors into irreducible pieces carries over to tensors as well. Examples are given by $\mathcal{F}_{ab} = \phi_{AB}\epsilon_{A'B'}$, a complex anti-self-dual 2-form, and ${}^{-}C_{abcd} = \Psi_{ABCD}\epsilon_{A'B'}\epsilon_{C'D'}$, a complex anti-self-dual tensor with the symmetries of the Weyl tensor. Here, ϕ_{AB} and Ψ_{ABCD} are symmetric.

[8] It is conventional to refer to spin-tensors, e.g., of the form $x^{AA'}$ or $\psi_{ABA'}$ simply as spinors.

[9] The ordering between primed and unprimed indices is irrelevant.

4.2 Spinors on Spacetime

Let now $(\mathcal{M}, g_{ab})$ be a Lorentzian 3 + 1 dimensional spin manifold with metric of signature $+ - - -$. The spacetimes we are interested in here are spin, in particular any orientable, globally hyperbolic 3 + 1 dimensional spacetime is spin, cf. [35, page 346]. If $\mathcal{M}$ is spin, then the orthonormal frame bundle SO($\mathcal{M}$) admits a lift to Spin($\mathcal{M}$), a principal SL(2, $\mathbb{C}$)-bundle. The associated bundle construction now gives vector bundles over $\mathcal{M}$ corresponding to the representations of SL(2, $\mathbb{C}$), in particular we have bundles of valence (k, l) spinors with sections $\phi_{A\cdots DA'\cdots D'}$. The Levi-Civita connection lifts to act on sections of the spinor bundles,

$$\nabla_{AA'} : \varphi_{B\cdots DB'\cdots D'} \to \nabla_{AA'}\varphi_{B\cdots DB'\cdots D'}, \tag{31}$$

where we have used the tensor-spinor correspondence to replace the index a by AA'. We shall denote the totally symmetric spinor bundles by $S_{k,l}$ and their spaces of sections by $\mathcal{S}_{k,l}$.

The above mentioned correspondence between spinors and tensors, and the decomposition into irreducible pieces, can be applied to the Riemann curvature tensor. In this case, the irreducible pieces correspond to the scalar curvature, traceless Ricci tensor, and the Weyl tensor, denoted by R, S_{ab}, and C_{abcd}, respectively. The Riemann tensor then takes the form

$$\begin{aligned} R_{abcd} = & -\tfrac{1}{12} g_{ad} g_{bc} R + \tfrac{1}{12} g_{ac} g_{bd} R + \tfrac{1}{2} g_{bd} S_{ac} - \tfrac{1}{2} g_{bc} S_{ad} \\ & - \tfrac{1}{2} g_{ad} S_{bc} + \tfrac{1}{2} g_{ac} S_{bd} + C_{abcd}. \end{aligned} \tag{32}$$

The spinor equivalents of these tensors are

$$C_{abcd} = \Psi_{ABCD}\bar{\epsilon}_{A'B'}\bar{\epsilon}_{C'D'} + \bar{\Psi}_{A'B'C'D'}\epsilon_{AB}\epsilon_{CD}, \tag{33a}$$

$$S_{ab} = -2\Phi_{ABA'B'}, \tag{33b}$$

$$R = 24\Lambda. \tag{33c}$$

4.3 Fundamental Operators

Projecting (31) on its irreducible pieces gives the following four *fundamental operators*, introduced in [4].

Definition 4.1 The differential operators

$$\mathscr{D}_{k,l} : \mathcal{S}_{k,l} \to \mathcal{S}_{k-1,l-1}, \quad \mathscr{C}_{k,l} : \mathcal{S}_{k,l} \to \mathcal{S}_{k+1,l-1},$$

$$\mathscr{C}^{\dagger}_{k,l} : \mathcal{S}_{k,l} \to \mathcal{S}_{k-1,l+1}, \quad \mathscr{T}_{k,l} : \mathcal{S}_{k,l} \to \mathcal{S}_{k+1,l+1}$$

are defined as

$$(\mathscr{D}_{k,l}\varphi)_{A_1\ldots A_{k-1}}{}^{A'_1\ldots A'_{l-1}} \equiv \nabla^{BB'}\varphi_{A_1\ldots A_{k-1}B}{}^{A'_1\ldots A'_{l-1}}{}_{B'}, \tag{34a}$$

$$(\mathscr{C}_{k,l}\varphi)_{A_1\ldots A_{k+1}}{}^{A'_1\ldots A'_{l-1}} \equiv \nabla_{(A_1}{}^{B'}\varphi_{A_2\ldots A_{k+1})}{}^{A'_1\ldots A'_{l-1}}{}_{B'}, \tag{34b}$$

$$(\mathscr{C}^\dagger_{k,l}\varphi)_{A_1\ldots A_{k-1}}{}^{A'_1\ldots A'_{l+1}} \equiv \nabla^{B(A'_1}\varphi_{A_1\ldots A_{k-1}B}{}^{A'_2\ldots A'_{l+1})}, \tag{34c}$$

$$(\mathscr{T}_{k,l}\varphi)_{A_1\ldots A_{k+1}}{}^{A'_1\ldots A'_{l+1}} \equiv \nabla_{(A_1}{}^{(A'_1}\varphi_{A_2\ldots A_{k+1})}{}^{A'_2\ldots A'_{l+1})}. \tag{34d}$$

The operators are called, respectively, the divergence, curl, curl-dagger, and twistor operators.

As we will see in Sect. 4.4, the kernels of $\mathscr{C}^\dagger_{2s,0}$ and $\mathscr{C}_{0,2s}$ are the massless spin-s fields. The kernels of $\mathscr{T}_{k,l}$, are the valence (k, l) Killing spinors, which we will discuss further in Sects. 4.5 and 4.7. A complete set of commutator properties of these operators can be found in [4].

4.4 Massless Spin-s Fields

For $s \in \frac{1}{2}\mathbb{N}$, $\varphi_{A\cdots D} \in \ker \mathscr{C}^\dagger_{2s,0}$ is a totally symmetric spinor $\varphi_{A\cdots D} = \varphi_{(A\cdots D)}$ of valence $(2s, 0)$ which solves the massless spin-s equation

$$(\mathscr{C}^\dagger_{2s,0}\varphi)_{A\cdots BD'} = 0.$$

For $s = 1/2$, this is the Dirac-Weyl equation $\nabla_{A'}{}^A\varphi_A = 0$, for $s = 1$, we have the left and right Maxwell equation $\nabla_{A'}{}^B\phi_{AB} = 0$ and $\nabla_A{}^{B'}\varphi_{A'B'} = 0$, i.e., $(\mathscr{C}^\dagger_{2,0}\phi)_{AA'} = 0$, $(\mathscr{C}_{0,2}\varphi)_{AA'} = 0$.

An important example is the Coulomb Maxwell field on Kerr,

$$\phi_{AB} = -\frac{2}{(r - ia\cos\theta)^2} o_{(A}\iota_{B)}. \tag{35}$$

This is a non-trivial sourceless solution of the Maxwell equation on the Kerr background. We note that the scalars components, see Sect. 4.8 below, of the Coulomb field $\phi_1 = (r - ia\cos\theta)^{-2}$ while $\phi_0 = \phi_2 = 0$.

For $s > 1$, the existence of a non-trivial solution to the spin-s equation implies curvature conditions, a fact known as the Buchdahl constraint [20],

$$0 = \Psi_{(A}{}^{DEF}\phi_{B\ldots C)DEF}. \tag{36}$$

This is easily obtained by commuting the operators in

$$0 = (\mathscr{D}_{2s-1,1}\mathscr{C}^\dagger_{2s,0}\phi)_{A\ldots C}. \tag{37}$$

For the case $s = 2$, the equation $\nabla_{A'}{}^{D}\Psi_{ABCD} = 0$ is the Bianchi equation, which holds for the Weyl spinor in any vacuum spacetime. Due to the Buchdahl constraint, it holds that in any sufficiently general spacetime, a solution of the spin-2 equation is proportional to the Weyl spinor of the spacetime.

4.5 Killing Spinors

Spinors $\varkappa_{A_1\cdots A_k}{}^{A'_1\cdots A'_l} \in \mathcal{S}_{k,l}$ satisfying

$$(\mathcal{T}_{k,l}\varkappa)_{A_1\cdots A_{k+1}}{}^{A'_1\cdots A'_{l+1}} = 0$$

are called Killing spinors of valence (k, l). We denote the space of Killing spinors of valence (k, l) by $\mathcal{KS}_{k,l}$. The Killing spinor equation is an over-determined system. The space of Killing spinors is a finite dimensional space, and the existence of Killing spinors imposes strong restrictions on $\mathcal{M}$, see Sect. 4.7 below. Killing spinors $\nu_{AA'} \in \mathcal{KS}_{1,1}$ are simply conformal Killing vector fields, satisfying $\nabla_{(a}\nu_{b)} - \frac{1}{2}\nabla^c\nu_c g_{ab}$. A Killing spinor $\kappa_{AB} \in \mathcal{KS}_{2,0}$ corresponds to a complex anti-selfdual conformal Killing-Yano 2-form $\mathcal{Y}_{ABA'B'} = \kappa_{AB}\epsilon_{A'B'}$ satisfying the equation

$$\nabla_{(a}\mathcal{Y}_{b)c} - 2\zeta_c g_{ab} + \zeta_{(a}g_{b)c} = 0, \tag{38}$$

where in the 4-dimensional case, $\zeta_a = \frac{1}{3}\nabla_b\mathcal{Y}^b{}_a$.

4.6 Algebraically Special Spacetimes

Let $\varphi_{A\cdots D} \in \mathcal{S}_{k,0}$. A spinor α_A is a *principal spinor* of $\varphi_{A\cdots D}$ if

$$\varphi_{A\cdots D}\alpha^A \cdots \alpha^D = 0.$$

An application of the fundamental theorem of algebra shows that any $\varphi_{A\cdots D} \in \mathcal{S}_{k,0}$ has exactly k principal spinors $\alpha_A, \ldots, \delta_A$, and hence is of the form

$$\varphi_{A\cdots D} = \alpha_{(A} \cdots \delta_{D)}.$$

If $\varphi_{A\cdots D} \in \mathcal{S}_{k,0}$ has n distinct principal spinors $\alpha^{(i)}_A$, repeated m_i times, then $\varphi_{A\cdots D}$ is said to have algebraic type $\{m_1, \ldots, m_n\}$. Applying this to the Weyl tensor leads to the Petrov classification, see Table 1. We have the following list of algebraic, or Petrov, types.[10]

[10] The Petrov classification is exclusive, so a spacetime belongs at each point to exactly one Petrov class.

Table 1 The Petrov classification

I	$\{1, 1, 1, 1\}$	$\Psi_{ABCD} = \alpha_{(A}\beta_B\gamma_C\delta_{D)}$
II	$\{2, 1, 1\}$	$\Psi_{ABCD} = \alpha_{(A}\alpha_B\gamma_C\delta_{D)}$
D	$\{2, 2\}$	$\Psi_{ABCD} = \alpha_{(A}\alpha_B\beta_C\beta_{D)}$
III	$\{3, 1\}$	$\Psi_{ABCD} = \alpha_{(A}\alpha_B\alpha_C\beta_{D)}$
N	$\{4\}$	$\Psi_{ABCD} = \alpha_A\alpha_B\alpha_C\alpha_D$
O	$\{-\}$	$\Psi_{ABCD} = 0$

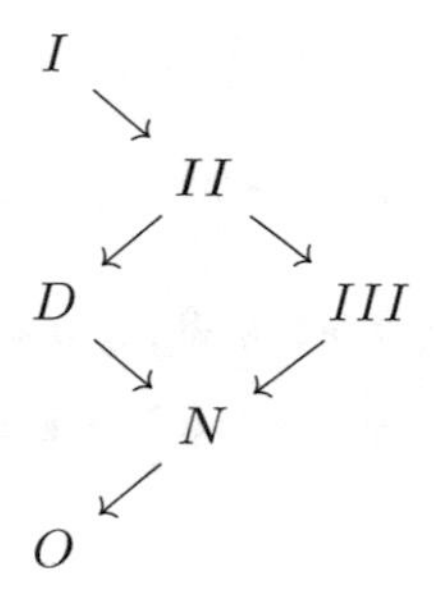

A principal spinor o_A determines a principal null direction $l_a = o_A\bar{o}_{A'}$. The Goldberg-Sachs theorem states that in a vacuum spacetime, the congruence generated by a null field l^a is geodetic and shear free[11] if and only if l_a is a repeated principal null direction of the Weyl tensor C_{abcd} (or equivalently o_A is a repeated principal spinor of the Weyl spinor Ψ_{ABCD}).

4.6.1 Petrov Type *D*

The Kerr metric is of Petrov type D, and many of its important properties follows from this fact. The vacuum type *D* spacetimes have been classified by Kinnersley [43], see also Edgar et al. [29]. The family of Petrov type D spacetimes includes the Kerr-NUT family and the boost-rotation symmetric C-metrics. The only Petrov type D vacuum spacetime which is asymptotically flat and has positive mass is the Kerr metric, see Theorem 5.1 below.

A Petrov type *D* spacetime has two repeated principal spinors o_A, ι_A, and correspondingly there are two repeated principal null directions l^a, n^a, for the Weyl tensor. We can without loss of generality assume that $l^a n_a = 1$, and define a null tetrad by adding complex null vectors $m^a, \bar{m}^a$ normalized such that $m^a\bar{m}_a = -1$. By the Goldberg-Sachs theorem both l^a, n^a are geodetic and shear free, and only one of the 5 independent complex Weyl scalars is non-zero, namely

$$\Psi_2 = -\, l^a m^b \bar{m}^d n^c C_{abcd}. \tag{39}$$

[11] If l^a is geodetic and shear then the spin coefficients σ, κ, cf. (50) below, satisfy $\sigma = \kappa = 0$.

In this case, the Weyl spinor takes the form

$$\Psi_{ABCD} = \frac{1}{6}\Psi_2 o_{(A} o_B \iota_C \iota_{D)}.$$

See (53) below for the explicit form of Ψ_2 in the Kerr spacetime.

The following result is a consequence of the Bianchi identity.

Theorem 4.2 ([70]) *Assume* $(\mathcal{M}, g_{ab})$ *is a vacuum spacetime of Petrov type D. Then* $(\mathcal{M}, g_{ab})$ *admits a one-dimensional space of Killing spinors* κ_{AB} *of the form*

$$\kappa_{AB} = -2\kappa_1 o_{(A} \iota_{B)}, \tag{40}$$

where o_A, ι_A *are the principal spinors of* Ψ_{ABCD} *and* $\kappa_1 \propto \Psi_2^{-1/3}$.

Remark 4.3 Since the Petrov classes are exclusive, we have that $\Psi_2 \neq 0$ for a Petrov type D space.

4.7 Spacetimes Admitting a Killing Spinor

Differentiating the Killing spinor equation $(\mathscr{T}_{k,l}\phi)_{A\cdots DA'\cdots D'} = 0$, and commuting derivatives yields an algebraic relation between the curvature, Killing spinor, and their covariant derivatives which restrict the curvature spinor, see [4, §2.3], see also [5, §3.2]. In particular, for a Killing spinor $\kappa_{A\cdots D}$ of valence $(k, 0)$, $k \geq 1$, the condition

$$\Psi_{(ABC}{}^F \kappa_{D\cdots E)F} = 0 \tag{41}$$

must hold, which restricts the algebraic type of the Weyl spinor. For a valence $(2, 0)$ Killing spinor κ_{AB}, the condition takes the form

$$\Psi_{(ABC}{}^E \kappa_{D)E} = 0. \tag{42}$$

It follows from (42) that a spacetime admitting a valence $(2, 0)$ Killing spinor is of type D, N, or O. The space of Killing spinors of valence $(2, 0)$ on Minkowski space (or any space of Petrov type O) has complex dimension 10. The explicit form in Cartesian coordinates $x^{AA'}$ is

$$\kappa^{AB} = U^{AB} + 2x^{A'(A} V^{B)}{}_{A'} + x^{AA'} x^{BB'} W_{A'B'},$$

where U^{AB}, $V^B{}_{A'}$, $W^{A'B'}$ are arbitrary constant symmetric spinors, see [1, Eq. (4.5)]. One of these corresponds to the spinor in (40), in spheroidal coordinates it takes the form given in (54) below.

A further application of the commutation properties of the fundamental operators yields that the 1-form

$$\xi_{AA'} = (\mathscr{C}^\dagger_{2,0}\kappa)_{AA'} \tag{43}$$

is a Killing field, $\nabla_{(a}\xi_{b)} = 0$, provided $\mathcal{M}$ is vacuum. Clearly the real and imaginary parts of ξ_a are also Killing fields. If ξ_a is proportional to a real Killing field,[12] we can without loss of generality assume that ξ_a is real. In this case, the 2-form

$$Y_{ab} = \tfrac{3}{2}i(\kappa_{AB}\bar{\epsilon}_{A'B'} - \bar{\kappa}_{A'B'}\epsilon_{AB}) \tag{44}$$

is a Killing-Yano tensor, $\nabla_{(a}Y_{b)c} = 0$, and the symmetric 2-tensor

$$K_{ab} = Y_a{}^c Y_{cb} \tag{45}$$

is a Killing tensor,

$$\nabla_{(a}K_{bc)} = 0. \tag{46}$$

Further, in this case,

$$\zeta_a = \xi^b K_{ab} \tag{47}$$

is a Killing field, see [24, 40]. Recall that the quantity $L_{ab}\dot{\gamma}^a\dot{\gamma}^b$ is conserved along null geodesics if L_{ab} is a conformal Killing tensor. For Killing tensors, this fact extends to all geodesics, so that if K_{ab} is a Killing tensor, then $K_{ab}\dot{\gamma}^a\dot{\gamma}^b$ is conserved along a geodesic γ^a. See [5] for further details and references.

4.8 GHP Formalism

Taking the point of view that the null tetrad components of tensors are sections of complex line bundles with action of the non-vanishing complex scalars corresponding to the rescalings of the tetrad, respecting the normalization, leads to the GHP formalism [36].

Given a null tetrad $l^a, n^a, m^a, \bar{m}^a$ we have a spin dyad o_A, ι_A as discussed above. For a spinor $\varphi_{A\cdots D} \in \mathcal{S}_{k,0}$, it is convenient to introduce the Newman-Penrose scalars

$$\varphi_i = \varphi_{A_1\cdots A_i A_{i+1}\cdots A_k}\iota^{A_1}\cdots\iota^{A_i}o^{A_{i+1}}\cdots o^{A_k}. \tag{48}$$

---

[12] We say that such spacetimes are of the generalized Kerr-NUT class, see [12] and the references therein.

In particular, Ψ_{ABCD} corresponds to the five complex Weyl scalars $\Psi_i, i = 0, \ldots 4$. The definition φ_i extends in a natural way to the scalar components of spinors of valence (k, l).

The normalization (30) is left invariant under rescalings $o_A \to \lambda o_A, \iota_A \to \lambda^{-1}\iota_A$ where λ is a non-vanishing complex scalar field on $\mathcal{M}$. Under such rescalings, the scalars defined by projecting on the dyad, such as φ_i given by (48) transform as sections of complex line bundles. A scalar φ is said to have type $\{p, q\}$ if $\varphi \to \lambda^p \bar{\lambda}^q \varphi$ under such a rescaling. Such fields are called properly weighted. The lift of the Levi-Civita connection $\nabla_{AA'}$ to these bundles gives a covariant derivative denoted Θ_a. Projecting on the null tetrad $l^a, n^a, m^a, \bar{m}^a$ gives the GHP operators

$$þ = l^a \Theta_a, \quad þ' = n^a \Theta_a, \quad ð = m^a \Theta_a, \quad ð' = \bar{m}^a \Theta_a.$$

The GHP operators are properly weighted, in the sense that they take properly weighted fields to properly weighted fields, for example, if φ has type $\{p, q\}$, then $þ\varphi$ has type $\{p + 1, q + 1\}$. This can be seen from the fact that $l^a = o^A \bar{o}^{A'}$ has type $\{1, 1\}$. There are 12 connection coefficients in a null frame, up to complex conjugation. Of these, 8 are properly weighted, the GHP spin coefficients. The other connection coefficients enter in the connection 1-form for the connection Θ_a.

The following formal operations take weighted quantities to weighted quantities,

$$\begin{aligned} &{}^-(\text{bar}) : l^a \to l^a,\ n^a \to n^a,\ m^a \to \bar{m}^a,\ \bar{m}^a \to m^a, && \{p, q\} \to \{q, p\}, \\ &{}'(\text{prime}) : l^a \to n^a,\ n^a \to l^a,\ m^a \to \bar{m}^a,\ \bar{m}^a \to m^a, && \{p, q\} \to \{-p, -q\}, \\ &{}^*(\text{star}) : l^a \to m^a,\ n^a \to -\bar{m}^a,\ m^a \to -l^a,\ \bar{m}^a \to n^a, && \{p, q\} \to \{p, -q\}. \end{aligned} \tag{49}$$

The properly weighted spin coefficients can be represented as

$$\kappa = m^b l^a \nabla_a l_b, \quad \sigma = m^b m^a \nabla_a l_b, \quad \rho = m^b \bar{m}^a \nabla_a l_b, \quad \tau = m^b n^a \nabla_a l_b, \tag{50}$$

together with their primes $\kappa', \sigma', \rho', \tau'$.

A systematic application of the above formalism allows one to write the tetrad projection of the geometric field equations in a compact form. For example, the Maxwell equation corresponds to the four scalar equations given by

$$(þ - 2\rho)\phi_1 - (ð' - \tau')\phi_0 = -\kappa\phi_2, \tag{51}$$

with its primed and starred versions.

Working in a spacetime of Petrov type D gives drastic simplifications, in view of the fact that choosing the null tetrad so that l^a, n^a are aligned with principal null directions of the Weyl tensor (or equivalently choosing the spin dyad so that o_A, ι_A are principal spinors of the Weyl spinor), as has already been mentioned, the Weyl scalars are zero with the exception of Ψ_2, and the only non-zero spin coefficients are ρ, τ and their primed versions.

5 The Kerr Spacetime

Taking into account the background material given in Sect. 4, we can now state some further properties of the Kerr spacetime. As mentioned above, the Kerr metric is algebraically special, of Petrov type D. An explicit principal null tetrad $(l^a, n^a, m^a, \bar{m}^a)$ is given by the Carter tetrad [74]

$$l^a = \frac{a(\partial_\phi)^a}{\sqrt{2}\Delta^{1/2}\Sigma^{1/2}} + \frac{(a^2+r^2)(\partial_t)^a}{\sqrt{2}\Delta^{1/2}\Sigma^{1/2}} + \frac{\Delta^{1/2}(\partial_r)^a}{\sqrt{2}\Sigma^{1/2}}, \tag{52a}$$

$$n^a = \frac{a(\partial_\phi)^a}{\sqrt{2}\Delta^{1/2}\Sigma^{1/2}} + \frac{(a^2+r^2)(\partial_t)^a}{\sqrt{2}\Delta^{1/2}\Sigma^{1/2}} - \frac{\Delta^{1/2}(\partial_r)^a}{\sqrt{2}\Sigma^{1/2}}, \tag{52b}$$

$$m^a = \frac{(\partial_\theta)^a}{\sqrt{2}\Sigma^{1/2}} + \frac{i\csc\theta(\partial_\phi)^a}{\sqrt{2}\Sigma^{1/2}} + \frac{ia\sin\theta(\partial_t)^a}{\sqrt{2}\Sigma^{1/2}}. \tag{52c}$$

In view of the normalization of the tetrad, the metric takes the form $g_{ab} = 2(l_{(a}n_{b)} - m_{(a}\bar{m}_{b)})$. We remark that the choice of l^a, n^a to be aligned with the principal null directions of the Weyl tensor, together with the normalization of the tetrad fixes the tetrad up to rescalings.

We have

$$\Psi_2 = -\frac{M}{(r - ia\cos\theta)^3}, \tag{53}$$

$$\kappa_{AB} = \tfrac{2}{3}(r - ia\cos\theta)o_{(A}\iota_{B)}. \tag{54}$$

With κ_{AB} as in (54), Eq. (43) yields

$$\xi^a = (\partial_t)^a, \tag{55}$$

and from (44) we get

$$Y_{ab} = a\cos\theta l_{[a}n_{b]} - irm_{[a}\bar{m}_{b]}. \tag{56}$$

With the normalizations above, the Killing tensor (45) takes the form

$$K_{ab} = \tfrac{1}{4}(2\Sigma l_{(a}n_{b)} - r^2 g_{ab}) \tag{57}$$

and (47) gives

$$\zeta^a = a^2(\partial_t)^a + a(\partial_\phi)^a. \tag{58}$$

Recall that for a geodesic γ, the quantity $\boldsymbol{k} = 4K_{ab}\dot{\gamma}^a\dot{\gamma}^b$, known as Carter's constant, is conserved. Explicitly,

$$\boldsymbol{k} = \dot{\gamma}_\theta^2 + a^2 \sin^2\theta \boldsymbol{e}^2 + 2a\boldsymbol{e}\boldsymbol{\ell}_z + a^2\cos^2\theta \boldsymbol{\mu}^2, \tag{59}$$

where $\dot{\gamma}_\theta = \dot{\gamma}^a(\partial_\theta)_a$. For $a \neq 0$, the tensor K_{ab} cannot be expressed as a tensor product of Killing fields [70], and similarly Carter's constant $\boldsymbol{k}$ cannot be expressed in terms of the constants of motion associated to Killing fields. In this sense K_{ab} and $\boldsymbol{k}$ manifest a *hidden symmetry* of the Kerr spacetime. As we shall see in Sect. 7, these structures are also related to symmetry operators and separability properties, as well as conservation laws, for field equations on Kerr, and more generally in spacetimes admitting Killing spinors satisfying certain auxiliary conditions.

5.1 Characterizations of Kerr

Consider a vacuum Cauchy data set (Σ, h_{ij}, k_{ij}). We say that (Σ, h_{ij}, k_{ij}) is asymptotically flat if Σ has an end $\mathbb{R}^3 \setminus B(0, R)$ with a coordinate system (x^i) such that

$$h_{ij} = \delta_{ij} + O_\infty(r^\alpha), \quad k_{ij} = O_\infty(r^{\alpha-1}) \tag{60}$$

for some $\alpha < -1/2$. The Cauchy data set (Σ, h_{ij}, k_{ij}) is asymptotically Schwarzschildean if

$$h_{ij} = -\left(1 + \frac{2A}{r}\right)\delta_{ij} - \frac{\alpha}{r}\left(\frac{2x_i x_j}{r^2} - \delta_{ij}\right) + o_\infty(r^{-3/2}), \tag{61a}$$

$$k_{ij} = \frac{\beta}{r^2}\left(\frac{2x_i x_j}{r^2} - \delta_{ij}\right) + o_\infty(r^{-5/2}), \tag{61b}$$

where A is a constant, and α, β are functions on S^2, see [13, §6.5] for details. Here, the symbols $o_\infty(r^\alpha)$ are defined in terms of weighted Sobolev spaces, see [13, §6.2] for details.

If $(\mathcal{M}, g_{ab})$ is vacuum and contains a Cauchy surface (Σ, h_{ij}, k_{ij}) satisfying (60) or (61), then $(\mathcal{M}, g_{ab})$ is asymptotically flat, respectively, asymptotically Schwarzschildean, at spatial infinity. In this case there is a spacetime coordinate system (x^α) such that $g_{\alpha\beta}$ is asymptotic to the Minkowski line element with asymptotic conditions compatible with (61). For such spacetimes, the ADM 4-momentum P^μ is well defined. The positive mass theorem states that P^μ is future directed causal $P^\mu P_\mu \geq 0$ (where the contraction is in the asymptotic Minkowski line element), $P^0 \geq 0$, and gives conditions under which P^μ is strictly timelike. This holds in particular if Σ contains an apparent horizon.

Mars [52] has given a characterization of the Kerr spacetime as an asymptotically flat vacuum spacetime with a Killing field ξ^a asymptotic to a time translation, positive mass, and an additional condition on the Killing form $F_{AB} = (\mathscr{C}_{1,1}\xi)_{AB}$,

$$\Psi_{ABCD}F^{CD} \propto F_{AB}.$$

A characterization in terms of algebraic invariants of the Weyl tensor has been given by Ferrando and Saez [30]. The just mentioned characterizations are in terms of spacetime quantities. It can be shown that Killing spinor initial data propagates, which can be used to formulate a characterization of Kerr in terms of Cauchy data. See [12–15].

We here give a characterization in terms spacetimes admitting a Killing spinor of valence $(2, 0)$.

Theorem 5.1 *Assume that* $(\mathcal{M}, g_{ab})$ *is vacuum, asymptotically Schwarzschildean at spacelike infinity, and contains a Cauchy slice bounded by an apparent horizon. Assume further* $(\mathcal{M}, g_{ab})$ *admits a non-vanishing Killing spinor* κ_{AB} *of valence* $(2, 0)$. *Then* $(\mathcal{M}, g_{ab})$ *is locally isometric to the Kerr spacetime.*

Proof Let P^μ be the ADM 4-momentum vector for $\mathcal{M}$. By the positive mass theorem, $P^\mu P_\mu \geq 0$. In the case where $\mathcal{M}$ contains a Cauchy surface bounded by an apparent horizon, then $P^\mu P_\mu > 0$ by [16, Remark 11.5].[13]

Recall that a spacetime with a Killing spinor of valence $(2, 0)$ is of Petrov type D, N, or O. From asymptotic flatness and the positive mass theorem, we have $C_{abcd}C^{abcd} = O(1/r^6)$, and hence there is a neighborhood of spatial infinity where $\mathcal{M}$ is Petrov type D. It follows that near spatial infinity, $\kappa_{AB} = -2\kappa_1 o_{(A}\iota_{B)}$, with $\kappa_1 \propto \Psi_2^{-1/3} = O(r)$. It follows from our asymptotic conditions that the Killing field $\xi_{AA'} = (\mathscr{C}^\dagger_{2,0}\kappa)_{AB}$ is $O(1)$ and hence asymptotic to a translation, $\xi^\mu \to A^\mu$ as $r \to \infty$, for some constant vector A^μ. It follows from the discussion in [2, §4] that A^μ is non-vanishing. Now, by [17, §III], it follows that in the case $P^\mu P_\mu > 0$, then A^μ is proportional to P^μ, see also [18]. We are now in the situation considered in the work by Bäckdahl and Valiente-Kroon, see [14, Theorem B.3], and hence we can conclude that $(\mathcal{M}, g_{ab})$ is locally isometric to the Kerr spacetime.

Remark 5.1

1. This result can be turned into a characterization in terms of Cauchy data along the lines in [13].
2. Theorem 5.1 can be viewed as a variation on the Kerr characterization given in [14, Theorem B.3]. In the version given here, the asymptotic conditions on the Killing spinor have been removed.

[13] Section 11 appears only in the ArXiv version of [16].

6 Monotonicity and Dispersion

The dispersive properties of fields, i.e., the tendency of the energy density contained within any stationary region to decrease asymptotically to the future is a crucial property for solutions of field equations on spacetimes, and any proof of stability must exploit this phenomenon. In view of the geometric optics approximation, the dispersive property of fields can be seen in an analogous dispersive property of null geodesics, i.e., the fact that null geodesics in the Kerr spacetime which do not orbit the black hole at a fixed radius must leave any stationary region in at least one of the past or future directions. In Sect. 6.1 we give an explanation for this fact using tools which can readily be adapted to the case of field equations, while in Sect. 6.2 we outline sketch how these ideas apply to fields.

We begin by a discussion of conservation laws. For a null geodesic γ^a, we define the energy associated with a vector field X and evaluated on a Cauchy hypersurface Σ to be

$$e_X[\gamma](\Sigma) = g_{ab} X^a \dot{\gamma}^b |_{\Sigma}.$$

Since $\dot{\gamma}^b \nabla_b \dot{\gamma}^a = 0$ for a geodesic, integrating the derivative of the energy gives

$$e_X[\gamma](\Sigma_2) - e_X[\gamma](\Sigma_1) = \int_{\lambda_1}^{\lambda_2} (\dot{\gamma}_a \dot{\gamma}_b) \nabla^{(a} X^{b)} \mathrm{d}\lambda, \tag{62}$$

where λ_i is the unique value of λ such that $\gamma(\lambda)$ is the intersection of γ with Σ_i. Formula (62) is particularly easy to work with, if one recalls that

$$\nabla^{(a} X^{b)} = -\frac{1}{2} \mathcal{L}_X g^{ab}.$$

The tensor $\nabla^{(a} X^{b)}$ is commonly called the "deformation tensor." In the following, unless there is room for confusion, we will drop reference to γ and Σ in referring to e_X.

Conserved quantities play a crucial role in understanding the behavior of geodesics as well as fields. By (62), the energy e_X is conserved if X^a is a Killing field. In the Kerr spacetime we have the Killing fields $\xi^a = (\partial_t)^a$, $\eta^a = (\partial_\phi)^a$ with the corresponding conserved quantities energy $\boldsymbol{e} = (\partial_t)^a \dot{\gamma}_a$ and azimuthal angular momentum $\boldsymbol{\ell}_z = (\partial_\phi)^a \dot{\gamma}_a$. In addition, the squared particle mass $\boldsymbol{\mu} = g_{ab} \dot{\gamma}^a \dot{\gamma}^b$, and the Carter constant $\boldsymbol{k} = K_{ab} \dot{\gamma}^a \dot{\gamma}^b$ are conserved along any geodesic γ^a in the Kerr spacetime. The presence of the extra conserved quantity allows one to integrate the equations of geodesic motion.[14]

For a covariant field equation derived from an action principle which depends on the background geometry only via the metric and its derivatives, the symmetric stress-energy tensor T_{ab} is conserved. As an example, we consider the wave equation

$$\nabla^a \nabla_a \psi = 0 \tag{63}$$

which has stress-energy tensor

$$T_{ab} = \nabla_{(a}\psi \nabla_{b)}\bar{\psi} - \tfrac{1}{2}\nabla^c \psi \nabla_c \bar{\psi} g_{ab}. \tag{64}$$

Let ψ be a solution to (63). Then T_{ab} is conserved, $\nabla^a T_{ab} = 0$. For a vector field X^a we have that $\nabla^a(T_{ab}X^b)$ is given in terms of the deformation tensor,

$$\nabla^a(T_{ab}X^b) = T_{ab}\nabla^{(a}X^{b)}.$$

Let $(J_X)_a = T_{ab}X^b$ be the current corresponding to X^a. By the above, we have conserved currents J_ξ and J_η corresponding to the Killing fields ξ^a, η^a.

An application of Gauss' law gives the analog of (62),

$$\int_{\Sigma_2} (J_X)_a d\sigma^a - \int_{\Sigma_1} (J_X)_a d\sigma^a = \int_\Omega T_{ab}\nabla^{(a}X^{b)},$$

where Ω is a spacetime region bounded by Σ_1, Σ_2.

6.1 Monotonicity for Null Geodesics

We shall consider only null geodesics, i.e., $\mu = 0$. In this case we have

$$\begin{aligned} k &= K_{ab}\dot{\gamma}^a\dot{\gamma}^b \\ &= 2\Sigma l_{(a}n_{b)}\dot{\gamma}^a\dot{\gamma}^b \\ &= 2\Sigma m_{(a}\bar{m}_{b)}\dot{\gamma}^a\dot{\gamma}^b. \end{aligned} \tag{65}$$

[14] In general, the geodesic equation in a 4-dimensional stationary and axisymmetric spacetime cannot be integrated, and the dynamics of particles may in fact be chaotic, see [34, 50] and the references therein. Note, however, that the geodesic equations are not *separable* in the Boyer-Lindquist coordinates. On the other hand, the Darboux coordinates have this property, cf. [33].

We note that the tensors $2\Sigma l_{(a}n_{b)}$ and $2\Sigma m_{(a}\bar{m}_{b)}$ are conformal Killing tensors, see Sect. 4.5. From (65) it is clear that $\boldsymbol{k}$ is non-negative. A calculation using (52) gives

$$
\begin{aligned}
2\Sigma l^{(a}n^{b)}\partial_a\partial_b &= \frac{1}{\Delta}[(r^2+a^2)\partial_t + a\partial_\phi]^2 - \Delta\partial_r^2 \\
2\Sigma m^{(a}\bar{m}^{b)}\partial_a\partial_b &= \partial_\theta^2 + \frac{1}{\sin^2\theta}\partial_\phi^2 + a^2\sin^2\theta\partial_t^2 + 2a\partial_t\partial_\phi.
\end{aligned}
$$

Let $Z = (r^2+a^2)\boldsymbol{e} + a\boldsymbol{\ell}_z$. Recall that $\dot{r} = \dot{\gamma}^r = g^{rr}\dot{\gamma}_r$ where $g^{rr} = -\Delta/\Sigma$. Now we can write $0 = g_{ab}\dot{\gamma}^a\dot{\gamma}^b$ in the form

$$
\Sigma^2\dot{r}^2 + \mathcal{R}(r; \boldsymbol{e}, \boldsymbol{\ell}_z, \boldsymbol{k}) = 0, \tag{66}
$$

where

$$
\mathcal{R} = -Z^2 + \Delta\boldsymbol{k}. \tag{67}
$$

Equation (66) allows one to make a qualitative analysis of the motion of null geodesics in the Kerr spacetime. In particular, we find that the location of orbiting null geodesics is determined by $\mathcal{R} = 0$, $\partial_r\mathcal{R} = 0$. Due to the form of $\mathcal{R}$, the location of orbiting null geodesics depends only on the ratios $\boldsymbol{k}/\boldsymbol{\ell}_z{}^2, \boldsymbol{e}/\boldsymbol{\ell}_z$. One finds that orbiting null geodesics exist for a range of radii $r_1 \le r \le r_2$, with $r_+ < r_1 < 3M < r_2$. Here r_1, r_2 depend on a, M and as $|a| \nearrow M$, $r_1 \searrow r_+$, and $r_2 \nearrow 4M$. The orbits at r_1, r_2 are restricted to the equatorial plane, those at r_1 are corotating, while those at r_2 are counterrotating. For $r_1 < r < r_2$, the range of θ depends on r. There is $r_3 = r_3(a, M)$, $r_1 < r_3 < r_2$ such that the orbits at r_3 reach the poles, i.e., $\theta = 0, \theta = \pi$, see Fig. 6. For such geodesics, it holds that $\boldsymbol{\ell}_z = 0$. Examples of null geodesics with non zero $\boldsymbol{\ell}_z$ are shown in Fig. 7.

For the following discussion, it is convenient to introduce

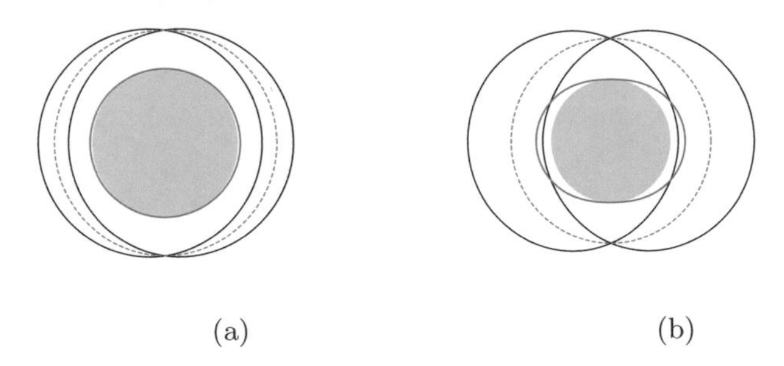

Fig. 6 The Kerr photon region. In (**a**), $|a| \ll M$ and the ergoregion, see Sect. 3.3, is well separated from the photon region (bordered in black). The radius r_3 where geodesics reach the poles is indicated by a grey, dashed line. In (**b**), $|a|$ is close to M and the ergoregion overlaps the photon region

Fig. 7 Examples of orbiting null geodesics in Kerr with $a = M/2$. In (**a**), the $\boldsymbol{k}/\boldsymbol{\ell}_z{}^2$ is small, while in (**b**), this constant is larger

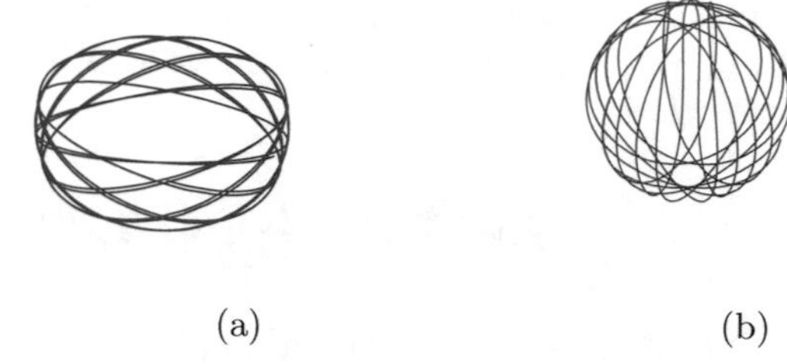

$$\boldsymbol{q} = \boldsymbol{k} - 2a\boldsymbol{e}\boldsymbol{\ell}_z - \boldsymbol{\ell}_z{}^2 = Q^{ab}\dot{\gamma}_a\dot{\gamma}_b,$$

where

$$Q^{ab} = (\partial_\theta)^a(\partial_\theta)^b + \frac{\cos^2\theta}{\sin^2\theta}(\partial_\phi)^a(\partial_\phi)^b + a^2\sin^2\theta(\partial_t)^a(\partial_t)^b. \tag{68}$$

By construction, $\boldsymbol{q}$ is a sum of conserved quantities and is, therefore, conserved. Further, it is non-negative, since it is a sum of non-negative terms. In the following we use $(\boldsymbol{e}, \boldsymbol{\ell}_z, \boldsymbol{q})$ as parameters for null geodesics. Since we are considering only null geodesics, there is no loss of generality compared to using $(\boldsymbol{e}, \boldsymbol{\ell}_z, \boldsymbol{k})$ as parameters.

For a null geodesic with given parameters $(\boldsymbol{e}, \boldsymbol{\ell}_z, \boldsymbol{q})$, a simple turning point analysis shows that there is a number $r_o \in (r_+, \infty)$ so that the quantity $(r - r_o)\dot{\gamma}^r$ increases overall. This quantity corresponds to the energy e_A for the vector field $A = -(r - r_o)\partial_r$. Following this idea, we may now look for a function $\mathcal{F}$ which will play the role of $-(r - r_o)$, so that for $A = \mathcal{F}\partial_r$, the energy e_A is non-decreasing for all λ and not merely non-decreasing overall. For $a \neq 0$, both r_o and $\mathcal{F}$ will necessarily depend on both the Kerr parameters (M, a) and the constants of motion $(\boldsymbol{e}, \boldsymbol{\ell}_z, \boldsymbol{q})$; the function $\mathcal{F}$ will also depend on r, but no other variables.

We define $A^a = \mathcal{F}(\partial_r)^a$ with

$$\mathcal{F} = \mathcal{F}(r; M, a, \boldsymbol{e}, \boldsymbol{\ell}_z, \boldsymbol{q}).$$

It is important to note that this is a map from the tangent bundle to the tangent bundle, and hence $A^a = \mathcal{F}(\partial_r)^a$ cannot be viewed as a standard vector field, which is a map from the manifold to the tangent bundle.

To derive a monotonicity formula, we wish to choose $\mathcal{F}$ so that e_A has a non-negative derivative. We define the covariant derivative of A by holding the values of $(\boldsymbol{e}, \boldsymbol{\ell}_z, \boldsymbol{q})$ fixed and computing the covariant derivative as if A were a regular vector field. Similarly, we define $\mathcal{L}_A g^{ab}$ by fixing the values of the constants of geodesic motion. Since the constants of motion have zero derivative along null geodesics, Eq. (62) remains valid.

Recall that null geodesics are conformally invariant up to reparameterization. Hence, it is sufficient to work with the conformally rescaled metric Σg^{ab}. Furthermore, since γ is a null geodesic, for any function q_{reduced}, we may subtract $q_{\text{reduced}}\Sigma g^{ab}\dot\gamma_a\dot\gamma_a$ wherever it is convenient. Thus, the change in e_A is given as the integral of

$$\Sigma\dot\gamma_a\dot\gamma_b\nabla^{(a}A^{b)} = \left(-\frac{1}{2}\mathcal{L}_A(\Sigma g^{ab}) - q_{\text{reduced}}\Sigma g^{ab}\right)\dot\gamma_a\dot\gamma_b.$$

The Kerr metric can be written as

$$\Sigma g^{ab} = -\Delta(\partial_r)^a(\partial_r)^b - \frac{1}{\Delta}\mathcal{R}^{ab}, \tag{69}$$

where the tensorial form of $\mathcal{R}^{ab}$ can be read off from the earlier definitions. We now calculate $-\mathcal{L}_A g^{ab}\dot\gamma_a\dot\gamma_b$ using (69). Ignoring distracting factors of Σ, Δ, the most important terms are

$$-2(\partial_r\mathcal{F})\dot\gamma_r\dot\gamma_r + \mathcal{F}(\partial_r\mathcal{R}^{ab})\dot\gamma_a\dot\gamma_b = -2(\partial_r\mathcal{F})\dot\gamma_r\dot\gamma_r + \mathcal{F}(\partial_r\mathcal{R}).$$

The second term in this sum will be non-negative if $\mathcal{F} = \partial_r\mathcal{R}(r; M, a; \boldsymbol{e}, \boldsymbol{\ell}_z, \boldsymbol{q})$. Recall that the vanishing of $\partial_r\mathcal{R}(r; M, a; \boldsymbol{e}, \boldsymbol{\ell}_z, \boldsymbol{q})$ is one of the two conditions for orbiting null geodesics. With this choice of $\mathcal{F}$, the instability of the null geodesic orbits ensures that, for these null geodesics, the coefficient in the first term, $-2(\partial_r\mathcal{F})$, will be positive. These observations motivate the form of $\mathcal{F}$ which yields non-negativity for all null geodesics.

It remains to make explicit choices of $\mathcal{F}$ and q_{reduced}. Once these choices are made, the necessary calculations are straight-forward but rather lengthy. Let z and w be smooth functions of r and the Kerr parameters (M, a). Let $\tilde{\mathcal{R}}'$ denote $\partial_r(\frac{z}{\Delta}\mathcal{R}(r; M, a; \boldsymbol{e}, \boldsymbol{\ell}_z, \boldsymbol{q}))$ and choose $\mathcal{F} = zw\tilde{\mathcal{R}}'$ and $q_{\text{reduced}} = (1/2)(\partial_r z)w\tilde{\mathcal{R}}'$. In terms of these functions,

$$\Sigma\dot\gamma_a\dot\gamma_b\nabla^{(a}A^{b)} = \tfrac{1}{2}w(\tilde{\mathcal{R}}')^2 - z^{1/2}\Delta^{3/2}\left(\partial_r\left(w\frac{z^{1/2}}{\Delta^{1/2}}\tilde{\mathcal{R}}'\right)\right)\dot\gamma_r^2. \tag{70}$$

If z and w are chosen to be positive, then the first term on the right-hand side of (70) which contains a square $(\tilde{\mathcal{R}}')^2$ is non-negative. If we now take $z = z_1 = \Delta(r^2 + a^2)^{-2}$ and $w = w_1 = (r^2 + a^2)^4/(3r^2 - a^2)$, then[15]

$$-\partial_r\left(w\frac{z^{1/2}}{\Delta^{1/2}}\tilde{\mathcal{R}}'\right) = 2\frac{3r^4 + a^4}{(3r^2 - a^2)^2}\boldsymbol{\ell}_z{}^2 + 2\frac{3r^4 - 6a^2r^2 - a^4}{(3r^2 - a^2)^2}\boldsymbol{q}. \tag{71}$$

The coefficient of $\boldsymbol{q}$ is positive for $r > r_+$ when $|a| < 3^{1/4}2^{-1/2}M \cong 0.93M$. Since $\boldsymbol{q}$ is non-negative, the right-hand side of (71) is non-negative, and hence also the

[15] Equation (71) corrects a misprint in [8, Eq. (1.15b)].

right-hand side of Eq. (70) is non-negative, for this range of a. Since Eq. (70) gives the rate of change, the energy e_A is monotone.

These calculations reveal useful information about the geodesic motion. The positivity of the term on the right-hand side of (71) shows that $\tilde{\mathcal{R}}'$ can have at most one root, which must be simple. In turn, this shows that $\mathcal{R}$ can have at most two roots. For orbiting null geodesics $\mathcal{R}$ must have a double root, which must coincide with the root of $\tilde{\mathcal{R}}'$. It is convenient to think of the corresponding value of r as being r_o.

The first term in (70) vanishes at the root of $\tilde{\mathcal{R}}'$, as it must so that e_A can be constantly zero on the orbiting null geodesics. When $a = 0$, the quantity $\tilde{\mathcal{R}}'$ reduces to $-2(r - 3M)r^{-4}(\boldsymbol{\ell}_z{}^2 + \boldsymbol{q})$, so that the orbits occur at $r = 3M$. The continuity in a of $\tilde{\mathcal{R}}'$ guarantees that its root converges to $3M$ as $a \to 0$ for fixed $(\boldsymbol{e}, \boldsymbol{\ell}_z, \boldsymbol{q})$.

From the geometrics optics approximation, it is natural to imagine that the monotone quantity constructed in this section for null geodesics might imply the existence of monotone quantities for fields, which would imply some form of dispersion. For the wave equation, this is true. In fact, the above discussion, when carried over to the case of the wave equation, closely parallels the proof of the Morawetz estimate for the wave equation given in [8], see Sect. 6.2 below. The quantity $(\dot{\gamma}_\alpha \dot{\gamma}_\beta)(\nabla^{(\alpha} X^{\beta)})$ corresponds to the Morawetz density, i.e., the divergence of the momentum corresponding to the Morawetz vector field. The role of the conserved quantities $(\boldsymbol{e}, \boldsymbol{\ell}_z, \boldsymbol{q})$ for geodesics is played, in the case of fields, by the energy fluxes defined via second order symmetry operators corresponding to these conserved quantities. The fact that the quantity $\mathcal{R}$ vanishes quadratically on the trapped orbits is reflected in the Morawetz estimate for fields, by a quadratic degeneracy of the Morawetz density at the trapped orbits.

6.2 Dispersive Estimates for Fields

As discussed in Sect. 6.1, one may construct a suitable function of the conserved quantities for null geodesics in the Kerr spacetime which is monotone along the geodesic flow. This function may be viewed as arising from a generalized vector field on phase space. The monotonicity property implies, as discussed there, that non-trapped null geodesics disperse, in the sense that they leave any stationary region in the Kerr space time. As mentioned in Sect. 6.1, in view of the geometric optics approximation for the wave equation, such a monotonicity property for null geodesics reflects the tendency for waves in the Kerr spacetime to disperse.

At the level of the wave equation, the analog of the just mentioned monotonicity estimate is called the Morawetz estimate. For the wave equation $\nabla^a \nabla_a \psi = 0$, a Morawetz estimate provides a current J_a defined in terms of ψ and some of its derivatives, with the property that $\nabla^a J_a$ has suitable positivity properties, and that the flux of J_a can be controlled by a suitable energy defined in terms of the field.

Let ψ be a solution of the wave equation $\nabla^a \nabla_a \psi = 0$. Define the current J_a by

$$J_a = T_{ab}A^b + \tfrac{1}{2}q(\bar{\psi}\nabla_a\psi + \psi\nabla_a\bar{\psi}) - \tfrac{1}{2}(\nabla_a q)\psi\bar{\psi},$$

where T_{ab} is the stress-energy tensor given by (64). We have

$$\nabla^a J_a = T_{ab}\nabla^{(a}A^{b)} + q\nabla^c\psi\nabla_c\bar{\psi} - \tfrac{1}{2}(\nabla^c\nabla_c q)\psi\bar{\psi}. \tag{72}$$

We now specialize to Minkowski space, with the line element $g_{ab}dx^a dx^b = dt^2 - dr^2 - d\theta^2 - r^2\sin^2\theta d\phi^2$. Let

$$E(\tau) = \int_{\{t=\tau\}} T_{tt}d^3x$$

be the energy of the field at time τ, where T_{tt} is the energy density. The energy is conserved, so that $E(t)$ is independent of t.

Setting $A^a = r(\partial_r)^a$, we have

$$\nabla^{(a}A^{b)} = g^{ab} - (\partial_t)^a(\partial_t)^b. \tag{73}$$

With $q = 1$, we get

$$\nabla^a J_a = -T_{tt}.$$

With the above choices, the bulk term $\nabla^a J_a$ has a sign. This method can be used to prove dispersion for solutions of the wave equation. In particular, by introducing suitable cutoffs, one finds that for any $R_0 > 0$, there is a constant C, so that

$$\int_{t_0}^{t_1}\int_{|r|\leq R_0} T_{tt}d^3x dt \leq C(E(t_0) + E(t_1)) \leq 2CE(t_0), \tag{74}$$

see [55]. The local energy, $\int_{|r|\leq R_0} T_{tt}d^3x$, is a function of time. By (74) it is integrable in t, and hence it must decay to zero as $t \to \infty$, at least sequentially. This shows that the field disperses. Estimates of this type are called Morawetz or integrated local energy decay estimates.

For a solution ϕ_{AB} of the Maxwell equation $(\mathscr{C}^\dagger_{2,0}\phi)_{AA'} = 0$, the stress-energy tensor T_{ab} given by

$$T_{ab} = \phi_{AB}\bar{\phi}_{A'B'}$$

is conserved, $\nabla^a T_{ab} = 0$. Further, T_{ab} has trace zero, with $T^a{}_a = 0$.

Restricting to Minkowski space and setting $J_a = T_{ab}A^b$, with $A^a = r(\partial_r)^a$ we have

$$\nabla^a J_a = -T_{tt}$$

which again gives local energy decay for the Maxwell field on Minkowski space.

For the wave equation on Schwarzschild we can choose

$$A^a = \frac{(r-3M)(r-2M)}{3r^2}(\partial_r)^a, \tag{75a}$$

$$q = \frac{6M^2 - 7Mr + 2r^2}{6r^3}. \tag{75b}$$

This gives

$$\begin{aligned} -\nabla^{(a}A^{b)} &= -\frac{Mg^{ab}(r-3M)}{3r^3} + \frac{M(r-2M)^2(\partial_r)^a(\partial_r)^b}{r^4} \\ &+ \frac{(r-3M)^2((\partial_\theta)^a(\partial_\theta)^b + \csc^2\theta(\partial_\phi)^a(\partial_\phi)^b)}{3r^5}, \end{aligned} \tag{76a}$$

$$\begin{aligned} -\nabla_a J^a &= \frac{M|\partial_r\psi|^2(r-2M)^2}{r^4} + \frac{\left(|\partial_\theta\psi|^2 + |\partial_\phi\psi|^2\csc^2\theta\right)(r-3M)^2}{3r^5} \\ &+ \frac{M|\psi|^2(54M^2 - 46Mr + 9r^2)}{6r^6}. \end{aligned} \tag{76b}$$

Here, A^a was chosen so that the last two terms (76a) have good signs. The form of q given here was chosen to eliminate the $|\partial_t\psi|^2$ term in (76b). The first terms in (76b) are clearly non-negative, while the last is of lower-order and can be estimated using a Hardy estimate [8]. The effect of trapping in Schwarzschild at $r = 3M$ is manifested in the fact that the angular derivative term vanishes at $r = 3M$.

In the case of the wave equation on Kerr, the above argument using a classical vector field cannot work due to the complicated structure of the trapping. However, making use of higher-order currents constructed using second order symmetry operators for the wave equation, and a generalized Morawetz vector field analogous to the vector field A^a as discussed in Sect. 6.1. This approach has been carried out in detail in [8].

If we apply the same idea for the Maxwell field on Schwarzschild, there is no reason to expect that local energy decay should hold, in view of the fact that the Coulomb solution is a time-independent solution of the Maxwell equation which does not disperse. In fact, with

$$A^a = \mathcal{F}(r)\left(1 - \frac{2M}{r}\right)(\partial_r)^a, \tag{77}$$

we have

$$\begin{aligned} -T_{ab}\nabla^{(a}A^{b)} &= -\phi^{AB}\bar{\phi}^{A'B'}(\mathscr{T}_{1,1}A)_{ABA'B'} \\ &= \left(|\phi_0|^2 + |\phi_2|^2\right)\frac{(r-2M)}{2r}\mathcal{F}'(r) \end{aligned} \tag{78}$$

$$-\frac{|\phi_1|^2\big(r(r-2M)\mathcal{F}'(r)-2\mathcal{F}(r)(r-3M)\big)}{r^2}. \tag{79}$$

If $\mathcal{F}'$ is chosen to be positive, then the coefficient of the extreme components in (79) is positive. However, at $r = 3M$, the coefficient of the middle component is necessarily of the opposite sign. It is possible to show that no choice of $\mathcal{F}$ will give positive coefficients for all components in (79).

The dominant energy condition that $T_{ab}V^aW^b \geq 0$ for all causal vectors V^a, W^a is a common and important condition on stress-energy tensors. In Riemannian geometry, a natural condition on a symmetric 2-tensor T_{ab} would be non-negativity, i.e., the condition that for all X^a, one has $T_{ab}X^aX^b \geq 0$.

However, in order to prove dispersive estimates for null geodesics and the wave equation, the dominant energy condition on its own is not sufficient and non-negativity cannot be expected for stress-energy tensors. Instead, a useful condition to consider is non-negativity modulo trace terms, i.e., the condition that for every X^a there is a q such that $T_{ab}X^aX^b + qT^a{}_a \geq 0$. For null geodesics and the wave equation, the tensors $\dot{\gamma}_a\dot{\gamma}_b$ and $\nabla_a u \nabla_b u = T_{ab} + T^\gamma{}_\gamma g_{ab}$ are both non-negative, so $\dot{\gamma}_a\dot{\gamma}_b$ and T_{ab} are non-negative modulo trace terms.

From Eq. (76a), we see that $-\nabla^{(a}A^{b)}$ is of the form $f_1 g^{ab} + f_2\partial_r^a\partial_r^b + f_3\partial_\theta^a\partial_\theta^b + f_4\partial_\phi^a\partial_\phi^b$ where f_2, f_3 and f_4 are non-negative functions. That is $-\nabla^{(a}A^{b)}$ is a sum of a multiple of the metric plus a sum of terms of the form of a non-negative coefficient times a vector tensored with itself. Thus, from the non-negativity modulo trace terms, for null geodesics and the wave equation, respectively, there are functions q such that $\dot{\gamma}_a\dot{\gamma}_b\nabla^aA^b = \dot{\gamma}_a\dot{\gamma}_b\nabla^aA^b + qg^{ab}\dot{\gamma}_a\dot{\gamma}_b \leq 0$ and $T_{ab}\nabla^aA^b + qT^a{}_a \leq 0$. For null geodesics, since $g^{ab}\dot{\gamma}_a\dot{\gamma}_b = 0$, the q term can be ignored. For the wave equation, one can use the terms involving q in Eqs. (72), to cancel the $T^a{}_a$ term in ∇^aJ_a. For the wave equation, this gives non-negativity for the first-order terms in $-\nabla^aJ_a$, and one can then hope to use a Hardy estimate to control the zeroth order terms.

If we now consider the Maxwell equation, we have the fact that the Maxwell stress-energy tensor is traceless, $T^a{}_a = 0$ and does not satisfy the non-negativity condition. Therefore, it also does not satisfy the condition of non-negativity modulo trace. This appears to be the fundamental underlying obstruction to proving a Morawetz estimate using T_{ab}. This can be seen as a manifestation of the fact that the Coulomb solution does not disperse.

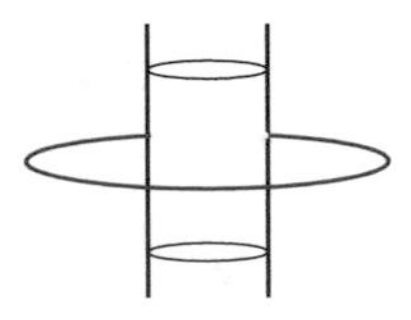

In fact, it is immediately clear that the Maxwell stress energy cannot be used directly to prove dispersive estimates since it does not vanish for the Coulomb field (35) on the Kerr spacetime. We remark that the existence of the Coulomb solution on the Kerr spacetime is a consequence of the fact that the exterior of the black hole contains non-trivial 2-spheres, and the existence of two conserved charge integrals $\int_S F_{ab} d\sigma^{ab}$, $\int_S (*F)_{ab} d\sigma^{ab}$. Hence this is valid also for dynamical black hole spacetimes.

7 Symmetry Operators

A symmetry operator for a field equation is an operator which takes solutions to solutions. In order to analyze higher spin fields on the Kerr spacetime, it is important to gain an understanding of the symmetry operators for this case. In the paper [4] we have given a complete characterization of those spacetimes admitting symmetry operators of second order for the field equations of spins $0, 1/2, 1$, i.e., the conformal wave equation, the Dirac-Weyl equation and the Maxwell equation, respectively, and given the general form of the symmetry operators, up to equivalence. In order to simplify the presentation here, we shall discuss only the spin-1 case, and restrict to spacetimes admitting a valence $(2, 0)$ Killing spinor κ_{AB}. We first give some background on the wave equation.

7.1 Symmetry Operators for the Kerr Wave Equation

As shown by Carter [21], if K_{ab} is a Killing tensor in a Ricci flat spacetime, the operator

$$K = \nabla_a K^{ab} \nabla_b \tag{80}$$

is a commuting symmetry operator for the d'Alembertian,

$$[\nabla^a \nabla_a, K] = 0.$$

In particular there is a second order symmetry operator for the wave equation, i.e., an operator which maps solutions to solutions,

$$\nabla^a \nabla_a \psi = 0 \quad \Rightarrow \quad \nabla^a \nabla_a K \psi = 0.$$

Due to the form of the Carter Killing tensor, K_{ab}, cf. (57), the operator K defined by (80) contains derivatives with respect to all coordinates.

Recall that $\nabla^a \nabla_a = \frac{1}{\mu_g} \partial_a \mu_g g^{ab} \partial_b$, where $\mu_g = \sqrt{\det(g_{ab})}$ is the volume element. For Kerr in Boyer-Lindquist coordinates, we have from (25) that $\mu_g = \Sigma \mu$, with $\mu = \sin\theta$. After rescaling the d'Alembertian by Σ, and using the just mentioned facts, one finds

$$\Sigma \nabla^a \nabla_a = -\partial_r \Delta \partial_r + \frac{\mathcal{R}(r; \partial_t, \partial_\phi, Q)}{\Delta}, \tag{81}$$

where

$$Q = \frac{1}{\mu} \partial_a \mu Q^{ab} \partial_b. \tag{82}$$

In view of the form of Q^{ab} given in (68), we see that Q contains derivatives only with respect to θ, ϕ, t, but not with respect to r. Thus, it is clear from (81) that Q is a commuting symmetry operator for the rescaled d'Alembertian $\Sigma \nabla^a \nabla_a$,

$$[\Sigma \nabla^a \nabla_a, Q] = 0.$$

In addition to the symmetry operator Q related to the Carter constant, we have the second order symmetry operators generated by the Killing fields $\xi^a \nabla_a = \partial_t$, $\eta^a \nabla_a = \partial_\phi$. The operator Q can be termed a hidden symmetry, since it cannot be represented in terms of operators generated by the Killing fields.

The above shows that we can write

$$\Sigma \nabla^a \nabla_a = \mathrm{R} + \mathrm{S},$$

where the operators R, S commute, $[\mathrm{R}, \mathrm{S}] = 0$, and R contains derivatives with respect to the non-symmetry coordinate r, and the two symmetry coordinates t, ϕ, while S contains derivatives with respect to the non-symmetry coordinate θ, and with respect to t, ϕ.

By making a separated ansatz

$$\psi_{\omega,\ell,m}(t, r, \theta, \phi) = e^{-i\omega t} e^{im\phi} R_{\omega,\ell,m}(r) S_{\omega,\ell,m}(\theta)$$

the equation $\nabla^a \nabla_a \psi = 0$ becomes a pair of scalar ordinary differential equations

$$\mathrm{R}R + \lambda R = 0 \tag{83a}$$

$$\mathrm{S}S = \lambda S, \tag{83b}$$

where $\lambda = \lambda_{\omega,\ell,m}$. Here it should be noted that Eq. (83b) is to be considered as a boundary value problem on $[0, \pi]$ with boundary conditions determined by the requirement that ϕ be smooth. In the Schwarzschild case $a = 0$, we can take $\mathrm{S} = \not{\Delta}$, the angular Laplacian. The eigenfunctions of $\not{\Delta}$ are the spherical harmonics $Y_{\ell,m}(\theta, \phi) = e^{im\phi} Y_\ell(\theta)$. The eigenvalues of $\not{\Delta}$ are $\lambda_{\ell,m} = -\ell(\ell+1)$.

The solutions to the eigenproblem $\mathrm{S}S = \lambda S$ are the spheroidal harmonics, the eigenvalues in this case are not known in closed form, depend on the time frequency ω, and are indexed by ℓ, m. For real ω, it is known that the eigensystem is complete, but for general ω this is not known.

One may now apply a Fourier transform and represent a typical solution ψ to the wave equation in the form

$$\psi = \int d\omega \sum_{\ell,m} e^{-i\omega t} e^{im\phi} R_{\omega,\ell,m} S_{\omega,\ell,m},$$

analyze the behavior of the separated modes $\psi_{\omega,\ell,m}$, and recover estimates for ψ after inverting the Fourier transform. In order to do this, one must show a priori that the Fourier transform can be applied. This can be done by applying cutoffs, and removing these after estimates have been proved using Fourier techniques. This approach has been followed in, e.g., [9, 10, 27]. In recent work by Dafermos, Rodnianski and Shlapentokh-Rothman, see [28], proving boundedness and decay for the wave equation on Kerr for the whole range $|a| < M$, makes use of the technical condition of time integrability, i.e., that the solution to the wave equation and its derivatives to a sufficiently high order is bounded in L^2 on time lines,

$$\int_{-\infty}^{\infty} dt |\partial^\alpha \psi(t, r, \theta, \phi)|.$$

This condition is consistent with integrated local energy decay and is removed at the end of the argument.

However, by working directly with currents defined in terms of second order symmetry operators, one may prove a Morawetz estimate directly for the wave equation on the Kerr spacetime. This was carried out for the case $|a| \ll M$ in [8]. This involves introducing a generalization of the vector field method to allow for currents defined in terms of generalized, operator valued, vector fields. These are operator analogs of the generalized vector field A^a introduced in Sect. 6.1.

Fundamental for either of the above mentioned approaches is that the analysis of the wave equation on the Kerr spacetime is based on the hidden symmetry manifested in the existence of the Carter constant, or the conserved quantity $\boldsymbol{q}$, and its corresponding symmetry operator Q.

8 Outlook

I will end by briefly discuss some further developments and give some references. Analysis on black hole spacetimes, and the black hole stability problem is the subject of intense work. Here I will give some brief, and partial remarks and references on the current (Sept 2021) state of the problem. Recall that the subextreme range of Kerr is $|a|/M < 1$, while slowly rotating Kerr black holes are characterized by $|a|/M \ll 1$.

8.1 Teukolsky

Linearized gravity on Petrov type D spacetimes, and Kerr in particular, is governed by the Teukolsky equation, a spin- and boost-weighted wave equation for suitably rescaled linearized extreme Weyl scalars. The Teukolsky Master Equation (TME) is admits a symmetry operator generalizing the symmetry operator Q discussed above, and as a consequence is separable. A generalized Morawetz vector field for the Teukolsky equation on slowly rotating Kerr spacetimes can be constructed following the approach of [8], cf. [51].

Similar results can be achieved using Fourier techniques, cf. [28] for the spin-0 case for subextreme Kerr. For the case of non-zero spin, there is currently[16] no complete proof of a Morawetz estimate or decay for the TME for the full subextreme range, see, however, [66].

For the Fourier based approaches to estimates for Teukolsky, mode stability, i.e., the absence of mode solutions is an essential step. By mode solutions is meant solutions of the radial Teukolsky equation with boundary conditions at the horizon and infinity that imply that no radiation is entering the black hole exterior, either across the horizon of from infinity [11, 65, 71].

8.2 Stability for Linearized Gravity on Kerr

Stability for linearized gravity on slowly rotating Kerr backgrounds is known [6, 37]. The approach in [6] relies on the outgoing radiation gauge (ORG) condition, adapted to one of the principal null directions of the Kerr geometry. In addition to the gauge invariant Teukolsky equation, the linearized Einstein equations in ORG imply a transport system. Provided a Morawetz estimate for the TME is given, as is provided by [51], improved decay estimates are shown using a hierarchy of equations derived by commuting the TME with suitable vector fields. The Teukolsky-Starobinsky Identities are used to get improved control near null infinity. Transport estimates are then used to close the estimates. The approach in [37], on the other hand, relies on treating the linearized Einstein equation on the Kerr spacetime as a perturbation of the Schwarzschild case.

8.3 Nonlinear Stability for Schwarzschild

As discussed in Sect. 2.7, angular momentum is quasilocally conserved on an axisymmetric spacetime. In particular, the black hole stability conjecture for this

[16] Sept. 2021.

case states that Cauchy data that are axially symmetric with vanishing angular momentum and close to Schwarzschild must evolve to a member of the Schwarzschild family. This has been proved for the polarized case [47]. See also [26] for related results.

8.4 Nonlinear Stability for Kerr

The black hole stability conjecture is open. An approach to this problem has been presented by Klainerman and Szeftel, cf. [48] and the references therein. A complete proof is not available as of this writing.[17]

The nonlinear version of the outgoing radiation gauge for deformations of Kerr has been introduced [7]. The reduced Einstein equation can be written in first-order symmetric hyperbolic form and is hence locally well-posed. The system implies a nonlinear Teukolsky equation as well as a transport system, which is analogous to the one used in [6].

Acknowledgments I am grateful for the organizers for the invitation to lecture at the 2019 Domodossola summer school and for their hospitality during this enjoyable event. I thank Steffen Aksteiner, Thomas Bäckdahl, Pieter Blue, Siyuan Ma, Marc Mars, and Claudio Paganini for helpful remarks.

References

1. S. Aksteiner, *Geometry and Analysis in Black Hole Spacetimes*. PhD thesis, Gottfried Wilhelm Leibniz Universität Hannover (2014). http://d-nb.info/1057896721/34
2. S. Aksteiner, L. Andersson, Charges for linearized gravity. Classical Quantum Gravity **30**(15), 155016 (2013). arXiv.org:1301.2674
3. L. Andersson, The global existence problem in general relativity, in *The Einstein Equations and the Large Scale Behavior of Gravitational Fields* (Birkhäuser, Basel, 2004), pp. 71–120
4. L. Andersson, T. Bäckdahl, P. Blue, Second order symmetry operators. Classical Quantum Gravity **31**(13), 135015 (2014). arXiv.org:1402.6252
5. L. Andersson, T. Bäckdahl, P. Blue, Spin geometry and conservation laws in the Kerr spacetime, in *One Hundred Years of General Relativity*, eds. by L. Bieri, S.-T. Yau (International Press, Boston, 2015), pp. 183–226. arXiv.org:1504.02069.
6. L. Andersson, T. Bäckdahl, P. Blue, S. Ma, Stability for linearized gravity on the Kerr spacetime. arXiv e-prints (2019)
7. L. Andersson, T. Bäckdahl, P. Blue, S. Ma, Nonlinear radiation gauge for near Kerr spacetimes. arXiv e-prints, page arXiv:2108.03148 (2021)
8. L. Andersson, P. Blue, Hidden symmetries and decay for the wave equation on the Kerr spacetime. Ann. Math. (2) **182**(3), 787–853 (2015)

[17] A proof of non-linear stability for slowly rotating black holes has appeared on arxiv in 2022.

9. L. Andersson, P. Blue, Uniform energy bound and asymptotics for the Maxwell field on a slowly rotating Kerr black hole exterior. J. Hyperbolic Differ. Equations **12**(04), 689–743 (2015)
10. L. Andersson, P. Blue, J.-P. Nicolas, A decay estimate for a wave equation with trapping and a complex potential. Int. Math. Res. Not. IMRN **2013**(3), 548–561 (2013)
11. L. Andersson, S. Ma, C. Paganini, B.F. Whiting, Mode stability on the real axis. J. Math. Phys. **58**(7), 072501 (2017)
12. T. Bäckdahl, J.A. Valiente Kroon, Geometric Invariant Measuring the Deviation from Kerr Data. Phys. Rev. Lett. **104**(23), 231102 (2010)
13. T. Bäckdahl, J.A. Valiente Kroon, On the construction of a geometric invariant measuring the deviation from Kerr data. Ann. Henri Poincaré **11**(7), 1225–1271 (2010)
14. T. Bäckdahl, J.A. Valiente Kroon, The 'non-Kerrness' of domains of outer communication of black holes and exteriors of stars. Royal Society of London Proceedings Series A **467**, 1701–1718 (2011). arXiv.org:1010.2421
15. T. Bäckdahl, J.A. Valiente Kroon, Constructing "non-Kerrness" on compact domains. J. Math. Phys. **53**(4), 042503 (2012)
16. R.A. Bartnik, P.T. Chruściel, Boundary value problems for Dirac-type equations. J. Reine Angew. Math. **579**, 13–73 (2005). arXiv.org:math/0307278
17. R. Beig, P.T. Chruściel, Killing vectors in asymptotically flat space-times. I. Asymptotically translational Killing vectors and the rigid positive energy theorem. J. Math. Phys. **37**, 1939–1961 (1996)
18. R. Beig, N. Ó Murchadha, The Poincaré group as the symmetry group of canonical general relativity. Ann. Physics **174**(2), 463–498 (1987)
19. A.N. Bernal, M. Sánchez, Further Results on the Smoothability of Cauchy Hypersurfaces and Cauchy Time Functions. Lett. Math. Phys. **77**, 183–197 (2006)
20. H. Buchdahl, On the compatibility of relativistic wave equations for particles of higher spin in the presence of a gravitational field. Il Nuovo Cimento **10**(1), 96–103 (1958)
21. B. Carter, Killing tensor quantum numbers and conserved currents in curved space. Phys. Rev. D **16**, 3395–3414 (1977)
22. Y. Choquet-Bruhat, R. Geroch, Global aspects of the Cauchy problem in general relativity. Commun. Math. Phys. **14**, 329–335 (1969)
23. D. Christodoulou, N. O'Murchadha, The boost problem in general relativity. Commun. Math. Phys. **80**, 271–300 (1981)
24. C.D. Collinson, P.N. Smith, A comment on the symmetries of Kerr black holes. Commun. Math. Phys. **56**, 277–279 (1977)
25. R.M. Corless, G.H. Gonnet, D.E.G. Hare, D.J. Jeffrey, D.E. Knuth, On the LambertW function. Adv. Comput. Math. **5**(1), 329–359 (1996)
26. M. Dafermos, G. Holzegel, I. Rodnianski, M. Taylor, The non-linear stability of the Schwarzschild family of black holes. arXiv e-prints, page arXiv:2104.08222 (2021)
27. M. Dafermos, I. Rodnianski, A proof of the uniform boundedness of solutions to the wave equation on slowly rotating Kerr backgrounds. Invent. Math. **185**(3), 467–559 (2011)
28. M. Dafermos, I. Rodnianski, Y. Shlapentokh-Rothman, Decay for solutions of the wave equation on Kerr exterior spacetimes III: The full subextremal case $|a| < M$ (2014). arXiv.org:1402.7034
29. S.B. Edgar, A.G.-P. Gómez-Lobo, J.M. Martín-García, Petrov D vacuum spaces revisited: identities and invariant classification. Classical Quantum Gravity **26**(10), 105022 (2009). arXiv.org:0812.1232
30. J.J. Ferrando, J.A. Sáez, An intrinsic characterization of the Kerr metric. Classical Quantum Gravity **26**(7), 075013 (2009). arXiv.org:0812.3310
31. F. Finster, N. Kamran, J. Smoller, S.-T. Yau, A Rigorous treatment of energy extraction from a rotating black hole. Commun. Math. Phys. **287**, 829–847 (2009)
32. Y. Fourès-Bruhat, Théorème d'existence pour certains systèmes d'équations aux dérivées partielles non linéaires. Acta Mathematica **88**(1), 141–225 (1952)
33. V. Frolov, A. Zelnikov, *Introduction to Black Hole Physics*. OUP Oxford, New York (2011)

34. J.R. Gair, C. Li, I. Mandel, Observable properties of orbits in exact bumpy spacetimes. Phys. Rev. D **77**(2), 024035 (2008). arXiv.org:0708.0628
35. R. Geroch, Spinor structure of space-times in general relativity. II. J. Math. Phys. **11**(1), 343–348 (1970)
36. R. Geroch, A. Held, R. Penrose, A space-time calculus based on pairs of null directions. J. Math. Phys. **14**, 874–881 (1973)
37. D. Häfner, P. Hintz, A. Vasy, Linear stability of slowly rotating Kerr black holes. Inventiones mathematicae **223**(3), 1227–1406 (2021)
38. J. Haláček, T. Ledvinka, The analytic conformal compactification of the Schwarzschild spacetime. Classical Quantum Gravity **31**(1), 015007 (2014)
39. L.-H. Huang, On the center of mass of isolated systems with general asymptotics. Classical Quantum Gravity **26**(1), 015012 (2009)
40. L.P. Hughston, P. Sommers, The symmetries of Kerr black holes. Commun. Math. Phys. **33**, 129–133 (1973)
41. V. Iyer, R.M. Wald, Some properties of the Noether charge and a proposal for dynamical black hole entropy. Phys. Rev. D **50**, 846–864 (1994)
42. R.P. Kerr, Gravitational field of a spinning mass as an example of algebraically special metrics. Phys. Rev. Lett. **11**, 237–238 (1963)
43. W. Kinnersley, Type D Vacuum Metrics. J. Math. Phys. **10**, 1195–1203 (1969)
44. S. Klainerman, I. Rodnianski, Rough solutions of the Einstein-vacuum equations. Ann. Math. (2) **161**(3), 1143–1193 (2005)
45. S. Klainerman, I. Rodnianski, J. Szeftel, Overview of the proof of the Bounded L^2 Curvature Conjecture (2012). arXiv.org:1204.1772
46. S. Klainerman, I. Rodnianski, J. Szeftel, The bounded L^2 curvature conjecture. Invent. Math. **202**(1), 91–216 (2015)
47. S. Klainerman, J. Szeftel, Global Nonlinear Stability of Schwarzschild Spacetime under Polarized Perturbations. arXiv:1711.07597 (2017)
48. S. Klainerman, J. Szeftel, Kerr stability for small angular momentum. arXiv e-prints (2021)
49. A. László, I. Rácz, Superradiance or total reflection? Springer Proc. Phys. **157**, 119–127 (2014). arXiv.org:1212.4847
50. G. Lukes-Gerakopoulos, T.A. Apostolatos, G. Contopoulos, Observable signature of a background deviating from the Kerr metric. Phys. Rev. D **81**(12), 124005 (2010). arXiv.org:1003.3120
51. S. Ma, Uniform energy bound and Morawetz estimate for extreme components of spin fields in the exterior of a slowly rotating Kerr black hole II: linearized gravity. Commun. Math. Phys. **377**(3), 2489–2551 (2020)
52. M. Mars, Uniqueness properties of the Kerr metric. Classical Quantum Gravity **17**, 3353–3373 (2000)
53. M. Mars, T.-T. Paetz, J.M.M. Senovilla, W. Simon, Characterization of (asymptotically) Kerr-de Sitter-like spacetimes at null infinity. Classical Quantum Gravity **33**(15), 155001 (2016). arXiv.org:1603.05839
54. B. Michel, Geometric invariance of mass-like asymptotic invariants. J. Math. Phys. **52**(5), 052504–052504 (2011)
55. C.S. Morawetz, Time decay for the nonlinear Klein-Gordon equations. Proc. Roy. Soc. Ser. A **306**, 291–296 (1968)
56. I.D. Novikov, V.P. Frolov, *Physics of black holes* (Fizika chernykh dyr, Moscow, Izdatel'stvo Nauka, 1986), 328 p (Dordrecht, Netherlands, Kluwer Academic Publishers, 1989), 351 p. Translation. Previously cited in issue 19, p. 3128, Accession no. A87-44677, 1989
57. B. O'Neill, *The geometry of Kerr black holes*. (A K Peters Ltd., Wellesley, MA, 1995)
58. D. Parlongue, Geometric uniqueness for non-vacuum Einstein equations and applications. arXiv:1109.0644 (2011). arXiv.org:1109.0644
59. R. Penrose, W. Rindler, Spinors and Space-time I & II, in *Cambridge Monographs on Mathematical Physics*. (Cambridge University, Cambridge, 1986)

60. E. Poisson, *A relativist's toolkit* (Cambridge University, Cambridge, 2004). The mathematics of black-hole mechanics
61. T. Regge, C. Teitelboim, Role of surface integrals in the Hamiltonian formulation of general relativity. Ann. Phys. **88**, 286–318 (1974)
62. H. Ringström, Cosmic censorship for Gowdy spacetimes. Living Rev. Relativ. **13**(2), 1–59 (2010)
63. H. Ringström, Origins and development of the Cauchy problem in general relativity. Classical Quantum Gravity **32**(12), 124003 (2015)
64. J. Sbierski, On the existence of a maximal Cauchy development for the Einstein equations: a dezornification. Ann. Henri Poincaré **17**(2), 301–329 (2016)
65. Y. Shlapentokh-Rothman, Quantitative Mode Stability for the Wave Equation on the Kerr Spacetime. Ann. Henri Poincaré **16**, 289–345 (2015)
66. Y. Shlapentokh-Rothman, R. Teixeira da Costa, Boundedness and decay for the Teukolsky equation on Kerr in the full subextremal range $|a| < M$: frequency space analysis. arXiv:2007.07211 (2020)
67. S.A. Teukolsky, The Kerr metric. Classical Quantum Gravity **32**(12), 124006 (2015)
68. R.M. Wald, *General relativity* (University of Chicago Press, Chicago, 1984)
69. R.M. Wald, Gravitational Collapse and Cosmic Censorship (1997). arXiv.org:gr-qc/9710068
70. M. Walker, R. Penrose, On quadratic first integrals of the geodesic equations for type {2,2} spacetimes. Commun. Math. Phys. **18**, 265–274 (1970)
71. B.F. Whiting, Mode stability of the Kerr black hole. J. Math. Phys. **30**, 1301–1305 (1989)
72. J. Winicour, L. Tamburino, Lorentz-Covariant Gravitational Energy-Momentum Linkages. Phys. Rev. Lett. **15**, 601–605 (1965)
73. W.W.-Y. Wong, A comment on the construction of the maximal globally hyperbolic Cauchy development. J. Math. Phys. **54**(11), 113511–113511 (2013)
74. R.L. Znajek, Black hole electrodynamics and the Carter tetrad. Mon. Not. R. Astron. Soc. **179**, 457–472 (1977)

Study of Fundamental Laws with Antimatter

Marco Giammarchi

Mathematics Subject Classification (2000) Primary 81V99; Secondary 81V72

Contents

1 Introduction

Antimatter entered the scene of Modern Physics with the Dirac equation [1] and the discovery of the positron in 1932 [2], offering a new view of particles (and the Universe) as well as an interesting ground to test physical laws at a fundamental level. The question naturally arose whether particles and antiparticles behaved in identical ways with respect to the laws of Physics.

The theoretical basis for testing particle–antiparticle asymmetries was set by the CPT theorem, devised in 1957 in the frame of Lagrangian quantum field theory in a flat spacetime [3]. This theorem relates (at the quantum level) the properties of particle and antiparticles, by explicitly constructing the relevant CPT operators in the Lagrangian formalism: for any spin 0, 1/2, and 1 field a particle

M. Giammarchi (✉)
Istituto Nazionale di Fisica Nucleare, Sezione di Milano, Milano, Italy
e-mail: marco.giammarchi@mi.infn.it

S. L. Cacciatori, A. Kamenshchik (eds.), *Einstein Equations: Local Energy, Self-Force, and Fields in General Relativity*, Tutorials, Schools, and Workshops in the Mathematical Sciences, https://doi.org/10.1007/978-3-031-21845-3_4

and its antiparticle would have the same mass and lifetimes and opposite electric charges (and magnetic moments). In addition, the Weak Equivalence Principle (WEP) of General Relativity (given its validity for all forms of energies) relates the (gravitational) mass of particles and antiparticles at the classical (non-quantum) level. In its most general form, including all forms of self-energies, this is called the Einstein Equivalence Principle.

From the observational viewpoint, it is important to note that the effect of a CPT violation and a violation of the WEP can concur in creating an experimental evidence of different behavior of a particle and its antiparticle. Let us show this with an elementary example, assuming that CPT could be applied to the macroscopic level and using Newton's second law. In this case, CPT invariance would lead to the equality of *inertial* masses:

$$m_i a = \overline{m}_i a \tag{1}$$

between a particle and its antiparticle. While the separate validity of the WEP for a particle and an antiparticle would imply

$$m_i = m_g \tag{2}$$

$$\overline{m}_i = \overline{m}_g \tag{3}$$

Supposing then that a particle/antiparticle different behavior is observed, say, in the Earth's gravitational field, this could mean either a breakdown of CPT invariance, through the chain:

$$\overline{m}_g = \overline{m}_i \neq m_i = m_g \tag{4}$$

or a violation of the WEP for antimatter because of

$$\overline{m}_g \neq \overline{m}_i = m_i = m_g \tag{5}$$

It is important to note that any level of CPT violation would necessarily imply Lorentz violation (see [4]). However, in this chapter, I will only focus on CPT and WEP measurements through the behavior of low-energy (T $\leq$ MeV) antimatter. To this goal, a variety of fundamental tests can be done or are being considered by making use of the following neutral systems:

- Anti-hydrogen, the bound state of an antiproton and a positron
- Positronium, the (unstable) bound state of electron and positron
- Mu-atom, the (unstable) bound state of a muon and an electron

In this work, I will shortly summarize the developments in the test for fundamental laws, using these systems.

2 Testing CPT

Tests of CPT can be done in several ways: they are generally based on the fact that the Standard Model satisfies CPT symmetry and contains the C operator relating a particle to its own antiparticle. Therefore, in any quantum relativistic system, some properties of an antiparticle can be deduced by the ones of the particle by means of the CPT symmetry [5].

As already mentioned above, under mild general assumptions, CPT symmetry implies the equality of charges, masses, and lifetimes of a particle and its antiparticle. Moreover, in any antimatter bound system (like anti-hydrogen), the transition frequencies should be the same as for the matter system, the hydrogen atom.

Tests of CPT have been made in various configurations and are summarized in Fig. 1. Many of those measurements refer to the charge and mass of the particle and the antiparticle as measured in Penning trap systems: the electron [6, 7], the proton [8], and the antiproton [9].

For what concern the electron–positron mass difference, the most stringent limit comes from cosmology and makes use of the null mass of the photon [10, 11], while

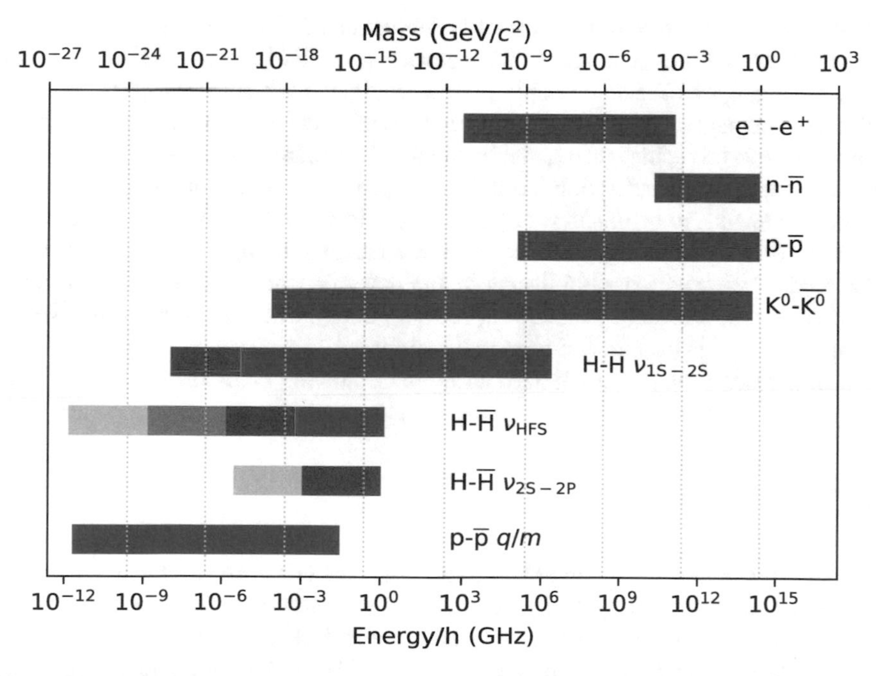

Fig. 1 Several CPT tests are compared. Bar's right-hand side: measured quantity. Length of bar: relative precision of the test. Let-hand side of the bar: sensitivity on an absolute energy scale. Blue: performed measurements. Orange: predicted sensitivity if existing precision on hydrogen is reached for anti-hydrogen. In the case of the Hyperfine Splitting (HFS) and the Lamb shift, the color refers to different future stages of experimental development

in laboratory the obtained relative limit is at the 10^{-9} level [12]. On the other hand, the most stringent limit on the antiproton/proton mass ratio comes from antiprotonic helium [13] and the antiproton charge-to-mass ratio [14, 19].

3 Testing the Weak Equivalence Principle

The weak equivalence principle is a cornerstone of General Relativity and has been tested for matter in a variety of configurations, reaching a level of about a part on 10^{15} for matter systems [15]. This is in striking contrast with the information we have for antimatter, with only a limit of about 100 on the ratio of inertial to gravitational mass for anti-hydrogen [16]. This is one of the most fundamental aspects of the research currently being done on antimatter.

4 Anti-hydrogen at CERN

Anti-hydrogen has a long tradition of being studied for testing fundamental laws. First produced at high energies in laboratories at CERN [17] and at Fermilab [18], anti-hydrogen was made available at low energies starting from 2002 [20, 21]. The development of the CERN Antiproton Decelerator machine was of paramount importance to this achievement, making available $\overline{p}$ beams with energies at the MeV scale. Subsequent deceleration has led to antiprotons confined at energies of a few kiloelectronvolts, to be mixed with positrons, obtained with radioactive sources.

The results of this experimentation have recently brought amazing results to the field. They were possible thanks to the development of magnetic and electric confinement and cooling techniques (Penning traps). Energies down to the Kelvin temperature range have in fact been reached for anti-hydrogen.

These results can be summarized along the following main lines.

4.1 Confinement

The number of anti-hydrogen atoms actually available has always been a matter of concern to experiments, typically ranging in the thousands at best per bunch of the Antiproton Decelerator. This fundamental problem has been successfully addressed by the ALPHA collaboration, making use of a special configuration for confinement.

An octupolar magnetic system, featuring the capability of confining both charged particles (antiprotons and positrons) and the neutral anti-hydrogen atom, has been developed. Using this technique, the ALPHA collaboration has been able to confine anti-atoms for as long as 1000 s [22]. A decisive proof of the confinement mechanism was provided by the ability to release anti-atoms from the trap by means of microwave-induced quantum transitions between trappable and untrappable states [23].

4.2 Anti-hydrogen Beams

The capability of obtaining antiparticle beams is also of decisive importance for experimentation on the WEP, since in principle a tiny fall during propagation can be measured by interferometric methods. As a reference, a neutral particle flying at 10^3 m/s in the Earth gravitational field will fall by about 4 μm along a meter length.

The Asacusa collaboration has been the first experiment capable of producing anti-hydrogen atoms to be detected at almost 3 m distance from the production point [24].

4.3 The Beginning of Anti-hydrogen Spectroscopy

The capability of confining anti-atoms has been instrumental to a variety of studies made by the ALPHA collaboration, actually opening up the field of spectroscopy of anti-atoms.

Basically, the goal is to study the configuration and radiative transitions analogous to that of hydrogen, see Fig. 2.

The first step in this direction has been the observation of the 1S-2S two-photon transition [25], soon followed by the observation of the hyperfine transition [26] and the first observation of the dipole-allowed 1S-2P transition, by means of a laser system tuned on the Lyman-α frequency [27]. These remarkable series of observation have *de facto* opened the field of anti-atoms spectroscopy.

Finally, one has to note the remarkable improvement constituted by the first extension of the laser cooling technique to antimatter [28], which is of fundamental importance for the next steps.

5 Positrons and Positronium

Positrons and positronium are also used to test fundamental laws in a variety of ways, from tests of masses and charges (see, e.g., [12]) to the recent observation of positron interferometry by the QUPLAS group [29].

In addition to the CPT tests made on positrons in traps of various kinds (see contributions already shown in Fig. 1), positronium is also being used to test fundamental laws.

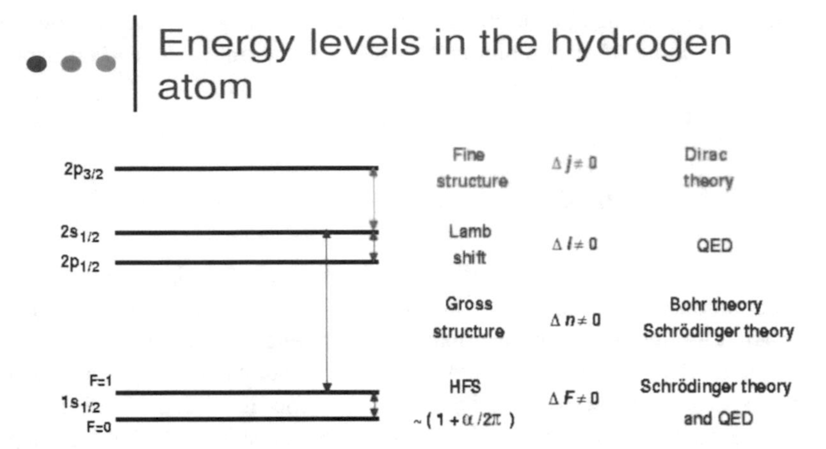

Fig. 2 The energy levels of the hydrogen atom are now being studied for the case of the antimatter system

Positronium (Ps) comes in the two varieties of para-Ps (anti-parallel spins, 124 *ps* lifetime) and ortho-Ps (parallel spins, 142 *ns* lifetime). Those states have been extensively used to test QED radiative corrections and are now being considered (mainly ortho-Ps) for the study of gravitation.

A neutral and well-collimated Ps beam is in fact being developed by several groups. This is especially challenging because of the small lifetime of ortho-Ps and involves the recently demonstrated excitation of Ps to Rydberg states (see, e.g., [30]). Once this goal is achieved, it will be possible to apply interferometric techniques to a Ps beam, similar to what was done for positrons.

6 The Mu-Atom

The muonium atom is a bound system composed of a μ^+ and an electron.[1] Its interest in fundamental physical studies stems from being relatively ($\sim$2 μs) long-lived and from being the leptonic system with the smallest possible Bohr radius. The research underway to prepare Mu-atom experiments is based first on high-energy beams of positive muons at the PSI Institute in Switzerland and at the J-PARC complex in Japan.

Mu-atom spectroscopy is an already well-studied problem for its interest in QED studies [31], and Mu-atom gravitation is being now considered by the current research programs. See [32] for a review of the activities at PSI.

[1] This system is often called—somehow improperly—*muonium*.

7 Conclusion

Antimatter is a very active research field, motivated by the study of fundamental physical symmetries like CPT and the Weak Equivalence Principle. While traditionally it was studied only for a few charged systems like antiproton or the positron, it is now becoming available in neutral systems like anti-hydrogen and positronium. These systems hold the tantalizing promise of the possibility of antimatter gravitational studies.

Acknowledgments I would like to thank Luca Venturelli for several useful discussions.

References

1. P.A.M. Dirac, The quantum theory of the electron. Proc. R. Soc. Lond. **A117**, 610 (1928)
2. C.D. Anderson, The apparent existence of easily deflectable positives. Science **76**, 238 (1932)
3. G. Lüders, Proof of the TCP theorem. Ann. Phys. **2**, 1 (1957)
4. O.W. Greenberg, CPT violation implies violation of Lorentz invariance. Phys. Rev. Lett. **89**, 231602 (2002)
5. R. Lehnert, CPT symmetry and its violation. Symmetry **8**, 114 (2016)
6. S. Sturm et al., High precision measurement of the atomic mass of the electron. Nature **506**, 467 (2014)
7. F. Koehler et al., The electron mass from g-factor measurements on hydrogen-like carbon $^{12}C^{5+}$. J. Phys. B At. Mol. Opt. Phys. **48**, 144032 (2015)
8. F. Heisse et al., High-Precision measurement of the Proton's Atomic Mass. Phys. Rev. Lett. **119**, 033001 (2017)
9. M. Hori et al., Buffer-gas cooling of antiprotonic helium to 1.5 to 1.7 K, and antiproton-to-electron mass ratio. Science **354**, 610 (2016)
10. A.D. Dolgov, V.A. Novikov, A cosmological bound on $e^+ - e^-$ mass difference. Phys. Lett. B **732**, 244 (2014)
11. P.A. Zyla et al., Review of particle physics. Prog. Theor. Exp. Phys. **2020**, 083C01 (2020)
12. M.S. Fee et al., Measurement of the positronium $1^3S_1 - 2^3S_1$ interval by continuous-wave two-photon excitation. Phys. Rev. A **48**, 192 (1993)
13. M. Hori et al., Two-photon laser spectroscopy of antiprotonic helium and the antiproton-to-electron mass ratio. Nature **475**, 484 (2011)
14. S. Ulmer et al., High-precision comparison of the antiproton-to-proton charge-to-mass ratio. Nature **524**, 196 (2015)
15. P. Touboul et al., Space test of the equivalence principle: first results of the MICROSCOPE mission. Classical and Quantum Gravity **36**, 225006 (2019)
16. C. Amole et al., Description and first application of a new technique to measure the gravitational mass of antihydrogen. Nat. Commun. **4**, 1785 (2013)
17. G. Baur et al., Production of antihydrogen. Phys. Lett. B **368**, 251 (1996)
18. G. Blanford et al., Observation of atomic antihydrogen. Phys. Rev. Lett. **80**, 3037 (1998)
19. G. Gabrielse et al., Precision mass spectroscopy of the antiproton and proton using simultaneously trapped particles. Phys. Rev. Lett. **82**, 3198 (1999)
20. M. Amoretti et al., Production and detection of cold antihydrogen atoms. Nature **419**, 456 (2002)
21. G. Gabrielse et al., Background-free observation of cold antihydrogen with field-ionization analysis of its states. Phys. Rev. Lett. **89**, 213401 (2002)

22. G.B. Andresen et al., Confinement of antihydrogen for 1000 seconds. Nature **7**, 558 (2011)
23. C. Amole et al., Resonant quantum transitions in trapped antihydrogen atoms. Nature **483**, 439 (2012)
24. N. Kuroda et al., A source of antihydrogen for in-flight hyperfine spectroscopy. Nat. Commun. **5**, 3089 (2014)
25. M. Ahmadi et al., Observation of the 1S-2S transition in trapped antihydrogen. Nature **541**, 506 (2017)
26. M. Ahmadi et al., Observation of the hyperfine spectrum of antihydrogen. Nature **548**, 66 (2017)
27. M. Ahmadi et al., Observation of the 1S-2P Lyman-α transition in antihydrogen. Nature **561**, 211 (2018)
28. C.J. Baker et al., Laser cooling of antihydrogen atoms. Nature **592**, 35 (2021)
29. S. Sala et al., First demonstration of antimatter interferometry. Sc. Adv. **5**, eaav7610 (2019)
30. S. Aghion et al., Laser excitation of the n=3 level of positroniuim for antihydrogen production. Phys. Rev. A **94**, 012507 (2016)
31. S. Kanda et al., New precise spectroscopy of the hyperfine structure in muonium with a high-intensity pulsed muon beam. Phys. Lett. B **815**, 136154 (2021)
32. P. Crivelli, Current status and prospects of muonium spectroscopy at PSI. SciPost Phys. Proc. **5**, 029 (2021)

Part II
Proceedings

Quantum Ergosphere and Brick Wall Entropy

Lennart Brocki, Michele Arzano, Jerzy Kowalski-Glikman, Marco Letizia, and Josua Unger

Contents

1 Introduction

The discovery that black holes carry an entropy proportional to their horizon area A divided by the Planck length squared ℓ^2 according to the celebrated Bekenstein–Hawking formula

$$S = \frac{A}{4\ell^2} \tag{1}$$

L. Brocki (✉) · J. Kowalski-Glikman · J. Unger
Institute for Theoretical Physics, University of Wrocław, Wroclaw, Poland
e-mail: lennart.brocki@uwr.edu.pl; jerzy.kowalski-glikman@uwr.edu.pl; unger.josua@uwr.edu.pl

M. Arzano
Dipartimento di Fisica "E. Pancini" and INFN, Università degli studi di Napoli "Federico II", Napoli, Italy
e-mail: michele.arzano@na.infn.it

M. Letizia
Perimeter Institute for Theoretical Physics, Waterloo, ON, Canada
e-mail: mletizia@uwaterloo.ca

S. L. Cacciatori, A. Kamenshchik (eds.), *Einstein Equations: Local Energy, Self-Force, and Fields in General Relativity*, Tutorials, Schools, and Workshops in the Mathematical Sciences, https://doi.org/10.1007/978-3-031-21845-3_5

is now more than 40 years old [1, 2]. Despite the numerous derivations of the entropy-area relation (1) existing in a variety of approaches to quantum gravity (see [3] for a comprehensive listing), the fundamental question concerning the nature of the degrees of freedom responsible for such entropy has not found yet a conclusive answer. Since due to quantum effects black holes radiate thermally [2], one of the earliest attempts at addressing this question focused on the quanta of a field in thermal equilibrium at the Hawking temperature near the horizon [4, 5] as possible candidates for the origin the Bekenstein–Hawking entropy. As it turns out the counting of modes needed for deriving the thermodynamic partition function of the field yields a divergent result due to an infinite contribution coming from the black hole horizon. 't Hooft noticed that introducing a crude regulator by requiring the vanishing of the field at a small radial distance from the horizon one can obtain a finite horizon contribution to the entropy proportional to the area. Appropriately tuning the distance of such "brick wall" from the horizon one can exactly reproduce the Bekenstein–Hawking formula (1). This result, albeit suggestive, replaces the question about the origin of the Bekenstein–Hawking entropy with a question about the nature of the brick wall boundary. In [6] the authors suggested that backreaction of the Hawking radiation can excite the quasinormal modes of the black hole thus effectively creating a "wall" of oscillations in the geometry close to the horizon. The Bekenstein–Hawking entropy can be then seen as emerging from an interplay between the degrees of freedom of the geometry and those of the field.

In the present letter we adopt a philosophy similar to [6] and study the effect of backreaction on the field propagating in the vicinity of the black hole horizon. We do this by replacing the usual Schwarzschild metric by a dynamic, "evaporating" metric first proposed in [7], in which the effects of backreaction are parametrized by the luminosity of the radiating black hole. After solving the field equations in such metric we proceed to the usual mode counting for the field. The key feature of our model is that the small luminosity creates a "quantum ergosphere," a region between the apparent horizon and the event horizon which effectively acts as a brick wall providing a finite horizon contribution to the entropy. As we show below, within the small luminosity and quasi-static approximations we use, we are able to reproduce the Bekenstein–Hawking result within very good accuracy.

2 Geometry of an Evaporating Black Hole

Our starting point is the result by Bardeen [7] (see also [8] and [9]) that the metric of a spherically symmetric black hole slowly emitting Hawking radiation has the following form

$$ds^2 = -e^{2\psi}\left(1 - \frac{2m}{r}\right)dv^2 + 2e^{\psi}dvdr + r^2 d\Omega, \tag{2}$$

where ψ and m are functions of the advanced time v and the radial coordinate r. For m constant and $\psi = 0$ this metric reduces to the Schwarzschild one, while for $\psi = 0$ and $m = m(v)$ it becomes the Vaidya metric. Following [8] we define the mass of the black hole at a given time to be $M(v) = m(v, r = 2m)$ and its luminosity to be $L = -\frac{dM}{dv}$. In this letter we work in the regime of small luminosity[1] $L \ll 1$ and up to first order in perturbation theory so that we can write

$$m(v,r) = M(v) \simeq M_0 - Lv\,. \tag{3}$$

In what follows we will focus on the near-horizon features of the metric (2). To this end we introduce a new "comoving" radial coordinate $\rho = r - 2M = r - 2M_0 + 2Lv$ and assume that ρ is small, of the same order as Lv, so that in our computations we will only keep terms which are at most linear in ρ and L. Further, we use the residual coordinate invariance to set $\psi(r = 2M) = 0$, which in our approximation conveniently makes the function ψ disappear from all the linearized expressions. Indeed

$$\psi(r) = \psi(r = 2M) + \left.\frac{\partial \psi}{\partial r}\right|_{r=2M} \rho \tag{4}$$

and it follows from Einstein equations that $\partial\psi/\partial r \sim L$ at $r = 2M$, so that the first term in (4) vanishes, while the second is of higher order and can be neglected. In terms of the comoving radial coordinate the metric near $r = 2M$ takes the form

$$ds^2 = -\left(\frac{\rho}{\rho + 2M} + 4L\right) dv^2 + 2dvd\rho + (\rho + 2M)^2 d\Omega. \tag{5}$$

The metric (2) has several horizon-like structures. We first consider the apparent horizon (AH), defined as the outermost trapped surface, i.e., the surface from which no light ray can move outwards. One characterizes this feature with the help of the expansion Θ of a congruence of null geodesics, which describes the fractional change of the congruence's area. The apparent horizon is defined as a surface for which $\Theta = 0$.

We define (see, e.g., [10]) null geodesics by their tangent vectors l^μ, with $l_\mu l^\mu = 0$, and introduce an auxiliary vector β^μ, an affine tangent vector for ingoing radial null geodesics, with normalization $l_\mu \beta^\mu = -1$. This auxiliary vector is needed because l^μ, parametrized by advanced time v, is not an affine tangent vector and the expansion is defined as the divergence of an affine tangent vector. The quantity $\kappa = -\beta^\mu l^\nu l_{\mu;\nu}$ measures to which extent l^μ is non-affine and the expansion is then given by

[1] For a black hole with solar mass the value would be $L \sim 10^{-38}$ in Planck units where we set the Newton constant G to 1.

$$\Theta = l^{\mu}{}_{;\mu} - \kappa \,. \tag{6}$$

Choosing the null vectors

$$l^{\mu} = (l^{v}, l^{r}) = \left(1, \frac{1}{2}\left(1 - \frac{2M}{r}\right)\right), \quad \beta^{\mu} = (0, -1), \tag{7}$$

we obtain the following expressions for the expansion in Bardeen coordinates (r, v) and comoving coordinates (ρ, v)

$$\Theta(r) = \frac{M}{r^2} - \frac{1}{4M}, \quad \Theta(\rho) = \frac{M}{(\rho + 2M)^2} - \frac{1}{4M}, \tag{8}$$

showing that the apparent horizon is located in the two coordinate systems at $r_{AH} = 2M$ and at $\rho_{AH} = 0$.

In order to capture another horizon-like structure present in the problem, York [8] gives a working definition of what we will call York event horizon (YEH), which lacks the teleological property of the event horizon and is instead based on the local condition

$$\frac{d^2 r}{dv^2} = 0, \tag{9}$$

i.e., it characterizes the YEH as the surface imprisoning photons for times long compared to the dynamical scale $4M$ of the black hole. According to this definition the YEH in the Bardeen and comoving coordinates lies at $r_{YEH} = 2M - 8ML$ and $\rho_{YEH} = -8ML$.

The region between York event horizon and apparent horizon was dubbed by York *quantum ergosphere* [8], and he argues that its presence is an irreducible property of an evaporating black hole.

We will use this observation to shed new light on the brick wall calculation of 't Hooft [5] by including the contributions due to the backreaction, here modelled by a small luminosity L.

The original result of 't Hooft is that the free energy of a scalar field living outside the Schwarzschild black hole has a horizon contribution given by

$$F = -\frac{2\pi^3}{45h}\left(\frac{2M}{\beta}\right)^4 + \dots \tag{10}$$

where β is the inverse Bekenstein–Hawking temperature and h is a small cut-off parameter with dimensions of length. From (10), using standard manipulations, one can calculate the thermodynamic entropy associated with the field and the resulting contribution from the horizon term above is proportional to the area of the black hole thus qualitatively reproducing the Bekenstein–Hawking entropy-area relation. As we will see, it is a consequence of finite luminosity that the brick wall thickness

h, which is arbitrary in the original 't Hooft calculation, can be now naturally identified with the distance between the event and apparent horizon. Therefore the *quantum ergosphere* [8], the region between the AH and the YEH, plays the role of a physically motivated brick wall.

3 Mode Counting and Calculation of Entropy

In order to proceed with the counting of modes of the field we start by solving the field equation in the vicinity of the black hole horizon in the comoving coordinates introduced earlier. A massless scalar field ϕ in this geometry with metric (5) obeys the Klein-Gordon equation

$$\left(4L+\frac{\rho}{\rho+2M}\right)\partial_\rho^2\phi+2\frac{\partial_\rho\phi}{\rho+2M}\left(1-\frac{M}{\rho+2M}+2L\right)+2\partial_\rho\partial_v\phi+ \\ \frac{2}{\rho+2M}\partial_v\phi-\frac{l(l+1)}{(\rho+2M)^2}\phi=0. \tag{11}$$

Since we are interested only in the contribution coming from the vicinity of the horizon in the case of small luminosity assuming $\rho/M_0 \ll 1$, $L \ll 1$ and using the quasi-static approximation $\frac{Lv}{2M_0} \ll 1$, we can write

$$\frac{2}{\rho+2M}\approx\frac{1}{M_0}\left(1-\frac{\rho}{2M_0}+\frac{Lv}{M_0}\right) \tag{12}$$

$$\frac{2}{\rho+2M}\left(1-\frac{M}{\rho+2M}+2L\right)\approx\frac{1}{2M_0}\left(1+4L+\frac{Lv}{M_0}\right) \tag{13}$$

and the Klein-Gordon equation (11) becomes

$$\left(4L+\frac{\rho}{2M_0}\right)\partial_\rho^2\phi+\frac{1}{2M_0}\left(1+4L+\frac{Lv}{M_0}\right)\partial_\rho\phi+2\partial_\rho\partial_v\phi \\ +\frac{1}{M_0}\left(1-\frac{\rho}{2M_0}+\frac{Lv}{M_0}\right)\partial_v\phi-\frac{l(l+1)}{(2M_0)^2}\phi=0. \tag{14}$$

We now make use of the standard WKB ansatz

$$\phi(\rho,v)=U(\rho)e^{-i\omega v}e^{i\int^\rho k(\rho')d\rho'}, \tag{15}$$

and find that the real part of (14) takes the form

$$\left(4L+\frac{\rho}{2M_0}\right)(U''-k^2U)+\frac{U'}{2M_0}+2\omega kU-\frac{l(l+1)}{(2M_0)^2}U=0, \tag{16}$$

which, as a consequence of our approximation scheme, is v-independent. This part is sufficient for obtaining the wavenumber k. The imaginary part could be used to compute the amplitude but since we are only interested in counting the modes of the field with the help of the wavenumber we can ignore it. Moreover the v-independence of (16) shows that for the purpose of the entropy computation, to be presented below, the geometry is static. In the WKB approximation one assumes that the amplitude $U(\rho)$ varies slowly compared to the wave number

$$\frac{U'}{U} \ll k, \quad \frac{U''}{U} \ll k^2, \tag{17}$$

and therefore (16) becomes

$$-\left(4L + \frac{\rho}{2M_0}\right)k^2 + 2\omega k - \frac{l(l+1)}{(2M_0)^2} = 0, \tag{18}$$

which can be solved for k giving

$$k^{\pm} \approx \frac{\omega \pm \sqrt{\omega^2 - \left(4L + \frac{\rho}{2M_0}\right)\frac{l(l+1)}{(2M_0)^2}}}{4L + \frac{\rho}{2M_0}}, \tag{19}$$

where again we neglected the terms which are of higher order in our approximation scheme. These two solutions correspond to incoming and outgoing modes, respectively, and can be used to calculate the thermodynamic entropy associated with the field via a count of its number of modes and the derivation of the statistical partition function.

By approximating the sum over those l that render the square root real with an integral, the number of modes with frequency up to ω is given by

$$g(\omega) = \int_0^{l_{max}} \nu(l,\omega)(2l+1)dl, \tag{20}$$

where $\nu(l,\omega)$ is the number of nodes in the mode with (l,ω) [11]. Such quantity can be explicitly calculated by considering the modes (19) in the box of the radial length Λ, which acts as an infrared regulator

$$\Lambda = \nu\frac{\lambda}{2} = \nu\frac{\pi}{k} \rightarrow \pi\nu = \Lambda k, \quad k = \frac{2\pi}{\lambda}, \tag{21}$$

where λ is the wavelength of the mode.

In the original brick wall calculation [5] it is assumed that the scalar field, whose entropy we are going to compute, vanishes beyond the brick wall, situated at a small distance h from the Schwarzschild black hole horizon at $r_{\rm Sch}$, so that all the relevant integrals have the lower limit at $r_{\rm Sch} + h$. In the case of the Schwarzschild black hole considered in [5] the apparent and event horizon coincide, $r_{\text{Sch}} = r_{EH} = r_{AH}$; however, in our case they are different and we must decide at which of the two we impose the scalar field boundary conditions. Our argument relies on the observation that in the brick wall picture the scalar field is to be in thermal equilibrium at temperature T which is identified with the temperature of Hawking radiation. However, by invoking the so-called tunneling picture, it can be argued that the Hawking radiation originates at the vicinity of the apparent, not the event horizon (see [12] and the references therein). Thus, remembering that the apparent horizon corresponds to $\rho = 0$, we choose the integration range in the formula above to go from 0 to Λ, where Λ is the infrared cut-off introduced before, whose explicit value will not interest us here, since the expression for the area contribution to the entropy does not depend on it. The number of nodes is thus given by the integral

$$2\pi\nu(l,\omega) = \int_0^\Lambda k^+ d\rho = \int_0^\Lambda \frac{\omega + \sqrt{\omega^2 - \left(4L + \frac{\rho}{2M_0}\right)\frac{l(l+1)}{(2M_0)^2}}}{4L + \frac{\rho}{2M_0}} d\rho, \tag{22}$$

where we used Eq. (19) and only considered the contribution from the outgoing modes. We notice that the ingoing solutions close to the apparent horizon are moving towards the singularity and one can argue that they cannot contribute to the entropy. As we will find below, this choice is also justified a posteriori, by the remarkable agreement of our final result with the Bekenstein–Hawking entropy relation.

Let us notice that the equation for the number of nodes above differs from the one obtained previously in the literature in [11] in two aspects. First, due to the approximations we made there is no dependence on the advanced time v and second, we do not have to introduce a cut-off close to the horizon, since the finite luminosity prevents the integrand from diverging at $\rho = 0$. The integration with respect to l in Eq. (20) is taken over those values for which the square root is real and yields

$$g(\omega) = \int_0^\Lambda \frac{5(2GM_0 + \rho)^4\omega^3}{6\pi(8M_0L + \rho(1 + 4L))^2} d\rho. \tag{23}$$

The leading contributions in the integral in (23) are thus given by

$$g(\omega) = \frac{5\omega^3 M_0^3}{3\pi L} + \frac{5\omega^3\Lambda^3}{18\pi(1 + 8L)}, \tag{24}$$

where the second term is the usual volume contribution and has no relevance for our discussion. The thermodynamic partition function of the field is given by

$$Z = e^{-\beta F}, \tag{25}$$

where F is the free energy

$$\pi \beta F = \int dg(\omega) \ln \left(1 - e^{-\beta\omega}\right). \tag{26}$$

Using (24) and neglecting the volume contribution to $g(\omega)$ we have

$$F = \frac{1}{\beta} \int_0^\infty \ln \left(1 - e^{-\beta\omega}\right) \frac{dg(\omega)}{d\omega} d\omega = -\frac{M_0^3 \pi^3}{9L\beta^4}, \tag{27}$$

from which we can calculate the entropy of the field associated with the horizon boundary

$$S = \beta^2 \frac{\partial}{\partial \beta} F = \frac{4M_0^3 \pi^3}{9L\beta^3}. \tag{28}$$

Comparing our result for the free energy (27) with the standard result obtained from the brick wall calculation (10) we see that the brick wall width parameter h introduced by 't Hooft can be expressed in terms of the luminosity of the black hole as

$$h = \frac{32}{5} L M_0, \tag{29}$$

and thus the backreaction of the quantum radiance on the horizon structures of the black hole naturally provides the regulator needed for a finite horizon contribution to the field entropy. It should be noticed that if we had counted also the ingoing modes in (22) the resulting brick wall size would have coincided with the quantum ergosphere, i.e., $h = 8LM_0$.

4 Calculation of Luminosity

In order to have an expression for the entropy (28) to be compared to the Bekenstein–Hawking relation (1) we now have to spell out the explicit form of the luminosity L in terms of the black hole mass M_0. In the first order approximation used in our calculation the luminosity L is a small quantity so that we can identify it with the luminosity of Hawking radiation in the case of a Schwarzschild black hole. To find it, one considers [13] a flux X of radiation with energy ω_k

$$X(\omega_k) = \frac{\Gamma(\omega_k)}{2\pi \left(e^{8\pi M_0 \omega_k} - 1\right)}, \tag{30}$$

where the factor Γ models the backscattering. Integrating the flux times the energy ω we find the luminosity that escapes to infinity

$$rL = \int_0^\infty d\omega\, \omega X(\omega). \tag{31}$$

The factor Γ can be approximated by DeWitt [14]

$$\Gamma \approx 27\pi M_0^2 \omega^2 \tag{32}$$

and integration over ω yields

$$L \approx \frac{1.69}{7680\pi M_0^2} \,. \tag{33}$$

Plugging the expression (33) in (28) we finally obtain

$$S = 0.987 \cdot 4\pi M_0^2 = 0.987 S_{BH}, \tag{34}$$

where S_{BH} is the Bekenstein–Hawking entropy. We thus see that our model reproduces the exact result of Bekenstein–Hawking with an accuracy close to 99%, which is a remarkable result given the rather crude approximations that we used.

There is an important comment to be made at this point. In our calculation of the entropy we assumed that the luminosity of black hole results from a single massless scalar mode, the same that we used to compute entropy. Since we do not know any massless scalar field it might be argued that one should use instead in our computations the massless fields that we know about, namely photons and gravitons, i.e., four massless degrees of freedom. Each degree of freedom will contribute the amount (28) to the entropy. As for the luminosity one can use the numerical results of Page [15], to see that the contribution to luminosity of photons and gravitons is of order of order of $3 \times 10^{-5}\, 1/M_0^2$ as compared to $7 \times 10^{-5}\, 1/M_0^2$ given by (33). This means that the final entropy will be by factor 8 larger in the case of photons and gravitons that it is in the case of a single scalar. On the other hand it is believed that the Bekenstein–Hawking entropy is fundamental, capturing some essential features of space-time and from that perspective it is hard to imagine that it could depend on the number of massless degrees of freedom in nature, which seems to be rather contingent. The fact that employing a single massless degree of freedom reproduces in our reasoning reproduces the correct value, with a small error, indicates that there might be something special about the single massless scalar field model.

5 Conclusion

In this letter we showed how small backreaction effects can be introduced in the derivation of the thermodynamic entropy of a field in thermal equilibrium in the proximity of a black hole horizon. The resulting changes due to a small but non-vanishing luminosity on the horizon structure of the black hole provide a natural brick wall regulator for the near-horizon modes of the field. Using the small luminosity and quasi-static approximations we were able to solve the field equations in the evaporating metric to find an explicit expression for the field modes, the degrees of freedom contributing to the thermodynamic partition function of the field. We showed that once the width of the quantum ergosphere is set by the Hawking luminosity the horizon contribution to the entropy of the field is in very good agreement with the Bekenstein–Hawking relation for the black hole entropy. In the original brick wall calculation the width of the brick wall had to be adjusted by hand in order to have the correct proportionality factor between entropy and the black hole area. From this point of view we find our result particularly suggestive: the non-trivial horizon geometry determined by the backreaction of the Hawking flux creates a "covariant" brick wall and leaves no arbitrary parameter to be tuned to obtain the desired result.

Acknowledgments This work is based on [16], which is available under the terms of the Creative Commons Attribution License (CC BY) https://creativecommons.org/licenses/by/4.0/. For LB, JKG, and JU this work is supported by funds provided by the National Science Center, projects number 2017/27/B/ST2/01902. ML acknowledges the Fondazione Angelo della Riccia and the Foundation BLANCEFLOR Boncompagni-Ludovisi, née Bildt for financial support. MA acknowledges support from the COST Action MP1405 "QSpace" for a Short Term Scientific Mission Grant supporting a visit to the University of Wroclaw where part of this work was carried out.

References

1. J.D. Bekenstein, Black holes and entropy. Phys. Rev. D **7**, 2333 (1973)
2. S.W. Hawking, Particle Creation by Black Holes. Commun. Math. Phys. **43**, 199 (1975). Erratum: [Commun. Math. Phys. **46**, 206 (1976)]
3. S. Carlip, Black Hole Entropy and the Problem of Universality. arXiv:0807.4192 [gr-qc]
4. W.H. Zurek, K.S. Thorne, Statistical mechanical origin of the entropy of a rotating, charged black hole. Phys. Rev. Lett. **54**, 2171 (1985)
5. G. 't Hooft, On the Quantum Structure of a Black Hole. Nucl. Phys. B **256**, 727 (1985)
6. M. Arzano, S. Bianco, O. Dreyer, From bricks to quasinormal modes: A new perspective on black hole entropy. Int. J. Mod. Phys. D **22**, 1342027 (2013)
7. J.M. Bardeen, Black Holes Do Evaporate Thermally. Phys. Rev. Lett. **46**, 382 (1981)
8. J.W. York, Jr., What happens to the horizon when a black hole radiates?, in *Quantum Theory Of Gravity**, ed. by *Christensen, S.m., pp. 135–147
9. J.W. York, Jr., Dynamical origin of black hole radiance. Phys. Rev. D **28**, 2929 (1983)
10. E. Poisson, *A Relativist's Toolkit: The Mathematics of Black-Hole Mechanics* (Cambridge University, Cambridge, 2009)
11. X. Li, Z. Zhao, Entropy of a Vaidya black hole. Phys. Rev. D **62**, 104001 (2000)
12. L. Vanzo, G. Acquaviva, R. Di Criscienzo, Class. Quant. Grav. **28**, 183001 (2011). https://doi.org/10.1088/0264-9381/28/18/183001
13. A. Fabbri, J. Navarro-Salas, Modeling black hole evaporation (Imp. Coll. Pr., London, UK, 2005)
14. B.S. DeWitt, Quantum field theory in curved spacetime. Phys. Rep. **19**, 295–357 (1975)
15. D.N. Page, Particle Emission Rates from a Black Hole: Massless Particles from an Uncharged, Nonrotating Hole. Phys. Rev. D **13**, 198 (1976). https://doi.org/10.1103/PhysRevD.13.198
16. M. Arzano, L. Brocki, J. Kowalski-Glikman, M. Letizia, J. Unger, Quantum ergosphere and brick wall entropy. Phys. Lett. B **797**, 134887 (2019). https://doi.org/10.1016/j.physletb.2019.134887

Geodesic Structure and Linear Instability of Some Wormholes

Francesco Cremona

Mathematics Subject Classification (2000) 83C10, 83C15, 83C20, 83C25

Contents

1 Introduction

We consider the metric

$$-\left(1+k^2(x^2+b^2)\right)dt^2+\frac{1}{1+k^2(x^2+b^2)}dx^2+(x^2+b^2)d\Omega^2 \quad (-\infty < t < x < +\infty). \tag{1}$$

where $d\Omega^2 = d\theta^2+\sin^2\theta\, d\varphi^2$ ($0<\theta<\pi$, $0<\varphi<2\pi$) is the usual line element of the spherical surface S^2 and b and k are two positive constants, respectively,

F. Cremona (✉)
Dipartimento di Matematica, Universitá di Milano, Milano, Italy
e-mail: francesco.cremona92@gmail.com

S. L. Cacciatori, A. Kamenshchik (eds.), *Einstein Equations: Local Energy, Self-Force, and Fields in General Relativity*, Tutorials, Schools, and Workshops in the Mathematical Sciences, https://doi.org/10.1007/978-3-031-21845-3_6

with the dimension of a length and dimensionless. In [3] (and, more recently, in the survey [2]), Bronnikov finds this metric as a solution to Einstein's equations for a gravitational field minimally coupled to a self-interacting scalar field whose action functional has the kinetic part artificially multiplied by -1; the scalar field and its self-interacting potential found by Bronnikov have the following expressions:

$$\phi(x) = \sqrt{\frac{2}{\kappa}} \arctan \frac{x}{b}, \tag{2}$$

$$V(\phi) = -\frac{k^2}{\kappa}\left[3 - 2\cos^2\left(\sqrt{\frac{\kappa}{2}}\phi\right)\right]. \tag{3}$$

In the limit case $k = 0$, the potential $V(\phi)$ vanishes, and the metric (1) reduces to the well-known metric

$$-dt^2 + dx^2 + (x^2 + b^2)d\Omega^2; \tag{4}$$

this was first introduced almost simultaneously by Ellis [7] and Bronnikov [1] and only at a later time by Morris and Thorne in the famous paper [12] and describes the configuration of a static spherically symmetric wormhole connecting two asymptotically flat universes. In order to take into account all the previously mentioned authors, using the initials of their names, we will refer to the spacetime structure (4) as "EBMT wormhole" and to Eqs. (4), (2) as "EBMT solution".

In the first part of this chapter, we focus on the general metric (1) and build an embedding diagram in a suitable three-dimensional ambient space, justifying the interpretation of this spacetime as a four-dimensional, static spherically symmetric wormhole with a throat of size b connecting two separate universes with asymptotic constant negative scalar curvature $-12k^2$ (Anti-de Sitter universes); for these reasons, throughout the chapter, we will refer to the metric (1) as "AdS-AdS wormhole", or more simply, "AdS wormhole", and to Eqs. (1), (2), and (3) as "AdS-AdS wormhole solution", or more simply, "AdS wormhole solution". From this general construction, in the limit case $k = 0$, we recover exactly the same embedding diagram for the EBMT wormhole (1) which was proposed by Hartle in his textbook [9].

In the second part of the chapter, we recall a general strategy to studying the geodesic motion in a four-dimensional, static spherically symmetric spacetime and specialize it to the case of the AdS wormhole (1); the motion of some timelike and null geodesics is computed numerically and plotted in the embedding diagram of the wormhole.

The last part of the chapter is dedicated to the linear stability analysis of the EMBT wormhole (4). Although the linear instability of this configuration has already been stated in other previous works [2, 4, 8, 14], we will summarize the new derivation proposed in [5]; in this latter paper, the linearized Einstein equations for a time-dependent spherically symmetric perturbation of the EBMT solution (4) and (2) is reduced to a wave-type equation for the perturbed wormhole radius without encountering any singularities in its derivation; we will make a brief comparison between this approach and the one of [2, 4, 8, 14] (for more details about this comparison, see the Introduction of [5]); some words will also be spent on a very recent work [6] in which the linear instability result of the EBMT wormhole is extended to the AdS wormhole solution (1), (2), and (3) in a general gauge-invariant setting.

The chapter is organized as follows: in Sect. 2, we build the metric of a static spherically symmetric spacetime wormhole, starting from its "tunnel-shaped" embedding in an ambient space with constant negative curvature; in Sect. 3, we introduce the EBMT wormhole and the AdS wormhole as solutions to the phantom scalar field Einstein's equation; in Sect. 4, taking advantage of the considerations of Sect. 2, we plot the embedding diagram for the AdS wormhole; in Sect. 5, the qualitative features of the timelike and null geodesics of the AdS wormhole are studied; finally, Sect. 6 contains a brief overview on the linear stability analysis of the EBMT wormhole performed in [5] and a comparison with other previous works on the same subject [2, 4, 8, 14].

Throughout this chapter, we use the signature convention $(-,+,+,+,)$, and we stipulate

$$c = 1\,, \qquad \hbar = 1\,, \qquad \kappa := 8\pi G\,, \tag{5}$$

where c is the speed of light, $\hbar$ is the reduced Planck constant, and G is the gravitational constant.

2 Modelling the Metric of a Wormhole

In the coordinate system

$$(t, x, \theta, \varphi) \in \mathbb{R} \times \mathbb{R} \times (0, \pi) \times (0, 2\pi)\,, \tag{6}$$

we consider a four-dimensional spacetime $(M, \mathfrak{g})$ with metric

$$\mathfrak{g} = -h(x)^2 dt^2 + q(x)^2 dx^2 + r(x)^2 d\Omega^2\,, \tag{7}$$

where $\mathrm{d}\Omega^2 = d\theta^2 + \sin^2\theta\, d\varphi^2$ is (again) the usual line element of the spherical surface S^2 and $h, q, r : \mathbb{R} \to (0, +\infty)$ are smooth functions. Let us start with few elementary considerations about the metric (7):

(i) Since the coefficients of the metric $\mathfrak{g}$ do not depend on the angular coordinates θ and φ, the spacetime M has the property of being spherically symmetric.
(ii) Since the coefficients of the metric $\mathfrak{g}$ do not depend on the temporal coordinate t, the spacetime M is static: the coordinates (x, θ, φ) define a coordinate

system for the "space" of a static observer in this spacetime, while the temporal coordinate t represents its "time".

(iii) The first term $-h(x)^2dt^2$ of $\mathfrak{g}$ is the proper time (physical time) measured by someone which is at rest according to the static observer; note that the proper time depends on the spatial variable x and, therefore, the function h can be used to quantify the "gap" between the time lengths signed by two clocks at rest for the static observer depending on their relative positions.

(iv) The second term

$$q(x)^2dx^2 + r(x)^2d\Omega^2 \tag{8}$$

is the metric of each of the three-dimensional spherically symmetric manifolds $\mathbf{t} := \{\mathcal{A} \in M \mid t(\mathcal{A}) = const\} \subset M$, which represent the space seen by the static observer at a fixed time; for this reason, we can say that the functions q and r determine what we reasonably would call the "shape" of the spacetime $(M, \mathfrak{g})$.

A very useful tool for visualizing the geometrical properties of the spatial slice $\mathbf{t}$ is the embedding diagram. In general, a smooth map $\iota : M \to N$ between two differential manifolds M and N is an embedding if both ι and its differential $d\iota$ are everywhere injective and $\iota : M \to \iota(M)$ is a homeomorphism. We will see that it might be impossible to embed the three-dimensional slice $\mathbf{t}$ in a four-dimensional flat space. However, if it is the case, profiting from the spherical symmetry of (8), one can build up a picture of the embedded slice $\iota(\mathbf{t})$ fixing the value of an angle. In fact, the embedding ι transforms the spacetime slices $\mathbf{t}_{\theta_0} := \{\mathcal{A} \in M \mid t(\mathcal{A}) = const, \theta(\mathcal{A}) = \theta_0\} \subset M$ into two-dimensional surfaces in the three-dimensional Euclidean space: these representations of the spatial part of a spacetime are called *embedding diagrams*. Note that, as the value of θ_0 is immaterial, from now on, we fix $\theta_0 = \pi/2$.

After these preliminary considerations, we can propose a naive definition of a static spherically symmetric wormhole spacetime: the metric (7) represents a static spherically symmetric wormhole if each of its $\mathbf{t}_{\frac{\pi}{2}}$ slices (defined by the metric (8) with $\theta = \pi/2$), once embedded in a three-dimensional flat space, looks as a "tunnel-shaped" hypersurface, a form familiar from popular accounts of wormholes. However, this statement is too restrictive since, as we will see later, there are spacetimes whose $\mathbf{t}_{\frac{\pi}{2}}$ slices cannot be embedded in a flat space, but nevertheless, in some sense these slices have still the "shape of a tunnel" if embedded in a different suitable ambient space.

In the remaining part of this section, we will give a more precise mathematical setting to the consideration of the previous paragraph with an example: we will build up the metric of a spacetime whose $\mathbf{t}_{\frac{\pi}{2}}$ slice is embeddable as a "tunnel-shaped" surface in a three-dimensional space with constant negative curvature.

Let us start considering a four-dimensional Riemannian manifold M_k with constant scalar negative curvature $R_{M_k} := -12k^2$, $k > 0$; this is the ambient space in which the slice $\mathbf{t}$ will be embedded. The reasons why we are interested in M_k

(instead of a four-dimensional flat space) will be clear a little bit later. It is widely known that introducing the system of coordinates $(z, \rho, \hat{\theta}, \hat{\varphi}) \in \mathbb{R} \times (0, +\infty) \times (0, \pi) \times (0, 2\pi)$, the line element of the space M_k reads

$$d\sigma_k^2 = \alpha(\rho)^2 dz^2 + \frac{1}{\alpha(\rho)^2} d\rho^2 + \rho^2 d\hat{\Omega}^2\,, \qquad \alpha(\rho) = \sqrt{1 + k^2\rho^2} \tag{9}$$

(here, again, $d\hat{\Omega}^2 = d\hat{\theta}^2 + \sin^2\hat{\theta}\, d\hat{\varphi}^2$). A very easy way to define a "tunnel-shaped" hypersurface $\mathcal{S}$ in M_k is

$$\mathcal{S} := \{(z, \rho, \hat{\theta}, \hat{\varphi}) \,:\, \rho = F(z)\} \subset M_k,$$

where the function $F : \mathrm{dom}(F) \subseteq \mathbb{R} \to (0, +\infty)$ is smooth and possesses a positive minimum of size $b > 0$ at a certain point of its domain, let us say, without loss of generality, at $z_0 = 0 \in \mathrm{dom}(F)$; moreover, we prescribe that $z_0 = 0$ is the only minimum point of F. In short, we require that the function F, which can be effectively regarded as the "profile" of the tunnel, has the following properties:

$$F \in \mathcal{C}^\infty\,, \quad F(0) = b > 0\,, \quad F'(0) = 0\,, \quad F'(z)z > 0 \ \text{ for all } z \in \mathrm{dom}(F)/\{0\}\,. \tag{10}$$

With this position, it is clear that the minimum b of the function F is the size of the tunnel throat, while the large z limits represent the far ends of two separate hypersurfaces of M_k linked by the tunnel throat; these hypersurfaces are defined, respectively, by $\{(z, F(z), \hat{\theta}, \hat{\varphi}) \,|\, z > 0\}\}$ and $\{(z, F(z), \hat{\theta}, \hat{\varphi}) \,|\, z < 0\}\}$. Moreover, as $\rho = F(z)$ and $d\rho = F'(z)dz$ on $\mathcal{S}$, the metric of the whole hypersurface $\mathcal{S}$ reads

$$dS^2 = \left[\alpha(F(z))^2 + \frac{F'(z)^2}{\alpha(F(z))^2}\right] dz^2 + F(z)^2 d\hat{\Omega}^2\,. \tag{11}$$

Therefore, we are looking for a spacetime M with metric (7) such that, given a "profile function" F, it exists an embedding $\iota : M \to M_k$ such that the embedded slice $\mathcal{S} := \iota(\mathbf{t}) \subset M_k$ has the metric (11). Working in coordinates, the embedding ι is specified by four smooth bijections:

$$z = z(x, \theta, \varphi)\,, \quad \rho = \rho(x, \theta, \varphi)\,, \quad \hat{\theta} = \hat{\theta}(x, \theta, \varphi)\,, \quad \hat{\varphi} = \hat{\varphi}(x, \theta, \varphi)\,;$$

note that, as we are in a radially symmetric configuration, we can take $\hat{\theta} = \theta$, $\hat{\varphi} = \varphi$ and the functions $z(\cdot)$ and $\rho(\cdot)$ to be angles-independent, i.e. we can set

$$z = z(x)\,, \quad \rho = \rho(x)\,, \quad \hat{\theta} = \theta\,, \quad \hat{\varphi} = \varphi\,. \tag{12}$$

Provided $\mathrm{Im}(z) \subseteq \mathrm{dom}(F)$, one can insert the embedding functions (12) and their differentials into (11), obtaining the original form of the $\mathbf{t}$ slice metric:

$$d\mathbf{t}^2 = \alpha(F(z(x)))^2 z'(x)^2 \left[1 + \frac{F'(z(x))^2}{\alpha(F(z(x)))^4}\right] dx^2 + F(z(x))^2 d\hat{\Omega}^2 . \tag{13}$$

Without loss of generality, from now on, we also stipulate that $z(0) = 0$ and $z'(x) > 0$ for all $x \in \mathbb{R}$.

By comparing the two expressions (8) and (13) for the metric of the slice $\mathbf{t}$, we get the expressions of the functions q and r in dependence of F and $z = z(x)$:

$$r(x) = F(z(x)) , \tag{14}$$

$$q(x) = \alpha(F(z(x)))z'(x)\sqrt{1 + \frac{F'(z(x))^2}{\alpha(F(z(x)))^4}} . \tag{15}$$

Summing up, in this section, we have proved that, for all smooth functions $h : \mathbb{R} \to (0, +\infty)$, $z : \mathbb{R} \to \mathrm{Im}(z)$, $F : \mathrm{dom}(F) \subseteq \mathbb{R} \to [b, +\infty)$ such that

(i) z is a bijection with $z'(x) > 0$ for all $x \in \mathbb{R}$.
(ii) $\mathrm{Im}(z) \subseteq \mathrm{dom}(F)$.
(iii) $z(0) = 0 \in \mathrm{dom}(F)$.
(iv) F satisfies the requirements in (10).

The metric (7) with r and q as in (14), (15) and α as in (9) represents a static spherically symmetric wormhole spacetime; indeed, the embedded hypersurface $\mathcal{S} = \iota(\mathbf{t}) \subset M_k$ has a "tunnel-shaped" structure with a throat of size b located at $z = 0$. Thanks to the injectivity of the embedding function z, one can easily see that the slice $\mathbf{t}$ of the wormhole spacetime has the throat of size b at $x = 0$ and links together the two separate universes defined, respectively, by $\{\mathcal{A} \in M \mid x(\mathcal{A}) > 0\}$ and $\{\mathcal{A} \in M \mid x(\mathcal{A}) < 0\}$.

2.1 An Embedding for the EBMT Wormhole

To our knowledge, there is at least one example in the literature of the previous construction in the limit case $k = 0$, namely in the case of a static spherically symmetric spacetime whose spatial part is embeddable as a "tunnel-shaped" hypersurface in a flat ambient space (indeed, for $k = 0$, the space M_k has zero curvature): in his textbook [9], Hartle describes how to build the embedding of the $\mathbf{t}_{\frac{\pi}{2}}$ slice of the EBMT wormhole (4) (therein referred generically as "a Wormhole Spacetime") in the three-dimensional Euclidean space $\mathbb{R}^3$. The construction made by Hartle leads to an embedding diagram for this wormhole, described (in our notation) by the profile and the embedding functions

$$F_{\text{EBMT}}(z; b) := b \cosh(z/b) , \qquad z_{\text{EBMT}}(x; b) := b \,\mathrm{arcsinh}(x/b). \tag{16}$$

In the following section, we show that these functions can be generalized in the case $k \neq 0$ in order to describe the embedding of the AdS wormhole.

3 Deriving Wormholes from Einstein's Equation

In a four-dimensional spacetime $(M, \mathfrak{g})$, we consider a gravitational field minimally coupled to a real scalar field ϕ with a non-null self-interacting potential $V(\phi)$. This system is described by the action functional

$$S[\mathfrak{g}_{\mu\nu}, \phi] := \int \left(\frac{R}{2\kappa} - \frac{\sigma}{2} \partial^{\mu}\phi \, \partial_{\mu}\phi \right) dv, \tag{17}$$

where κ, R and dv are, respectively, the usual coupling constant defined in Eq. (5), the scalar curvature of the Lorentzian manifold $(M, \mathfrak{g})$ and the volume element corresponding to the metric $\mathfrak{g}$ ($dv = \sqrt{|\det(\mathfrak{g}_{\mu\nu})|} \prod_{\lambda} dx^{\lambda}$ in any spacetime coordinate system $(x^{\lambda})_{\lambda}$); σ is a constant which is normally set to 1. Both the metric and the scalar field are assumed to be smooth.

The first stationarity condition $\delta S / \delta \mathfrak{g}_{\mu\nu} = 0$ is equivalent to Einstein's equations

$$R_{\mu\nu} - \frac{1}{2} \mathfrak{g}_{\mu\nu} R = \kappa T_{\mu\nu} \,, \tag{18}$$

where stress–energy tensor is given by

$$T_{\mu\nu} := \sigma \left(\partial_{\mu}\phi \, \partial_{\nu}\phi - \frac{1}{2} \mathfrak{g}_{\mu\nu} \partial^{\lambda}\phi \, \partial_{\lambda}\phi \right) - \mathfrak{g}_{\mu\nu} V(\phi) \,, \tag{19}$$

The second stationarity condition $\delta S / \delta \phi = 0$ gives the Klein–Gordon equation for the scalar field ϕ

$$\Box \phi = \sigma V(\phi) \,, \tag{20}$$

where $\Box := \nabla_{\mu} \nabla^{\mu}$ and ∇_{μ} is the covariant derivative induced by the metric $\mathfrak{g}$.

As we have already mentioned, if ϕ is an ordinary field, the constant σ equals 1; however, a different and interesting class of scalar filed is defined by the position $\sigma = -1$: these are called *phantom scalar fields*. The latter anomalous choice of the value of σ corresponds to an artificial change of the sign of (the kinetic part of) the action functional (17) and reproduces, in the background of General Relativity, a surprising and well-known feature of quantum fields in their vacuum states [10, 15]. In their almost simultaneous papers, Ellis [7] and Bronnikov [1] introduced actually a phantom scalar field in order to prove that the EBMT wormhole

$$-\, dt^2 + dx^2 + (x^2 + b^2) d\Omega^2 \tag{21}$$

could be seen as a solution to Einstein's field equations (18), (19), and (20); the phantom scalar field ϕ which supports the metric (21) was found to be non-self-interactive and to depend only on the spatial coordinate x:

$$\phi(x) = \sqrt{\frac{2}{\kappa}} \arctan \frac{x}{b} . \tag{22}$$

More recently, in 2006, Bronnikov [3] proposed a generalization of the EBMT wormhole solution: with some reparameterization of the constants involved therein, this new general spacetime was found to have a metric of the form (7) with coefficients

$$h(x) = q^{-1}(x) = \sqrt{1+\left(x^2+b^2\right)\left(M \arctan \frac{x}{b} - K\right) + bMx}\,, \quad r(x) = \sqrt{x^2+b^2} \tag{23}$$

and to be supported by a phantom scalar field with a self-interacting potential given by

$$\phi = \sqrt{\frac{2}{\kappa}} \arctan \frac{x}{b}\,, \tag{24}$$

$$\begin{aligned} V(\phi) = {} & \frac{K}{\kappa}\left[3 - 2\cos^2\left(\sqrt{\frac{\kappa}{2}}\phi\right)\right] \\ & - \frac{M}{\kappa}\left\{3 \sin\left(\sqrt{\frac{\kappa}{2}}\phi\right) \cos\left(\sqrt{\frac{\kappa}{2}}\phi\right) + \sqrt{\frac{\kappa}{2}}\phi\left[3 - 2\cos^2\left(\sqrt{\frac{\kappa}{2}}\phi\right)\right]\right\}, \end{aligned} \tag{25}$$

where K, M and $b > 0$ are three constants. Obviously, the EBMT wormhole solution is recovered in the limit case $M, K = 0$.

Let us now focus on the particular choices $M = 0$ and $K = -k^2 < 0$ (with $k > 0$) so that the general solution (23), (24), and (25) reduces to

$$h(x) = q^{-1}(x) = \sqrt{1 + k^2(x^2+b^2)}\,, \quad r(x) = \sqrt{x^2+b^2}\,, \tag{26}$$

$$\phi = \sqrt{\frac{2}{\kappa}} \arctan \frac{x}{b}\,, \qquad V(\phi) = -\frac{k^2}{\kappa}\left[3 - 2\cos^2\left(\sqrt{\frac{\kappa}{2}}\phi\right)\right] . \tag{27}$$

In the limit case $b = 0$, the metric (26) is non-singular for $x > 0$ and $x < 0$ and describes in this two regions as many Anti-de Sitter (AdS) universes with constant negative scalar curvature $R_{\text{AdS}} = -12k^2$, while, in the general case $b > 0$, the functions h, q, r are smooth for all $x \in \mathbb{R}$; since $r(x) \sim |x|$ and $h(x) = q^{-1}(x) \sim \sqrt{1+k^2x^2}$ for $x \to \pm\infty$, the metric (26) approaches asymptotically to the AdS metric and, therefore, we reasonably would say that, for every $b > 0$, the metric (26) describes a wormhole with a throat of size b connecting two separate asymptotically AdS universes with negative scalar curvature $R_{\text{AdS}} = -12k^2$; for these reasons, as already mentioned in the Introduction section, we will refer to the metric (26) as "AdS-AdS wormhole", or more simply, "AdS wormhole".

4 An Embedding for the AdS Wormhole

In this subsection, we show how to find a profile function F and an embedding function z (introduced in Sect. 2) for the AdS wormhole. More precisely, we are looking for two functions F and z satisfying the requirements (i)–(iv) under Eq. (15) and such that the metric (7), (14), and (15) reduces to the AdS metric (26).

We start recalling that $\rho(x) = F(z(x))$ on the embedded hypersurface $\mathcal{S} = \iota(\mathbf{t})$, so that Eqs. (14), and (15) are equivalent to

$$r(x) = \rho(x)\,, \tag{28}$$

$$q(x) = \alpha(\rho(x))\sqrt{z'(x)^2 + \frac{\rho'(x)^2}{\alpha(\rho(x))^4}}\,, \tag{29}$$

$$F(z) := \rho(x(z))\,, \tag{30}$$

where $x = x(z)$ is the inverse function of $z = z(x)$ and α is as in (9).

If we know the metric coefficients q and r, Eqs. (28) and (29) become a system of two differential equations in the unknown $\rho = \rho(x)$, $z = z(x)$, which is a trivially solved setting

$$\rho(x) = r(x)\,, \quad z(x) = \int_0^x \frac{1}{\alpha(r(\tilde{x}))}\sqrt{q(\tilde{x})^2 - \frac{r'(\tilde{x})^2}{\alpha(r(\tilde{x}))^2}}d\tilde{x} \tag{31}$$

(note that in solving Eq. (29) we have required that $z(0) = 0$).

At this point, we think that it is necessary to spend some words about the use of the parameter k that we have made until now. In Sect. 2, $k \equiv k_{\mathrm{as}} = \sqrt{-R_{M_k}/12}$, where R_{M_k} is the constant negative scalar curvature of the ambient space of the embedding, while, in the present section, $k \equiv k_{\mathrm{w}} = \sqrt{-R_{\mathrm{AdS}}/12}$, where R_{AdS} is the asymptotic constant negative scalar curvature of the AdS wormhole. Note that the integral in Eq. (31) makes sense only if the functions q, r and α are such that $q(x)^2 - r'(x)^2/\alpha(r(x)^2 \geq 0$ for every $x \in \mathbb{R}$; in the case of the AdS wormhole, i.e. choosing the functions q and r as in Eq. (26) with $k = k_{\mathrm{w}}$ and recalling the definition of α in Eq. (9) with $k = k_{\mathrm{as}}$, this occurs whenever k_{as} is chosen such that $k_{\mathrm{as}} \geq k_{\mathrm{w}}$. We would like to stress the fact that in any other case it is not possible to embed the whole slice $\mathbf{t}$ of the AdS wormhole in an ambient space with constant curvature; in particular, it is impossible to embed $\mathbf{t}$ in a four-dimensional flat space (defined by the limit occurrence $k_{\mathrm{as}} = 0$) unless $k_{\mathrm{w}} = 0$ (which is the case of the EBMT wormhole). From now on, we make the easiest choice possible $k = k_{\mathrm{as}} = k_{\mathrm{w}}$, so that the second integral in Eq. (31) can be easily solved giving the following explicit expression for the function $z = z(x)$:

$$z(x) \equiv z_{\mathrm{AdS}}(x; k, b) = \frac{b}{\sqrt{1+b^2k^2}} \operatorname{arcsinh}\left(\frac{x}{b\sqrt{1+k^2(x^2+b^2)}}\right) ; \tag{32}$$

this can be inverted so that Eq. (30) reads

$$F(z) \equiv F_{\mathrm{AdS}}(z; k, b) = \frac{b \cosh\left(\frac{\sqrt{1+b^2k^2}z}{b}\right)}{\sqrt{1+b^2k^2\left(1-\cosh^2\left(\frac{\sqrt{1+b^2k^2}z}{b}\right)\right)}} . \tag{33}$$

An elementary computation shows that, for all $b > 0$ and $k > 0$, the functions z and F satisfy the conditions (i)–(iv) after Eq. (15). Note that Eqs. (32) and (33) actually generalize Eq. (16) to the case $k > 0$ as $z_{\mathrm{AdS}}(x; 0, b) = z_{\mathrm{EBMT}}(x; b)$ and $F_{\mathrm{AdS}}(x; 0, b) = F_{\mathrm{EBMT}}(x; b)$.

At this point, one might want to visualize the embedding of the AdS wormhole, namely a three-dimensional picture of the embedded slice $\iota(\mathbf{t}_{\frac{\pi}{2}})$. Obviously, this is not possible since the ambient space M_k is not flat, unless we settle for an approximation. For example, we observe that in the limit case $k \to 0$ or $\rho \to 0$ the metric of the ambient space (9) tends to become flat. This means that the bidimensional surface in $\mathbb{R}^3$ defined as

$$\mathcal{S}_{\mathrm{AdS}}(k, b) := \{(z, \rho, \varphi) : \rho = F_{\mathrm{AdS}}(z; k, b)\} \subset \mathbb{R}^3$$

approaches to $\iota(\mathbf{t}_{\frac{\pi}{2}}) \subset M_k$ in a region "suitably close to" the origin; this region can be very large if k is very small. In Fig. 1, one can see the plots of the profile function F_{AdS} and the embedding diagram $\mathcal{S}_{\mathrm{AdS}} \subset \mathbb{R}^3$ for a particular choice of the parameters b and k.

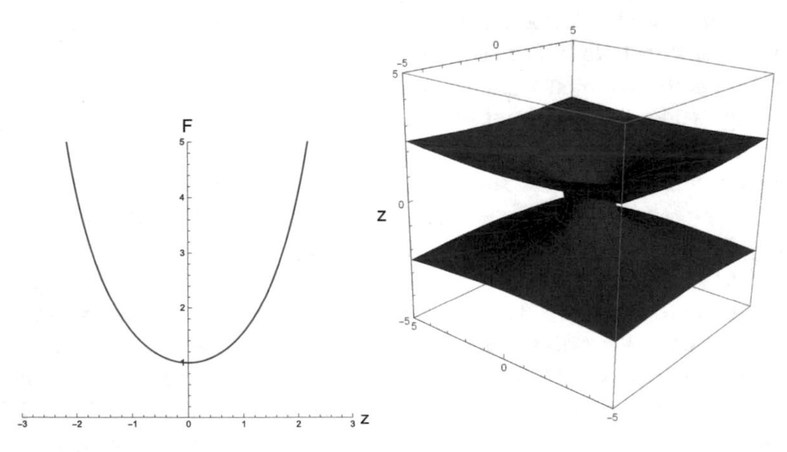

Fig. 1 The profile function F_{AdS} and the embedding diagram $\mathcal{S}_{\mathrm{AdS}}$ of the AdS wormhole with $b = 1$ and $k = 0.1$

5 The Timelike and Null Geodesics Motion

5.1 General Spherically Symmetric Case

Let us firstly recall that the trajectory described by a free-falling particle or by a light ray in a spacetime $(M, \mathfrak{g})$ is represented by a geodesic of M, i.e. a world line $\tau \mapsto \mathcal{P}(\tau)$ such that [13, 16]

$$\frac{\nabla}{d\tau}\frac{d\mathcal{P}}{d\tau} = 0, \tag{34}$$

where ∇ is the covariant derivative defined by the Levi-Civita connection of the metric $\mathfrak{g}$. Moreover, it is always possible to redefine the parameter τ in such a way that, for all τ,

$$\mathfrak{g}_{\mathcal{P}(\tau)}\left(\frac{d\mathcal{P}}{d\tau}, \frac{d\mathcal{P}}{d\tau}\right) = -\mathrm{k}, \qquad \mathrm{k} = \begin{cases} 1 \text{ for a timelike geodesic (free-falling particle)} \\ 0 \text{ for a null geodesic (light ray).} \end{cases} \tag{35}$$

In this section, we will study the geodesic motion in a four-dimensional spherically symmetric spacetime defined in coordinate system $(x^\mu) := (t, x, \theta, \varphi) \in \mathbb{R} \times \mathbb{R} \times (0, \pi) \times (0, 2\pi)$ by the metric

$$\mathfrak{g} = -h(x)^2 dt^2 + \frac{1}{h(x)^2} dx^2 + r(x)^2 d\Omega^2 ; \tag{36}$$

each geodesic $\mathcal{P}$ in this spacetime is described locally by four functions of the parameter τ

$$(x^\mu(\mathcal{P}(\tau))) =: (x^\mu(\tau)) =: (t(\tau), x(\tau), \theta(\tau), \varphi(\tau)) \tag{37}$$

satisfying the geodesic equation (34) which locally reads

$$\frac{d^2 x^\mu}{d\tau^2} + \Gamma^\mu_{\lambda\nu} \frac{dx^\lambda}{d\tau}\frac{dx^\nu}{d\tau} = 0 , \tag{38}$$

where $\Gamma^\mu_{\lambda\nu}$ are the Christoffel symbols of the Levi-Civita connection of $\mathfrak{g}$.

Moreover, it can be proved [11] that the geodesic equations (38) are equivalent to the Euler–Lagrange equations $\frac{d}{d\tau}\frac{\partial \mathcal{L}}{\partial \dot{x}^\mu} - \frac{\partial \mathcal{L}}{\partial x^\mu} = 0$, $\mu = 1, ..., 4$, for the Lagrangian

$$\mathcal{L}(x^\mu, \dot{x}^\mu) := \frac{1}{2}\mathfrak{g}_{\lambda\nu}(x^\mu)\dot{x}^\lambda \dot{x}^\nu = -\frac{h(x)^2}{2}\dot{t}^2 + \frac{1}{2h(x)^2}\dot{x}^2 + \frac{r^2}{2}\left(\dot{\theta}^2 + \sin^2\theta\, \dot{\varphi}^2\right) ; \tag{39}$$

these are satisfied if and only if the following system of four ordinary differential equations hold:

$$\frac{d}{d\tau}\left(h(x)^2\dot{t}\right) = 0\,, \tag{40}$$

$$\frac{d}{d\tau}\left(\frac{1}{h(x)^2}\dot{x}\right) = h(x)r(x)r'(x)\left(\dot{\theta}^2 + \sin^2\theta\,\dot{\varphi}^2\right) - h(x)^2h'(x)\dot{t}^2\,, \tag{41}$$

$$\frac{d}{d\tau}\left(r(x)^2\dot{\theta}\right) = r(x)^2\sin\theta\cos\theta\,\dot{\varphi}^2\,, \tag{42}$$

$$\frac{d}{d\tau}\left(r(x)^2\dot{\varphi}\right) = \frac{d}{d\tau}\left(r(x)^2\dot{\varphi}\cos^2\theta\right)\,. \tag{43}$$

Before starting with the study of the system (40)–(43), let us summarize some general and useful results about Lagrangian systems:

(i) In a time-independent n-dimensional Lagrangian system ($\mathcal{L}(q^i, \dot{q}^i)$, $q^i = q^i(t)$, $i = 1, ..., n$), the total energy function defined by $\mathcal{E} := \frac{\partial\mathcal{L}}{\partial\dot{q}^i}\dot{q}^i - \mathcal{L}$ is conserved.

(ii) In the hypothesis of (i), the system of the n Euler–Lagrange equations for the Lagrangian $\mathcal{L}$ is equivalent to the system made up of $n-1$ Euler–Lagrange equations and the conservation law $\mathcal{E} = const$;

(iii) In the hypothesis of (i) and if the Lagrangian $\mathcal{L}$ is a quadratic function in the generalized velocities $\dot{q}^i$, it follows that $\mathcal{E} = \mathcal{L}$ and the conservation law reads $\mathcal{L} = const$.

These results immediately apply to the Lagrangian (39); moreover, since for all τ

$$\mathcal{L}(x^\mu(\tau), \dot{x}^\mu(\tau)) = \frac{1}{2}\mathfrak{g}_{\lambda\nu}(x^\mu(\tau))\dot{x}^\lambda(\tau)\dot{x}^\nu(\tau) = \frac{1}{2}\mathfrak{g}_{\mathcal{P}(\tau)}\left(\frac{d\mathcal{P}}{d\tau}, \frac{d\mathcal{P}}{d\tau}\right)\,,$$

recalling the position (35), we have that the conservation law becomes $\mathcal{L} = -\mathtt{k}/2$.

We are now ready to study the system (40)–(43), starting from the third equation (42); obviously, this equation and the initial conditions

$$\tau_0 = 0\,, \quad \theta(\tau_0) = \frac{\pi}{2}\,, \quad \dot{\theta}(\tau_0) = 0 \tag{44}$$

imply that $\theta(\tau) = \frac{\pi}{2}$ for every τ. Since it is always possible to redefine the coordinates θ and φ and the parameter τ so that the previous conditions on θ are true, from now on we assume (44); in this way, the four-dimensional system (40)–(43) reduces to the three-dimensional system:

$$\frac{d}{d\tau}\left(h(x)^2\dot{t}\right) = 0\,, \tag{45}$$

$$\frac{d}{d\tau}\left(r(x)^2\dot{\varphi}\right) = 0\,, \tag{46}$$

$$-\frac{h(x)^2}{2}\dot{t}^2 + \frac{1}{2h(x)^2}\dot{x}^2 + \frac{r(x)^2}{2}\dot{\varphi}^2 = -\frac{\mathrm{k}}{2}\,, \tag{47}$$

where we have substituted the second equation (41) with the conservation law $\mathcal{L} = -\mathrm{k}/2$, thanks to (i)–(iii) after Eq. (43) and the forthcoming remark.

Let us start with Eqs. (45) and (46); note that, hopefully performing the parameter change $\tau \mapsto -\tau$, the vector $d\mathcal{P}/d\tau$ can be regarded as future-oriented, so that one can always suppose that $\dot{t} > 0$. Thanks to this remark, it is clear that Eqs. (45) and (46) hold if and only if there exist two constants $E \geq -\mathrm{k}/2$ and $L \in \mathbb{R}$ [16] such that

$$h(x)^2\dot{t} = \sqrt{\mathrm{k} + 2E}\,, \qquad r(x)^2\dot{\varphi} = L\,; \tag{48}$$

these two equations can be easily solved, leading to

$$t(\tau) = \int_0^\tau \frac{\sqrt{\mathrm{k}+2E}}{h(x(\tilde{\tau}))^2}d\tilde{\tau} + t(0)\,, \qquad \varphi(\tau) = \int_0^\tau \frac{L}{r(x(\tilde{\tau}))^2}d\tilde{\tau} + \varphi(0)\,. \tag{49}$$

Note that it results from Eq. (48) that the two constants E and L are fully determined by the initial data:

$$E := \frac{\dot{t}(0)^2 h(x(0))^4 - \mathrm{k}}{2}\,, \qquad L = \dot{\varphi}(0) r(x(0))^2\,.$$

It is easy to see that in the limit case of a particle moving slowly in a weak gravitational potential (i.e. $h(x) \simeq 1$) it is possible to prove that E and L approach, respectively, to the classical total energy and the angular momentum per unit rest mass of the particle [16]; therefore, in the timelike case, one can interpret L and E as a relativistic generalization of the total energy and the angular momentum per unit rest mass of a free-falling particle and, in the null case, $\hbar L$ and $\hbar E$ as the angular momentum and the total energy of a photon (recall however that in (5) we have stipulated $\hbar = 1$).

Inserting Eq. (49) into the conservation law (47), we have that the reduced Lagrangian system (45), (46), and (47) is equivalent to the dynamical system made up of Eq. (49) and

$$\frac{1}{2}\dot{x}^2 + V_{\mathrm{eff}}(x) = E \tag{50}$$

where we have defined the effective potential

$$V_{\text{eff}}(x) := \frac{L^2}{2} \frac{h(x)^2}{r(x)^2} + \frac{\mathrm{k}}{2} \left(h(x)^2 - 1 \right) . \tag{51}$$

Summing up, provided a suitable change of coordinates, the problem of finding the qualitative behaviour of a timelike ($\mathrm{k} = 1$) or a null ($\mathrm{k} = 0$) geodesic in a spacetime with a metric of the form (36) is reduced to studying its radial motion, which satisfies Eqs. (50) and (51); since this radial motion is the same as the motion of a unit mass particle of energy E in ordinary one-dimensional, nonrelativistic mechanics moving in the effective potential V_{eff}, in order to understand the qualitative features of the geodesic motion one has to investigate the analytical properties of V_{eff} in dependence of the values of the parameters appearing in its definition (51).

5.2 *The Case of the AdS Wormhole*

Let us consider the AdS wormhole (which has obviously a metric of the form (36)); in this case, we have that the effective potential (51) reads

$$V_{\text{eff}}(x) \equiv V_{\text{eff,AdS}}(x; b, k, \mathrm{k}, L) := \frac{L^2}{2} \left(\frac{1}{x^2 + b^2} + k^2 \right) + \frac{\mathrm{k}}{2} k^2 (x^2 + b^2) . \tag{52}$$

In the following, a complete analysis of the analytical properties of V_{eff} will be performed by varying the values of the parameters b, k, k, L in their respective ranges; in this way, we can deduce some information about the geodesic motion in the AdS wormhole.

We start from the limit case of the EBMT wormhole corresponding to the choice $k = 0$: depending on the value of the angular momentum L, we encounter only two qualitatively different situations: if $L = 0$, the potential V_{eff} is identically null, while, if $L \neq 0$, it possesses an asymptotically null "bell curve" shape with the maximum $L^2/(2b^2)$ located in $x = 0$. This means that in the EBMT wormhole:

(i) There is no difference between timelike and null geodesics.
(ii) If $E > L^2/(2b^2)$, both particles and light rays heading towards the centre of the wormhole will pass from one universe to the other and never come back (unless they accelerate or are deviated).

We now focus on the AdS case, so suppose $k > 0$. Considering firstly the motion of a light ray ($\mathrm{k} = 0$), we have a situation similar to that of the EBMT wormhole: if $L = 0$, the potential is again identically null, while if $L \neq 0$, the potential V_{eff} has once more a "bell curve" shape with the maximum of value $L^2(k^2 + 1/b^2)/2$ located in $x = 0$ and a non-null horizontal asymptote of value $L^2k^2/2$. In Fig. 2, the effective potential V_{eff} and the total energy E have been plotted in three different possible occurrences; in each case, the motion of a particular null geodesic $\mathcal{P}(\tau)$, $\tau \in [0, \tau_1]$ has been computed numerically and plotted in the embedding diagram of the AdS wormhole.

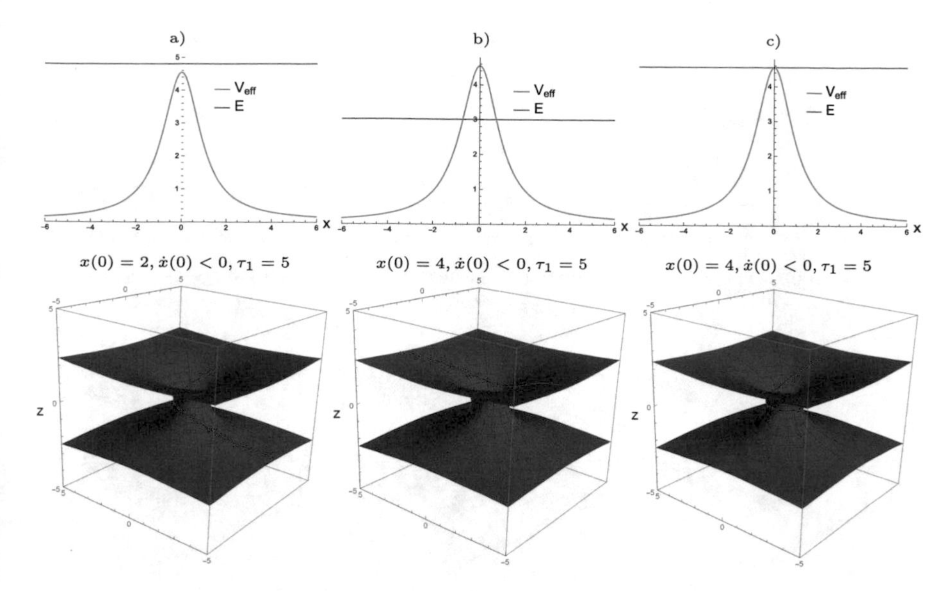

Fig. 2 Effective potential and embedding diagram of null geodesics in the AdS wormhole with $b = 1$ and $k = 0.1$. (**a**) $L = 3$, $E = 4.8$. (**b**) $L = 3$, $E = 3$. (**c**) $L = 3$, $E \simeq 4.545$

Secondly, we consider the timelike geodesic motion (k $= 1$). If $|L| \leq b^2k$, the effective potential is a convex function with the minimum $V_0 := b^2k^2/2 + L^2(k^2 + 1/b^2)/2$ at $x = 0$; if $|L| > b^2k$, the effective potential has a "Mexican hat" shape, with the local maximum of value V_0 located in $x = 0$, the two local minima in $x = \pm\sqrt{|L|/k - b^2}$ both of value $k|L|(1 + k|L|/2)$ and limits $V_{\text{eff}}(x) \to +\infty$ for $x \to \pm\infty$. Therefore, in the AdS wormhole, the timelike geodesics:

(i) Orbit in a bounded region of the spacetime which depends on b, k, E, L and is defined by

$$\left\{\mathcal{A} : |x(\mathcal{A})| \leq \frac{1}{\sqrt{2}k}\sqrt{2E - k^2(L^2 + 2b^2) + \sqrt{4E^2 + k^2L^2(k^2L^2 - 4(E+1))}}\right\}.$$

(ii) If $E > V_0$, the particles pass from one universe to the other and keep oscillating between them (unless they accelerate).

Figure 3 contains the plot of three possible effective potentials V_{eff} and values of the total energy E for the timelike geodesic motion; as in Fig. 2, for each possibility, the motion of a particular null geodesic $\mathcal{P}(\tau)$, $\tau \in [0, \tau_1]$, has been computed numerically and plotted in the embedding diagram of the AdS wormhole.

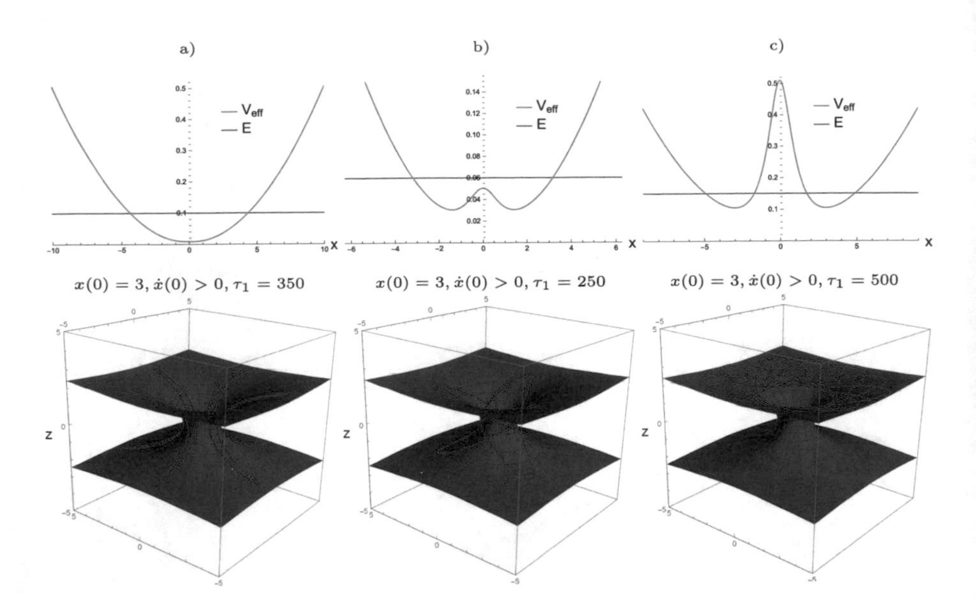

Fig. 3 Effective potential and embedding diagram of timelike geodesics in the AdS wormhole with $b = 1$ and $k = 0.1$. (**a**) $L = 0.09$, $E = 0.1$. (**b**) $L = 0.3$, $E = 0.06$. (**c**) $L = 1$, $E = 0.15$

6 Linear Instability of the EBMT Wormhole

6.1 Radial Perturbations of the EBMT Wormhole

In the first three subsections of this section, we focus on the EBMT wormhole solution defined by Eqs. (21) and (22) and briefly review the deduction of the linear instability of this configuration as it is performed in [5].

The starting point of this analysis is the introduction of a time-dependent spherically symmetric perturbation of the EBMT wormhole solution; this means that we consider a time-dependent spherically symmetric metric

$$\mathfrak{g} = -h(t,x)^2 dt^2 + q(t,x)^2 dx^2 + r(t,x)^2 d\Omega^2$$

defined by the positions

$$h(t,x) = 1 + \delta h(t,x)\,, \quad q(t,x) = 1 + \delta q(t,u)\,, \quad r(t,x) = \sqrt{x^2+b^2} + \delta r(t,x) \tag{53}$$

and a non-self-interactive phantom scalar field defined by

$$\phi(t,x) = \sqrt{\frac{2}{\kappa}}\arctan\frac{x}{b} + \delta\phi(t,x)\,, \qquad V(\phi) = 0\,. \tag{54}$$

With the purpose of simplifying the future computation, we stipulate that the perturbation functions δh, δq, δr and $\delta\phi$ have the expressions[1]

$$\delta h(t,x) := 0\,, \quad \delta q(s,u) := \varepsilon \mathcal{Q}\left(\frac{t}{b},\frac{x}{b}\right),$$

$$\delta r(t,x) := \frac{\varepsilon b^2}{x^2+b^2}\mathcal{R}\left(\frac{t}{b},\frac{x}{b}\right), \quad \delta\phi(t,x) := \varepsilon\sqrt{\frac{2}{\kappa}}\Phi\left(\frac{t}{b},\frac{x}{b}\right), \tag{55}$$

where ε is a small real parameter and $\mathcal{Q}, \mathcal{R}, \Phi : \mathbb{R}\times\mathbb{R} \to \mathbb{R}$ are smooth, dimensionless functions to be determined. Moreover, let us define the new variables

$$s := \frac{t}{b}\,, \qquad y := \frac{x}{b}\,.$$

on which the new perturbation functions $\mathcal{Q}$, $\mathcal{R}$ and Φ depend.

6.2 A Master Equation for $\mathcal{R}$

The system obtained from the linearization of Einstein's equations (18) and (19) about the perturbed solution (53), (54), and (55) with respect to the parameter ε can be not difficultly solved. Indeed, in [5], it is shown that this system implies that the perturbation functions $\mathcal{Q}$ and Φ have the following expression in dependence on $\mathcal{R}$:

$$\mathcal{Q}(s,y) = -\frac{2\mathcal{R}(s,y)}{(y^2+1)^{\frac{3}{2}}} + \mathcal{P}_0(y) + s\,\mathcal{P}_1(y)\,, \tag{56}$$

$$\Phi(s,y) = \frac{\mathcal{R}_y(s,y)}{\sqrt{1+y^2}} - y\left(\mathcal{P}_0(y) + s\,\mathcal{P}_1(y)\right) + \mathcal{C}\,, \tag{57}$$

where $\mathcal{C} \in \mathbb{R}$ is an immaterial constant, since Φ appears in the (linearized) field equations only through its derivatives, and $\mathcal{P}_0, \mathcal{P}_1 : \mathbb{R} \to \mathbb{R}$ are smooth functions which depend on the set of initial data

$$\mathcal{Q}_0(x) := \mathcal{Q}(0,x)\,, \quad \mathcal{R}_0(x) := \mathcal{R}(0,x)\,, \quad \mathcal{Q}_1(x) := \mathcal{Q}_s(0,x)\,, \quad \mathcal{R}_1(x) := \mathcal{R}_s(0,x)\,, \tag{58}$$

via the relation

[1] Note that the choice $\delta h(t,x) := 0$ is not restrictive at all since it is always possible to find a gauge in which this position holds, while the coefficient $b^2/(x^2+b^2)$ in front of the function $\mathcal{R}$ has been introduced in order to simplify the subsequent computation.

$$\mathcal{P}_i(x) = \mathcal{Q}_i(x) + \frac{2\mathcal{R}_i(x)}{(1+x^2)^{\frac{3}{2}}} \quad (i = 0, 1). \tag{59}$$

Moreover, the perturbation function $\mathcal{R}$ has to satisfy the following equation:

$$(\mathcal{R}_{ss} + \mathfrak{H}\,\mathcal{R})(s, y) = \mathcal{J}_0(y) + s\,\mathcal{J}_1(y), \tag{60}$$

where

$$\mathfrak{H} := -\frac{d^2}{dy^2} + \mathcal{V}, \quad \mathcal{V}(y) := -\frac{3}{(y^2+1)^2} \quad (y \in \mathbb{R}), \tag{61}$$

$$\mathcal{J}_i(y) := -\sqrt{y^2+1}\left(2\,\mathcal{P}_i(y) + y\,\mathcal{P}_{i,y}(y)\right) \quad \mathcal{P}_i \text{ as in Eq. (59)} \quad (i = 0, 1). \tag{62}$$

This wave-type equation for $\mathcal{R}$ is referred to as *master equation*; the source term $\mathcal{J}_0(y) + s\,\mathcal{J}_1(y)$ is fully determined by the functions $\mathcal{P}_0$ and $\mathcal{P}_1$ and therefore by the set of initial data on $\mathcal{Q}_i, \mathcal{R}_i$ $(i = 0, 1)$ (see Eq. (59)).

Taking advantage of some general considerations about Schrödinger-type operators in the framework of Hilbert spaces, one can prove[2] that the operator $\mathfrak{H}$ appearing in the master equation is self-adjoint once restricted to the domain of functions which are square-integrable together with their second derivatives; moreover, $\mathfrak{H}$ possesses a continuous spectrum which coincides with the interval $[0, +\infty)$ and a discrete spectrum consisting of exactly one eigenvalue $-E$; a numerical estimate for E is given in [8]:

$$E \simeq 1.40 .$$

From now on, we call $\mathcal{Y}_{-E}$ the normalized real eigenfunction for the eigenvalue $-E$.

6.3 *Linear Instability of the EBMT Wormhole*

The linear instability of the EBMT wormhole is proved finding at least one solution of the linearized Einstein equations such that one or more corresponding geometrical objects diverge in the large s limit. One can obtain the simplest solution of this kind choosing the initial data

$$\mathcal{R}_0(y) := \mathcal{Y}_{-E}(y), \quad \mathcal{R}_1(y) := 0, \quad \mathcal{Q}_0(y) = \frac{-2\mathcal{Y}_{-E}(y)}{(y^2+1)^{\frac{3}{2}}}, \quad \mathcal{Q}_1(y) := 0.$$

[2] For more details, see [5].

With these positions, from the general solution to the master equation (60) and from Eqs. (56) and (57), we get

$$\mathcal{R}(s,y) = \mathcal{Y}_{-E}(y)\cosh(\sqrt{E}s)\,, \quad \mathcal{Q}(s,y) = -\frac{2\mathcal{R}(s,y)}{(y^2+1)^{\frac{3}{2}}}\,, \quad \Phi(s,y) = \frac{\mathcal{R}_y(s,y)}{\sqrt{y^2+1}} + \mathcal{C}\,.$$

One can see that the linearized scalar curvature corresponding to this solution has the expression

$$R = -\frac{2b^2}{(x^2+b^2)^2} + \frac{4\varepsilon}{b^2}\mathcal{K}\left(\frac{x}{b}\right)\cosh\left(\sqrt{E}\,\frac{t}{b}\right) + O(\varepsilon^2)\,,$$

$$\mathcal{K}(y) := \left(\frac{1}{(y^2+1)^{7/2}} - \frac{E}{(1+y^2)^{3/2}}\right)\mathcal{Y}_{-E}(y) + \frac{x}{(1+y^2)^{5/2}}\mathcal{Y}_{-E,y}(y)\,. \tag{63}$$

The coefficient of ε of the scalar curvature $\mathcal{R}$ evaluated in $x = 0$ diverges exponentially for $t \to \pm\infty$; indeed, it can be proved that the above function $\mathcal{K}$ is not identically zero and in particular

$$\mathcal{K}(0) = (1-E)\mathcal{Y}_{-E}(0) \neq 0\,.$$

We would like to underline the fact that the above divergence does not depend on the gauge chosen. Indeed, any smooth coordinate change $(t,x) \mapsto (\tilde{t},\tilde{x})$, ε-close to the identity transforms the linearized scalar curvature $R = R(t,s)$ into a function $\tilde{R} = \tilde{R}(\tilde{t},\tilde{x})$ whose coefficient of the first order in ε, at $\tilde{x} = 0$, still diverges (again exponentially) for $\tilde{t} \to \pm\infty$.

6.4 A Comparison with [2, 4, 8, 14]

Admittedly, the linear instability of the EBMT wormhole has been previously stated in other works before [5], what is more in a gauge-invariant framework: in 2009, González, Guzmán and Sarbach proved in [8] the linear instability of a general wormhole solution of the Einstein-scalar field equations found by Bronnikov in [1], which includes the EBMT wormhole as a special case; some years later, in 2011, the linear instability result of Bronnikov's solution was then extended by Bronnikov himself, Fabris and Zhidenko in [4] to the whole class of static, spherically symmetric non-self-interactive scalar field solutions with throats; let us say that the same extension appears also in the recent survey [2]. However, in contrast to the previously sketched approach of [5], the main idea of these papers consists in fixing the radial coordinate and deriving a master equation for the perturbation function of the scalar field: this is a wave-type equation whose effective potential is singular at the throat (exactly where the derivative of the radial coordinate becomes null). In order to deal with this singularity, the authors of [8] have to introduce a suitable first order operator which, once applied to the function fulfilling the master equation, transforms the master equation into a regular one; this method was then generalized and called "S-deformation method" in [4] (and also in [2]). Let us add that the same approach is used also in [14] in order to prove the linear instability of a multidimensional generalization of the EBMT wormhole to spacetime dimension $d+1$ (with $d > 3$).

Summing up, we want to stress the fact that the derivation of the linear instability of the EBMT wormhole proposed by Cremona, Pirotta and Pizzocchero introduces a novelty in the study of the stability of wormhole configurations: indeed, the instability result is obtained without encountering any singularity and no S-deformation of the master equation is necessary.

However, one could object that Cremona et al. do not make use of gauge-invariant quantities from the very beginning of their derivation and have to prove *a posteriori* the gauge invariance of the instability result. In order to respond to this criticism, let us say that very recently, this problem has been overcome in [6], where a gauge-invariant strategy is introduced of studying the linear stability of static, spherically symmetric scalar field solutions with throats, obtaining a master equation which is non-singular everywhere (provided that the scalar field does not have critical points). Moreover, as this approach also includes the case of self-interactive scalar field, it is used to prove for the first time the linear instability of the AdS wormhole.

Acknowledgments This work was supported by INdAM, Gruppo Nazionale per la Fisica Matematica and Università degli Studi di Milano.

The author acknowledges prof. Livio Pizzocchero for his kind and professional support and prof. Sergio Cacciatori for the opportunity of contributing to this publication.

References

1. K.A. Bronnikov, Scalar-tensor theory and scalar charge. Acta Phys. Polon. **B4**, 251–266 (1973)
2. K.A. Bronnikov, Scalar fields as sources for wormholes and regular black holes. Particles **1**, 56–81 (2018). https://doi.org/10.3390/particles1010005
3. K.A. Bronnikov, J.C. Fabris, Regular phantom black holes. Phys. Rev. Lett. **96**, 251101 (2006)
4. K.A. Bronnikov, J.C. Fabris, A. Zhidenko, On the stability of scalar-vacuum space-times. Eur. Phys. J. C Particles Fields **71**, 1791 (12 pp.) (2011)
5. F. Cremona, F. Pirotta, L. Pizzocchero, On the linear instability of the Ellis-Bronnikov-Morris-Thorne wormhole. Gen. Relativ. Gravitat. **51**, 19 (2019)
6. F. Cremona, L. Pizzocchero, O. Sarbach, Gauge-invariant spherical linear perturbations of wormholes in Einstein gravity minimally coupled to a self-interacting phantom scalar field. arXiv:1911.13103 [gr-qc] (2019)
7. H. Ellis, Ether flow through a drainhole: A particle model in general relativity. J. Math. Phys. **14**, 104–118 (1973)
8. J.A. González, F.S. Guzmán, O. Sarbach, Instability of wormholes supported by a ghost scalar field. I: Linear stability analysis. Classical and Quantum Gravity **26**, 015010 (14 pp.) (2009)
9. J.B. Hartle, *Gravity: An Introduction to Einstein's General Relativity* (Addison-Wesley, San Francisco, 2003)
10. S.W. Hawking, G.F.R. Ellis, *The Large Scale Structure of Space-Time* (Cambridge University, Cambridge, 1973)
11. C.W. Misner, K.S. Thorne, J.A. Wheeler *Gravitation* (W. H. Freeman, San Francisco, 1973)

12. M.S. Morris, K.S. Thorne, Wormholes in spacetime and their use for interstellar travel: a tool for teaching general relativity. Am. J. Phys **56**, 395–412 (1988)
13. M.M. Postnikov, *The Variational Theory of Geodesics* (Dover Publications, New York, 1983)
14. T. Torii, H. Shinkai Wormholes in higher dimensional space-time: Exact solutions and their linear stability analysis. Phys. Rev. **D88**, 064027 (2013)
15. M. Visser, Lorentzian Wormholes: From Einstein to Hawking (Springer, New York, 1996)
16. M. Wald, *General Relativity* (The University of Chicago Press, Chicago and London, 1984)

New Trends in the General Relativistic Poynting–Robertson Effect Modeling

Vittorio De Falco

Contents

1 Introduction

The actual revolutionary discoveries occurred in the last years represented by the detection of gravitational waves first from a binary black holes (BHs) [1] and then from a neutron stars (NSs) [2] systems and the first imaging of the matter motion around the supermassive BH in M87 Galaxy [3] constitute a strong motivation to improve the actual theoretical models to validate Einstein theory or possible extension of it when benchmarked with the observations. The motion of relatively small-sized test particles, like dust grains or gas clouds, meteors, accretion disk matter elements, around radiating sources located outside massive compact objects is strongly affected by gravitational and radiation fields, and an important effect to be taken into account is the general relativistic PR effect [4, 5].

This phenomenon occurs each time the radiation field invests the test particle, raising up its temperature, which for the Stefan–Boltzmann law starts remitting radiation. This process of absorption and remission of radiation generates a recoil

V. De Falco (✉)
Department of Physics, Scuola Superiore Meridionale, Largo San Marcellino, Naples, Italy
e-mail: vittorio.defalco-ssm@unina.it

S. L. Cacciatori, A. Kamenshchik (eds.), *Einstein Equations: Local Energy, Self-Force, and Fields in General Relativity*, Tutorials, Schools, and Workshops in the Mathematical Sciences, https://doi.org/10.1007/978-3-031-21845-3_7

force opposite to the test body orbital motion. Such mechanism removes thus very efficiently angular momentum and energy from the test particle, forcing it to spiral inward or outward depending on the radiation field intensity. This effect has been extensively studied in Newtonian gravity within Classical Mechanics [4] and Special Relativity [5] and then applied in the Solar System [6]. Only in 2009–2011, this model has been proposed in General Relativity (GR) by Bini and collaborators within the equatorial plane of the Ker spacetime [7, 8]. Recently, it has been extended in the 3D space in Kerr metric [9–11]. One of the most evident implications of such effect is the formation of stable structures, termed critical hypersurfaces, around the compact object [12]. This phenomenon has been analyzed also under a Lagrangian formulation [13–15]. The novel aspect of such approach consists in the introduction of new techniques to deal with the nonlinearities in gravity patterns based on two new fundamental aspects: (1) use of an integrating factor to make closed differential forms [13] and (2) development of a new method termed *energy formalism*, which permits to analytically determine the Rayleigh potential associated with the radiation force [14, 15].

The chapter is structured as follows: in Sect. 2, the 3D model and its proprieties are described; in Sect. 3, the stability of the critical hypersurfaces is discussed within the Lyapunov theory; in Sect. 4, we analytically determine the Rayleigh dissipation function by using the energy formalism; finally, in Sect. 5, the conclusions are drawn.

2 General Relativistic 3D PR Effect Model

We consider a rotating compact object, whose geometry is described by the Kerr metric. Using the signature $(-,+,+,+)$ and geometrical units ($c = G = 1$), the metric line element, $ds^2 = g_{\alpha\beta}dx^\alpha dx^\beta$, in Boyer–Lindquist coordinates, parameterized by mass M and spin a, reads as [16]

$$\mathrm{d}s^2 = \left(\frac{2Mr}{\Sigma} - 1\right)\mathrm{d}t^2 - \frac{4Mra\sin^2\theta}{\Sigma}\mathrm{d}t\mathrm{d}\varphi + \frac{\Sigma}{\Delta}\mathrm{d}r^2 + \Sigma\mathrm{d}\theta^2 + \rho\sin^2\theta\mathrm{d}\varphi^2, \quad (1)$$

where $\Sigma \equiv r^2 + a^2\cos^2\theta$, $\Delta \equiv r^2 - 2Mr + a^2$, and $\rho \equiv r^2 + a^2 + 2Ma^2 r\sin^2\theta/\Sigma$. The determinant of the metric is $g = -\Sigma^2\sin^2\theta$. The orthonormal frame adapted to the zero angular momentum observers (ZAMOs) is [9, 10]

$$\boldsymbol{e}_{\hat{t}} \equiv \boldsymbol{n} = \frac{(\boldsymbol{\partial}_t - N^\varphi\boldsymbol{\partial}_\varphi)}{N}, \quad \boldsymbol{e}_{\hat{r}} = \frac{\boldsymbol{\partial}_r}{\sqrt{g_{rr}}}, \quad \boldsymbol{e}_{\hat{\theta}} = \frac{\boldsymbol{\partial}_\theta}{\sqrt{g_{\theta\theta}}}, \quad \boldsymbol{e}_{\hat{\varphi}} = \frac{\boldsymbol{\partial}_\varphi}{\sqrt{g_{\varphi\varphi}}}, \quad (2)$$

where $N = (-g^{tt})^{-1/2}$ and $N^\varphi = g_{t\varphi}/g_{\varphi\varphi}$. The nonzero ZAMO kinematical quantities in the decomposition of the ZAMO congruence are acceleration $\boldsymbol{a}(n) =$

$\nabla_{\boldsymbol{n}}\boldsymbol{n}$, expansion tensor along the $\hat{\varphi}$-direction $\boldsymbol{\theta}_{\hat{\varphi}}(n)$, and the relative Lie curvature vector $\boldsymbol{k}_{(\mathbf{Lie})}(n)$ (see Table 1 in [9], for their explicit expressions).

The radiation field is modeled as a coherent flux of photons traveling along null geodesics on the Kerr metric. The related stress–energy tensor is [9, 10]

$$T^{\mu\nu} = \mathcal{I}^2 k^\mu k^\nu, \qquad k^\mu k_\mu = 0, \qquad k^\mu \nabla_\mu k^\nu = 0, \tag{3}$$

where $\mathcal{I}$ is a parameter linked to the radiation field intensity and $\boldsymbol{k}$ is the photon four-momentum field. Splitting $\boldsymbol{k}$ with respect to the ZAMO frame, we obtain [10]

$$\boldsymbol{k} = E(n)[\boldsymbol{n} + \hat{\boldsymbol{\nu}}(k,n)], \tag{4}$$

$$\hat{\boldsymbol{\nu}}(k,n) = \sin\zeta\,\sin\beta\,\boldsymbol{e}_{\hat{r}} + \cos\zeta\,\boldsymbol{e}_{\hat{\theta}} + \sin\zeta\,\cos\beta\,\boldsymbol{e}_{\hat{\varphi}}, \tag{5}$$

where $E(n)$ is the photon energy measured in the ZAMO frame, $\hat{\boldsymbol{\nu}}(k,n)$ is the photon spatial unit relative velocity with respect to the ZAMOs, and β and ζ are the two angles measured in the ZAMO frame in the azimuthal and polar directions, respectively. The radiation field is governed by the two impact parameters (b, q), associated, respectively, with the two emission angles (β, ζ). The radiation field photons are emitted from a spherical rigid surface having a radius $R_\star$ centered at the origin of the Boyer–Lindquist coordinates and rotating with angular velocity $\Omega_\star$. The photon impact parameters are [10]

$$b = -\left[\frac{g_{t\varphi} + g_{\varphi\varphi}\Omega_\star}{g_{tt} + g_{t\varphi}\Omega_\star}\right]_{r=R_\star}, \qquad q = \left[b^2\cot^2\theta - a^2\cos^2\theta\right]_{r=R_\star}. \tag{6}$$

The related photon angles in the ZAMO frame are [10]

$$\cos\beta = \frac{bN}{\sqrt{g_{\varphi\varphi}}(1 + bN^\varphi)}, \qquad \zeta = \pi/2. \tag{7}$$

The parameter $\mathcal{I}$ has the following expression [10]:

$$\mathcal{I}^2 = \frac{\mathcal{I}_0^2}{\sqrt{\left(r^2 + a^2 - ab\right)^2 - \Delta\left[q + (b-a)^2\right]}}, \tag{8}$$

where $\mathcal{I}_0$ is $\mathcal{I}$ evaluated at the emitting surface.

A test particle moves with a timelike four-velocity $\boldsymbol{U}$ and a spatial three-velocity with respect to the ZAMO frames, $\boldsymbol{\nu}(U,n)$, which both read as [10]

$$\boldsymbol{U} = \gamma(U,n)[\boldsymbol{n} + \boldsymbol{\nu}(U,n)], \tag{9}$$

$$\boldsymbol{\nu} = \nu(\sin\psi\,\sin\alpha\boldsymbol{e}_{\hat{r}} + \cos\psi\boldsymbol{e}_{\hat{\theta}} + \sin\psi\,\cos\alpha\boldsymbol{e}_{\hat{\varphi}}), \tag{10}$$

where $\gamma(U,n) \equiv \gamma = 1/\sqrt{1-||\boldsymbol{\nu}(U,n)||^2}$ is the Lorentz factor, $\nu = ||\boldsymbol{\nu}(U,n)||$, and $\gamma(U,n) = \gamma$. We have that ν represents the magnitude of the test particle spatial velocity $\boldsymbol{\nu}(U,n)$, α is the azimuthal angle of the vector $\boldsymbol{\nu}(U,n)$ measured clockwise from the positive $\hat{\varphi}$ direction in the $\hat{r}-\hat{\varphi}$ tangent plane in the ZAMO frame, and ψ is the polar angle of the vector $\boldsymbol{\nu}(U,n)$ measured from the axis orthogonal to the $\hat{r}-\hat{\varphi}$ tangent plane in the ZAMO frame.

We assume that the radiation test particle interaction occurs through the Thomson scattering, characterized by a constant momentum transfer cross-section σ, independent from direction and frequency of the radiation field. We can split the photon four-momentum (4) in terms of the velocity $\boldsymbol{U}$ as [10]

$$\boldsymbol{k} = E(U)[\boldsymbol{U} + \hat{\boldsymbol{\mathcal{V}}}(k,U)], \tag{11}$$

where $E(U)$ is the photon energy measured by the test particle. The radiation force can be written as [10]

$$\mathcal{F}_{(\mathrm{rad})}(U)^{\hat{\alpha}} \equiv -\sigma\tilde{\mathcal{I}}^2(T^{\hat{\alpha}}{}_{\hat{\beta}}U^{\hat{\beta}} + U^{\hat{\alpha}}T^{\hat{\mu}}{}_{\hat{\beta}}U_{\hat{\mu}}U^{\hat{\beta}}) = \tilde{\sigma}\,[\mathcal{I}E(U)]^2\,\hat{\mathcal{V}}(k,U)^{\hat{\alpha}}, \tag{12}$$

where m is the test particle mass and the term $\tilde{\sigma}[\mathcal{I}E(U)]^2$ reads as [10]

$$\tilde{\sigma}[\mathcal{I}E(U)]^2 = \frac{A\gamma^2(1+bN^{\varphi})^2[1-\nu\sin\psi\cos(\alpha-\beta)]^2}{N^2\sqrt{\left(r^2+a^2-ab\right)^2-\Delta\left[q+(b-a)^2\right]}}, \tag{13}$$

with $A = \tilde{\sigma}[\mathcal{I}_0E_p]^2$ being the luminosity parameter, which can be equivalently written as $A/M = L/L_{\mathrm{EDD}} \in [0,1]$ with L the emitted luminosity at infinity and L_{EDD} the Eddington luminosity, and $E_p = -k_t$ is the conserved photon energy along the test particle trajectory. The terms $\hat{\mathcal{V}}(k,U)^{\hat{\alpha}}$ are the radiation field components, whose expressions are [10]

$$\begin{aligned}
\hat{\mathcal{V}}^{\hat{r}} &= \frac{\sin\beta}{\gamma[1-\nu\sin\psi\cos(\alpha-\beta)]} - \gamma\nu\sin\psi\sin\alpha, \quad \hat{\mathcal{V}}^{\hat{\theta}} = -\gamma\nu\cos\psi,\\
\hat{\mathcal{V}}^{\hat{\varphi}} &= \frac{\cos\beta}{\gamma[1-\nu\sin\psi\cos(\alpha-\beta)]} - \gamma\nu\sin\psi\cos\alpha, \quad \hat{\mathcal{V}}^{\hat{t}} = \gamma\nu\left[\frac{\sin\psi\cos(\alpha-\beta)-\nu}{1-\nu\sin\psi\cos(\alpha-\beta)}\right].
\end{aligned} \tag{14}$$

Collecting all the information together, it is possible to derive the resulting equations of motion for a test particle moving in a 3D space, which are [10]

$$\begin{aligned}
\frac{d\nu}{d\tau} = &-\frac{1}{\gamma}\left\{\sin\alpha\sin\psi\left[a(n)^{\hat{r}} + 2\nu\cos\alpha\sin\psi\,\theta(n)^{\hat{r}}{}_{\hat{\varphi}}\right]\right.\\
&\left.+\cos\psi\left[a(n)^{\hat{\theta}} + 2\nu\cos\alpha\sin\psi\,\theta(n)^{\hat{\theta}}{}_{\hat{\varphi}}\right]\right\} + \frac{\tilde{\sigma}[\Phi E(U)]^2}{\gamma^3\nu}\hat{\mathcal{V}}^{\hat{t}},
\end{aligned} \tag{15}$$

$$
\begin{aligned}
\frac{d\psi}{d\tau} &= \frac{\gamma}{\nu}\left\{\sin\psi\left[a(n)^{\hat\theta} + k_{(\mathrm{Lie})}(n)^{\hat\theta}\,\nu^2\cos^2\alpha + 2\nu\cos\alpha\sin^2\psi\,\theta(n)^{\hat\theta}{}_{\hat\varphi}\right]\right.\\
&\left. -\sin\alpha\cos\psi\left[a(n)^{\hat r} + k_{(\mathrm{Lie})}(n)^{\hat r}\,\nu^2 + 2\nu\cos\alpha\sin\psi\,\theta(n)^{\hat r}{}_{\hat\varphi}\right]\right\} \\
&+\frac{\tilde\sigma[\Phi E(U)]^2}{\gamma\nu^2\sin\psi}\left[\hat{\mathcal{V}}^{\hat t}\cos\psi - \hat{\mathcal{V}}^{\hat\theta}\nu\right],
\end{aligned}
\tag{16}
$$

$$
\begin{aligned}
\frac{d\alpha}{d\tau} &= -\frac{\gamma\cos\alpha}{\nu\sin\psi}\left[a(n)^{\hat r} + 2\theta(n)^{\hat r}{}_{\hat\varphi}\,\nu\cos\alpha\sin\psi + k_{(\mathrm{Lie})}(n)^{\hat r}\,\nu^2\right. \\
&\left. +k_{(\mathrm{Lie})}(n)^{\hat\theta}\,\nu^2\cos^2\psi\sin\alpha\right] + \frac{\tilde\sigma[\Phi E(U)]^2\cos\alpha}{\gamma\nu\sin\psi}\left[\hat{\mathcal{V}}^{\hat r} - \hat{\mathcal{V}}^{\hat\varphi}\tan\alpha\right],
\end{aligned}
\tag{17}
$$

$$
U^{\hat r} \equiv \frac{dr}{d\tau} = \frac{\gamma\nu\sin\alpha\sin\psi}{\sqrt{g_{rr}}},
\tag{18}
$$

$$
U^{\hat\theta} \equiv \frac{d\theta}{d\tau} = \frac{\gamma\nu\cos\psi}{\sqrt{g_{\theta\theta}}},
\tag{19}
$$

$$
U^{\hat\varphi} \equiv \frac{d\varphi}{d\tau} = \frac{\gamma\nu\cos\alpha\sin\psi}{\sqrt{g_{\varphi\varphi}}} - \frac{\gamma N^\varphi}{N},
\tag{20}
$$

$$
U^{\hat t} \equiv \frac{dt}{d\tau} = \frac{\gamma}{N},
\tag{21}
$$

where τ is the affine parameter along the test particle trajectory.

2.1 Critical Hypersurfaces

The dynamical system defined by Eqs. (15)–(20) exhibits a critical hypersurface outside around the compact object, where there exists a balance among gravitational and radiation forces, see Fig. 1. On such region, the test particle moves purely circular with constant velocity ($\nu = \mathrm{const}$) with respect to the ZAMO frame ($\alpha = 0, \pi$) and the polar axis orthogonal to the critical hypersurface ($\psi = \pm\pi/2$). These requirements entail $d\nu/d\tau = d\alpha/d\tau = 0$, from which we have [10]

$$
\nu = \cos\beta,
\tag{22}
$$

$$
\begin{aligned}
&a(n)^{\hat r} + 2\theta(n)^{\hat r}{}_{\hat\varphi}\nu + k_{(\mathrm{Lie})}(n)^{\hat r}\nu^2 \\
&= \frac{A(1+bN^\varphi)^2\sin^3\beta}{N^2\gamma\sqrt{\left(r^2_{(\mathrm{crit})}+a^2-ab\right)^2 - \Delta_{(\mathrm{crit})}\left[q+(b-a)^2\right]}},
\end{aligned}
\tag{23}
$$

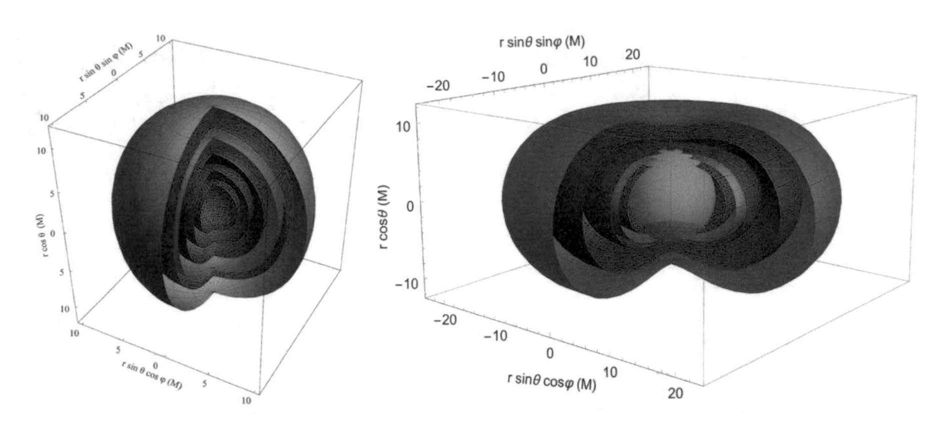

Fig. 1 Left panel: critical hypersurfaces for $\Omega_\star = 0$ and the luminosity parameters $A = 0.5, 0.7, 0.8, 0.85, 0.87, 0.9$ at a constant spin $a = 0.9995$. The respective critical radii in the equatorial plane are $r^{\rm eq}_{\rm (crit)} \sim 2.71M, 4.01M, 5.52M, 7.04M, 7.99M, 10.16M$, while at poles they are $r^{\rm pole}_{\rm (crit)} \sim 2.97M, 4.65M, 6.56M, 8.38M, 9.48M, 11.9M$. Right panel: critical hypersurfaces for an NS (gray sphere) with $\Omega_\star = 0.031$, $R_\star = 6M$, and luminosity parameters $A = 0.75, 0.78, 0.8, 0.85, 0.88$ at a constant spin $a = 0.41$. The respective critical radii in the equatorial plane are $r^{\rm eq}_{\rm (crit)} \sim 8.88M,\ 10.61M,\ 12.05M,\ 17.26M,\ 22.43M$, while at poles they are $r^{\rm pole}_{\rm (crit)} \sim 4.73M,\ 5.28M,\ 5.74M,\ 7.43M,\ 9.11M$. The red arrow is the polar axis

where the first condition means that the test particle moves on the critical hypersurface with constant velocity equal to the azimuthal photon velocity, whereas the second condition determines the critical radius $r_{\rm (crit)}$ as a function of the polar angle through an implicit equation, once A, a, $R_\star$, and $\Omega_\star$ are assigned.

In general, we have $d\psi/d\tau \neq 0$ because the ψ angle changes during the test particle motion on the critical hypersurface, having the so-called *latitudinal drift*. This effect, occurring for the interplay of gravitational and radiation actions in the polar direction, brings definitively the test particle on the equatorial plane [9, 10]. Only for $\psi = \theta = \pi/2$, we have $d\psi/d\tau = 0$, corresponding to the equatorial ring. However, we can have $d\psi/d\tau = 0$, also for a $\theta = \bar{\theta} \neq \pi/2$, having the so-called *suspended orbits*. The condition for this last configuration for $b \neq 0$ reads as [10]

$$
\begin{aligned}
& a(n)^{\hat{\theta}} + k_{\rm (Lie)}(n)^{\hat{\theta}}\, \nu^2 + 2\nu \sin^2\psi\, \theta(n)^{\hat{\theta}}{}_{\hat{\varphi}} \\
& + \frac{A(1+bN^{\varphi})^2(1-\cos^2\beta\sin\psi)\cos\beta}{\gamma N^2 \sqrt{\left(r^2_{\rm (crit)} + a^2 - ab\right)^2 - \Delta_{\rm (crit)}\left[q + (b-a)^2\right]\tan\psi}} = 0,
\end{aligned}
\tag{24}
$$

which permits to be solved in terms of ψ. Instead, for $b = 0$, we obtain $\psi = \pm\pi/2$ [9]. The critical points are either the suspended orbits or the equatorial ring, where the test particle ends its motion. In Fig. 2, we display some selected test particle trajectories to give an idea how the PR effect alters the matter motion surrounding a radiation source around a compact object [10].

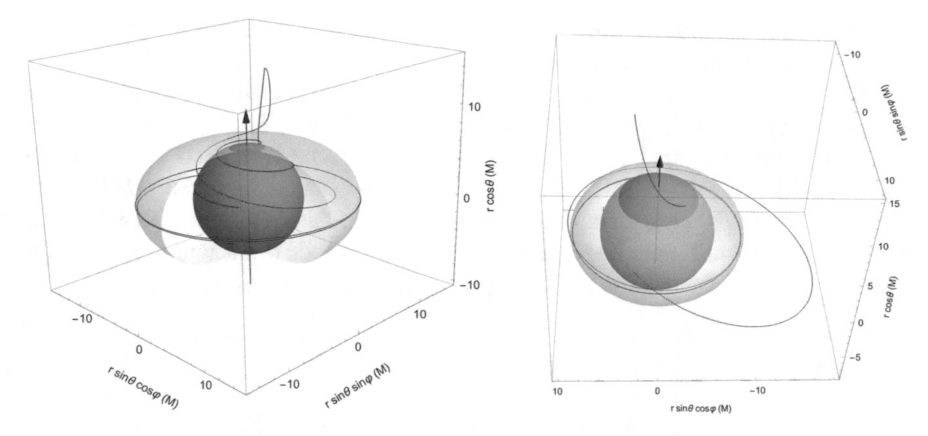

Fig. 2 Left panel: test particle trajectories around an NS of spin $a = 0.41$, radius $R_\star = 6M$, angular velocity $\Omega_\star = 0.031$, and luminosity parameter $A = 0.8$, starting at the position $(r_0, \theta_0) = (15M, 10°)$ with the initial velocity $\nu_0 = 0.01$ oriented in the azimuthal corotating direction (orange) and oriented radially toward the emitting surface (red). Right panel: test particle trajectories around an NS of spin $a = 0.07$, radius $R_\star = 6M$, angular velocity $\Omega_\star = 0.005$, and luminosity parameter $A = 0.85$, starting at the position $(r_0, \theta_0) = (15M, 10°)$ with the initial velocity $\nu_0 = 0.01$ oriented in the azimuthal corotating direction (orange) and oriented radially toward the emitting surface (red). The black sphere corresponds to the emitting surface of the NS. The blue–gray surface denotes the critical hypersurface

3 Stability of the Critical Hypersurfaces

To prove the stability of the critical hypersurfaces, we consider only those initial configurations, where the test particle ends its motion on them without escaping at infinity. Once the stability has been proven, it immediately follows that the critical equatorial ring is a stable attractor (region where the test particle is attracted for ending its motion), and the whole critical hypersurface is a basin of attraction [12].

Bini and collaborators have proved it only in the Schwarzschild case within the linear stability theory (see Appendix in Ref. [8]). This method consists in linearizing the dynamical system toward the critical points of the critical hypersurface and then looking at its eigenvalues. Theoretically, such method is simple, but practically it implies several calculations (especially in the Kerr case).

There is a simpler and more physical approach based on the Lyapunov theory. The dynamical system (15)–(20), $\dot{\boldsymbol{x}} = \boldsymbol{f}(\boldsymbol{x})$, is defined in the domain $\mathcal{D}$, while the critical hypersurface is defined by $\mathcal{H}$. Let $\Lambda = \Lambda(\boldsymbol{x})$ be a real-valued function, continuously differentiable in all points of $\mathcal{D}$, and then Λ is a Lyapunov function for $\dot{\boldsymbol{x}} = \boldsymbol{f}(\boldsymbol{x})$ if it fulfills the following conditions:

$$\text{(I)} \qquad \Lambda(\boldsymbol{x}) > 0, \quad \forall \boldsymbol{x} \in \mathcal{D} \setminus \mathcal{H}; \tag{25}$$

$$\text{(II)} \qquad \Lambda(\boldsymbol{x_0}) = 0, \quad \forall \boldsymbol{x_0} \in \mathcal{H}; \tag{26}$$

$$\text{(III)} \qquad \dot{\Lambda}(\boldsymbol{x}) \equiv \nabla\Lambda(\boldsymbol{x}) \cdot \boldsymbol{f}(\boldsymbol{x}) \leq 0, \quad \forall \boldsymbol{x} \in \mathcal{D}. \tag{27}$$

Once the Lyapunov function Λ has been found for all points belonging to the critical hypersurface $\mathcal{H}$, a theorem due to Lyapunov assures that $\mathcal{H}$ is stable [12].

The advantage to use this approach relies on easily studying the behavior of a dynamical system without knowing the analytical solution. The Lyapunov function is not unique, and there is no fixed rules to determine it, indeed several times one is guided by the physical intuitions. For the general relativistic PR effect, three different Lyapunov functions have been determined. The proof that they are Lyapunov function is based on expanding all the kinematic terms with respect to the radius estimating thus their magnitude (see Ref. [12] for further details).

- *The relative mechanical energy* of the test particle with respect to the critical hypersurface measured in the ZAMO frame is

$$\mathbb{K} = \frac{m}{2}\left|\nu^2 - \nu_{\rm crit}^2\right| + (A - M)\left(\frac{1}{r} - \frac{1}{r_{\rm crit}}\right), \tag{28}$$

  where $\nu_{\rm crit}(\theta) = [\cos\beta]_{r=r_{\rm crit}(\theta)}$, which includes as a particular case the velocity $\nu_{\rm eq} = [\cos\beta]_{r=r_{\rm crit}(\pi/2)}$ in the equatorial ring. Its derivative is

$$\dot{\mathbb{K}} = m\ \mathrm{sgn}\left(\nu^2 - \cos^2\beta\right)\left[\nu\frac{d\nu}{d\tau} - \cos\beta\frac{d(\cos\beta)}{d\tau}\right] - \frac{A - M}{r^2}\dot{r}, \tag{29}$$

  where $\mathrm{sgn}(x)$ is the signum function.
- *The angular momentum* of the test particle measured in the ZAMO frame is

$$\mathbb{L} = m(r\nu\sin\psi\cos\alpha - r_{\rm crit}\nu_{\rm crit}). \tag{30}$$

  Its derivative is given by

$$\begin{aligned}\dot{\mathbb{L}} = m\ \Big[&-\dot{r}_{\rm crit}\nu_{\rm crit} - r_{\rm crit}\frac{d(\nu_{\rm crit})}{d\tau} + r\frac{d\nu}{d\tau}\cos\alpha\sin\psi + \nu(\dot{r}\cos\alpha\sin\psi \\ &-r\sin\alpha\sin\psi\ \dot{\alpha} + r\sin\alpha\cos\psi\ \dot{\psi})\Big].\end{aligned} \tag{31}$$

- *The Rayleigh dissipation function* is (see Sect. 4 for its derivation and meaning)

$$\mathbb{F} = \tilde{\sigma}\mathcal{I}^2\left[\lg\left(\frac{\mathbb{E}_{\rm crit}}{E_p}\right) - \lg\left(\frac{\mathbb{E}}{E_p}\right)\right], \tag{32}$$

  where E_p is the photon energy and $\mathbb{E} \equiv E(U)$, defined as

$$\mathbb{E} \equiv -k_\alpha U^\alpha = \gamma\frac{E_p}{N}(1 + bN^\varphi)[1 - \nu\sin\psi\cos(\alpha - \beta)]. \tag{33}$$

$\mathbb{E}_{\text{crit}}$ is the energy $\mathbb{E}$ evaluated on the critical hypersurface, given by

$$\mathbb{E}_{\text{crit}} = [\mathbb{E}]_{r=R_\star,\alpha=0,\pi,\psi=\pm\pi/2,\nu=\nu_{\text{crit}}} = \frac{E_p|(\sin\beta)_{\text{crit}}|}{N_{\text{crit}}}(1+bN^\varphi_{\text{crit}}). \tag{34}$$

Its derivative is

$$\dot{\mathbb{F}} = \tilde{\sigma}(\dot{\mathcal{I}^2})\left[\lg\left(\frac{\mathbb{E}_{\text{crit}}}{E_p}\right) - \lg\left(\frac{\mathbb{E}}{E_p}\right)\right] + \tilde{\sigma}\mathcal{I}^2\left[\frac{\dot{\mathbb{E}}_{\text{crit}}}{\mathbb{E}_{\text{crit}}} - \frac{\dot{\mathbb{E}}}{\mathbb{E}}\right]. \tag{35}$$

In Fig. 3, we calculate a test particle orbit in the equatorial plane reaching the critical hypersurface, and in the other panels, we show the three proposed functions (i.e.,

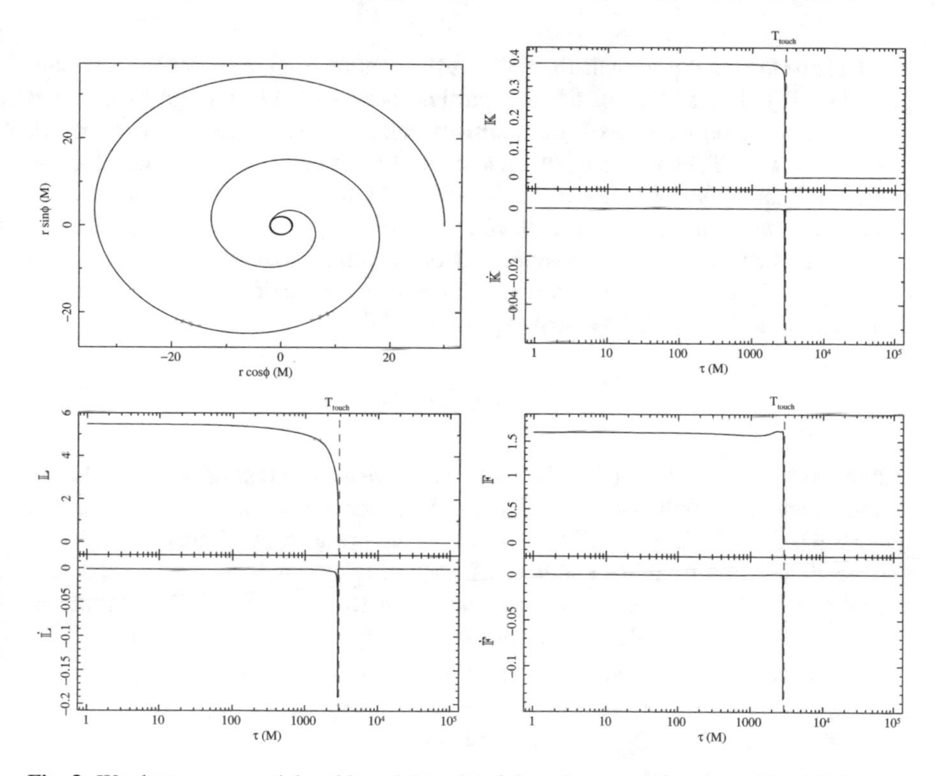

Fig. 3 We show a test particle orbit and the related three Lyapunov functions. *Upper left panel:* test particle moving around a rotating compact object with mass $M = 1$, spin $a = 0.3$, luminosity parameter $A = 0.2$, and photon impact parameter $b = 0$. The test particle starts its motion at the position $(r_0, \varphi_0) = (30M, 0)$ with velocity $(\nu_0, \alpha_0) = (\sqrt{M/r_0}, 0)$. The critical hypersurface is a circle with radius $r_{(\text{crit})} = 2.07M$. The energy (see Eqs. (28) and (29), and *upper right panel*), the angular momentum (see Eqs. (30) and (31), and *lower left panel*), and the Rayleigh potential (see Eqs. (33) and (35), and *lower right panel*) together with their τ-derivatives are all expressed in terms of the proper time τ. The dashed blue lines in all plots represent the proper time T_{touch} at which the test particle reaches the critical hypersurface, and it amounts to $T_{\text{touch}} = 2915M$

$\mathbb{K}$, $\mathbb{L}$, $\mathbb{F}$) together with their derivatives (i.e., $\dot{\mathbb{K}}$, $\dot{\mathbb{L}}$, $\dot{\mathbb{F}}$), to graphically prove that they verify the three proprieties to be Lyapunov functions.

It is important to note that the first two Lyapunov functions (energy and angular momentum) are written using the classical definition, and not the general relativistic expression, as instead it has been done with the third Lyapunov function. This is not in contradiction with the definition of Lyapunov function, rather they are very useful to carry out more easily the calculations. For example, even a mathematical function with no direct physical meaning with the system under study but verifying the conditions to be a Lyapunov function is a good candidate to prove the stability of the critical hypersurfaces.

4 Analytical form of the Rayleigh Dissipation Function

We describe the energy formalism, which is the method used to derive the Rayleigh potential [15]. The motion of the test particle occurs in $\mathcal{M}$, a simply connected domain (the region outside of the compact object including the event horizon). We denote with $T\mathcal{M}$ the tangent bundle of $\mathcal{M}$, whereas $T^*\mathcal{M}$ stands for the cotangent bundle over $\mathcal{M}$. Let $\boldsymbol{\omega} : T\mathcal{M} \rightarrow T^*\mathcal{M}$ be a smooth differential semi-basic one-form. Defined $\boldsymbol{X} = (t, r, \theta, \varphi)$ and $\boldsymbol{U} = (U^t, U^r, U^\theta, U^\varphi)$, the radiation force components (12) are the components of the differential semi-basic one-form $\boldsymbol{\omega}(\boldsymbol{X}, \boldsymbol{U}) = F_{(\mathrm{rad})}(\boldsymbol{X}, \boldsymbol{U})^\alpha \mathbf{d}X_\alpha$. We note that $\boldsymbol{\omega}$ is closed under the vertical exterior derivative $\mathbf{d}^{\mathbf{V}}$ if $\mathbf{d}^{\mathbf{V}}\boldsymbol{\omega} = 0$. The local expression of this operator is

$$\mathbf{d}^{\mathbf{V}} F = \frac{\partial F}{\partial U_\alpha} \mathbf{d}X_\alpha. \tag{36}$$

For the Poincaré lemma (generalized to the vertical differentiation), the closure condition and the simply connected domain $\mathcal{M}$ guarantee that $\boldsymbol{\omega}$ is exact. Therefore, it exists a 0-form $V(\boldsymbol{X}, \boldsymbol{U}) \in \mathcal{C}^\infty(T\mathcal{M}, \mathfrak{m})$, called primitive, such that $-\mathbf{d}^{\mathbf{V}} V = \boldsymbol{\omega}$.

Due to the nonlinear dependence of the radiation force on the test particle velocity field, the semi-basic one-form turns out to be not exact [13]. However, the PR phenomenon exhibits the peculiar propriety according to which $\boldsymbol{\omega}(\boldsymbol{X}, \boldsymbol{U})$ becomes exact through the introduction of the integrating factor $\mu = \left(E_{\mathrm{p}}/\mathbb{E}\right)^2$ [13]. Considering the energy $\mathbb{E} = -k_\beta U^\beta$ and substituting all the occurrences of $\mathbb{E}$ in $F_{(\mathrm{rad})}(\boldsymbol{X}, \boldsymbol{U})^\alpha$, see Eq. (12), we obtain [14, 15]

$$\mathbb{F}_{(\mathrm{rad})}(\boldsymbol{X}, \boldsymbol{U})^\alpha = -k^\alpha \mathbb{E}(\boldsymbol{X}, \boldsymbol{U}) + \mathbb{E}(\boldsymbol{X}, \boldsymbol{U})^2 U^\alpha. \tag{37}$$

Using the chain rule from the velocity to the energy derivative operator, we have

$$\frac{\partial\,(\,\cdot\,)}{\partial U_\alpha} = -k^\alpha\, \frac{\partial\,(\,\cdot\,)}{\partial \mathbb{E}}. \tag{38}$$

Therefore, the V function satisfies the usual primitive condition [14, 15]

$$\mu F^{\alpha}_{(\mathrm{rad})} = k^{\alpha}\frac{\partial V}{\partial \mathbb{E}}. \tag{39}$$

Such differential equation for V contains the k^{α} factor, which represents an obstacle for the integration process. To get rid of this term, we can consider the scalar product of both members of Eq. (39) by U_{α}, which permits to obtain a more manageable integral equation for V [14, 15], i.e.,

$$V = -\int \left(\frac{\mu F^{\alpha}}{\mathbb{E}}\right) \mathrm{d}\mathbb{E} + f(\boldsymbol{X}, \boldsymbol{U}), \tag{40}$$

where f is constant with respect to $\mathbb{E}$, i.e., $\partial f/\partial \mathbb{E} = 0$ and V is still a function of the local coordinates $(\boldsymbol{X}, \boldsymbol{U})$. Integrating Eq. (40), the final result is [14, 15]

$$V = \tilde{\sigma}\mathcal{I}^2 \left[\ln\left(\frac{\mathbb{E}}{E_{\mathrm{p}}}\right) + \frac{1}{2}\left(U_{\alpha}U^{\alpha} + 1\right)\right]. \tag{41}$$

Equation (41) is consistent with the classical description [4, 5]. The PR effect configures as the first dissipative nonlinear system in GR for which we know the analytical form of the Rayleigh potential.

4.1 Discussions of the Results

The energy function $\mathbb{E}$ and the chain rule both represent the fundamental aspects of the energy formalism, since they permit to simplify the demanding calculations for the V primitive. We are able to substantially reduce the coordinates involved in the calculations, passing from 4 initial parameters, represented by $\boldsymbol{U}$, to one only, i.e., the energy $\mathbb{E}$. In particular, the f function occurring in Eq. (40) embodies our ignorance about the analytic form of V as a function of the local coordinates $(\boldsymbol{X}, \boldsymbol{U})$. In some cases, as in our model, the f function can be determined by applying the integration process for an exact differential one-form. Such method has also the peculiar propriety, that it is independent from the considered metric, permitting to be applied to generic metric theories of gravity, and for its generality and simplicity, it can also be applied in different physical and mathematical fields.

The Rayleigh potential (40) is a valuable tool to investigate the proprieties of the general relativistic PR effect and more in general the radiation processes in high-energy astrophysics. This potential contains a logarithm of the energy, which physically is interpreted as the absorbed energy from the test particle. Therefore, it represents a new class of functions, never explored and discovered in the literature, used to describe the radiation absorption processes. Another important implication of the Rayleigh potential relies on the direct connection between theory and

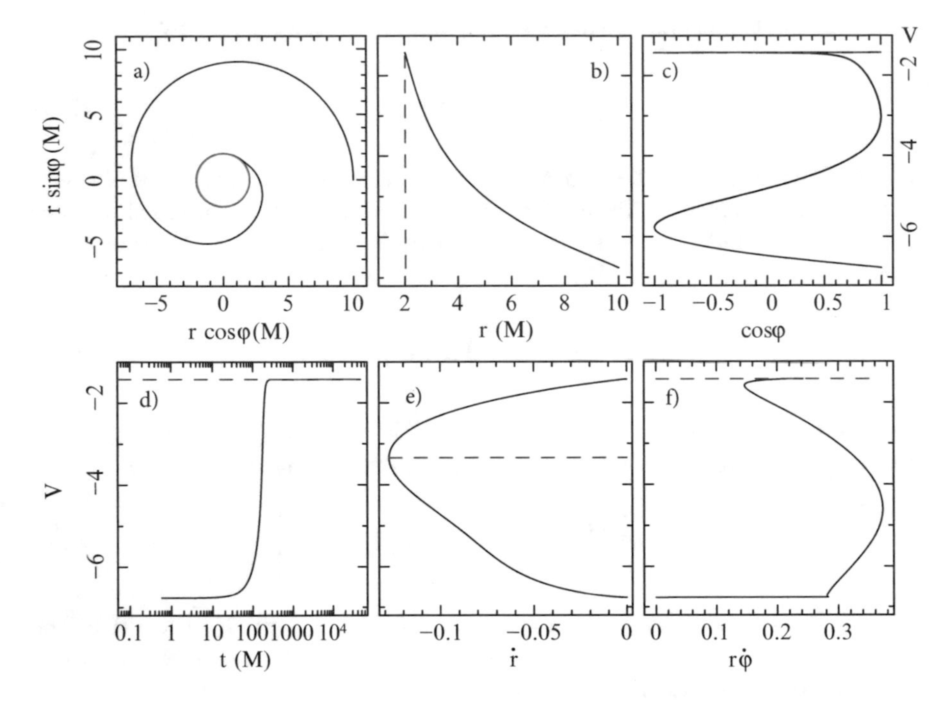

Fig. 4 Test particle trajectory with the Rayleigh potential V for mass $M = 1$ and spin $a = 0.1$, luminosity parameter $A = 0.1$, and photon impact parameter $b = 1$. The test particle moves in the spatial equatorial plane with initial position $(r_0, \varphi_0) = (10M, 0)$ and velocity $(\nu_0, \alpha_0) = (\sqrt{1/10M}, 0)$. (**a**) Test particle trajectory spiraling toward the BH and stopping on the critical radius (red dashed line) $r_{(\mathrm{crit})} = 2.02M$. The continuous green line is the event horizon radius $r^+_{(\mathrm{EH})} = 1.99M$. Rayleigh potential versus (**b**) radial coordinate, (**c**) azimuthal coordinate, (**d**) time coordinate, (**e**) radial velocity, and (**f**) azimuthal velocity. The blue dashed line in panel (**e**) marks the minimum value attained by the radial velocity, corresponding to $\dot{r} = -0.13$

observations. In Fig. 4, we show in panel (a) the test particle trajectory (what we can observe) and in panels $(b) - (f)$ the Rayleigh potential in terms of the coordinates $r, \varphi, t, \dot{r}, \dot{\varphi}$, respectively (what comes from the theory). Therefore, observing the test particle motion, it is possible to theoretically reconstruct the Rayleigh function; vice versa new Rayleigh functions can be proposed to study then the dynamics and see what we should observe (see Ref. [15] for details).

5 Conclusions

In this work, we have presented the fully general relativistic treatment of the 3D PR effect in the Kerr geometry, which extends the previous works framed in the 2D equatorial plane of relativistic compact objects. The radiation field comes from a rigidly rotating spherical source around the compact object. The emitted photons are parametrized by two impact parameters (b, q), where b can be variable and q depends on the value assumed by b and the polar angle θ, position occupied by the test particle in the 3D space. The resulting equations of motion represent a system of six coupled ordinary and highly nonlinear differential equations of first order. The motion of test particles is strongly affected by PR effect together with general relativistic effects. Such dynamical system admits the existence of critical hypersurfaces, regions where the gravitational attraction is balanced by the radiation forces.

We have presented the method to prove the stability of the critical hypersurfaces by employing the Lyapunov functions. Such strategy permits to simplify the calculations and to catch important physical aspects of the PR effect. Three different Lyapunov functions have been proposed, all with a different and precise meaning. The first two are deduced by the definition of the PR effect, which removes energy and angular momentum from the test particle. The third example is less intuitive because it is based on the Rayleigh dissipation function, determined by the use of an integrating factor and the introduction of the energy formalism.

Such method revealed to be very useful for two reasons: (1) a substantial reduction of the calculations from the 4 variables (i.e., the velocities $\boldsymbol{U}$) to only one (i.e., the energy $\mathbb{E}$) and (2) the obtained expression of the V potential as a function of $\mathbb{E}$ suffices for the description of the dynamics, being very important whenever the evaluation of $f(\boldsymbol{X}, \boldsymbol{U})$ turns out to be too laborious. In this way, we have obtained for the first time an analytical expression of the Rayleigh potential in GR, and we have discovered a new class of functions, represented by the logarithms, which physically describe the absorption processes in high-energy astrophysics.

As future projects, we plan to improve the actual theoretical assessments used to treat the radiation field in some ingenue aspects, like the momentum transfer cross-section will be not anymore constant, but it will depend on the angle and frequency of the incoming radiation field, and the radiation field is not emitted anymore by a point-like source, but from a finite extended source. We would like also to apply this theoretical model to some astrophysical situations in accretion physics, like accretion disk model, type-I X-ray burst, and photospheric radius expansion.

The new method to prove the stability of the critical hypersurfaces through Lyapunov functions can be easily applied to any possible extensions of the general relativistic PR effect model, naturally with the due modifications. Instead, the energy formalism opens up new frontiers in the study of the dissipative systems in metric theory of gravity and more broadly in other mathematical and physical research fields thanks to its general structure and versatile applicability. It permits to acquire more information on the mathematical structure and the physical meaning of the problem under study, because as discussed in Fig. 4, it is incredibly evident the profound connection between observations and theory.

Acknowledgments The author thanks the Silesian University in Opava and Gruppo Nazionale di Fisica Matematica of Istituto Nazionale di Alta Matematica for support. The results contained in the present paper have been partially presented at the summer school DOOMOSCHOOL 2019. The author acknowledges the support of INFN *sez. di Napoli, iniziative specifiche* TEONGRAV.

References

1. B.P. Abbott et al., Observation of gravitational waves from a binary black hole merger. Phys. Rev. Lett. **116**, 061102 (2016)
2. B.P. Abbott et al., GW170817: Observation of gravitational waves from a binary neutron star inspiral. Phys. Rev. Lett. **119**, 161101 (2017)
3. EHT Collaboration et al., First M87 event horizon telescope results. Astrophys. J. **875**, L1–L6 (2019)
4. J.H. Poynting, Radiation in the solar system: its effect on temperature and its pressure on small bodies. Mon. Not. R. Astron. Soc. **64**, 1 (1903)
5. H.P. Robertson, Dynamical effects of radiation in the solar system. Mon. Not. R. Astron. Soc. **97**, 423 (1937)
6. J.A. Burns, P.L. Lamy, S. Soter, Radiation forces on small particles in the solar system. Icarus **40**, 1–48 (1979)
7. D. Bini, R.T. Jantzen, L. Stella, The general relativistic Poynting Robertson effect. Classical and Quantum Gravity **26**, 5 (2009)
8. D. Bini, A. Geralico, R.T. Jantzen, O. Semerák, L. Stella, The general relativistic Poynting-Robertson effect: II. A photon flux with nonzero angular momentum. Classical and Quantum Gravity **28**, 3 (2011)
9. V. De Falco et al., Three-dimensional general relativistic Poynting-Robertson effect: Radial radiation field. Phys. Rev. D **99**, 023014 (2019)
10. P. Bakala et al., Three-dimensional general relativistic Poynting-Robertson effect II: Radiation field from a rigidly rotating spherical source. Phys. Rev. D **100**(10), 104053 (2019)
11. M. Wielgus, Optically thin outbursts of rotating neutron stars cannot be spherical. MNRAS **488**, 4937–4941 (2019)
12. V. De Falco et al., Stable attractors in the three-dimensional general relativistic Poynting-Robertson effect. Phys. Rev. D **101**(2), 024025 (2019)
13. V. De Falco, E. Battista, M. Falanga, Lagrangian formulation of the general relativistic Poynting-Robertson effect. Phys. Rev. D **97**, 8 (2019)
14. V. De Falco, E. Battista, Analytical Rayleigh potential for the general relativistic Poynting-Robertson effect. Europhys. Lett. **127**, 30006 (2019)
15. V. De Falco, E. Battista, Dissipative systems in metric theories of gravity. Foundations and applications of the energy formalism. Phys. Rev. D (2019), submitted
16. C.W. Misner, K.S. Thorne, J.A. Wheeler, *Gravitation* (W.H.Freeman and Co., San Francisco, 1973)

Brief Overview of Numerical Relativity

Mario L. Gutierrez Abed

To my parents, my wife Laura, and my soon-to-be-born baby boy Alessandro.

Contents

Mathematics Subject Classification (2000) Primary 83C05, 83F05; Secondary 83-01, 83-06

1 ADM Formalism of Numerical Relativity

We want to determine the dynamical evolution of a physical system governed by the field equations of general relativity (GR); this will essentially boil down to formulate the EFEs as a *Cauchy problem* with constraints. Analogous to classical dynamics, where the evolution of a system is uniquely determined by the initial positions and velocities of its constituents, we can evolve the gravitational field as determined by specifying the metric quantities g_{ab} and $\partial_t g_{ab}$ at a given (initial) instant of time t. This process can be simplified (for the sake of clarification) in two steps:

M. L. G. Abed (✉)
Newcastle University, Department of Mathematics, Statistics & Physics, Newcastle upon Tyne, UK
e-mail: mariogutierrezabed@gmail.com

S. L. Cacciatori, A. Kamenshchik (eds.), *Einstein Equations: Local Energy, Self-Force, and Fields in General Relativity*, Tutorials, Schools, and Workshops in the Mathematical Sciences, https://doi.org/10.1007/978-3-031-21845-3_8

(i) Specify g_{ab} and $\partial_t g_{ab}$ (actually, it will be related quantities) everywhere on some $3D$ spacelike hypersurface labelled by coordinate $x^0 = t =$ constant.
(ii) Provided that we can obtain expressions for second-time derivatives of the 4-metric g_{ab} at all points on the hypersurface from the EFEs, we then integrate forward in time the metric quantities from step (i).

However, even though this seems like a straightforward proposal, one must first define what is actually meant by "spatial hypersurface," since in GR there is no preferred timelike direction and, crucially, no global concept of time. This makes the problem of solving the Einstein equations numerically substantially different from other typical non-relativistic Cauchy problems. Moreover, it turns out that obtaining the appropriate expressions for $\partial_t^2 g_{ab}$ for such an integration is not so trivial. We require a total of ten second derivatives which, at first glance at least, appear to be achievable from the ten field equations ${}^{(4)}G^{ab} = 8\pi T^{ab}$.[1,2] However, note that by the (contracted) *Bianchi identities*,[3]

$$\begin{aligned} 0 &= \nabla_b {}^{(4)}G^{ab} \\ &= \partial_0 {}^{(4)}G^{a0} + \partial_i {}^{(4)}G^{ai} + {}^{(4)}G^{bc\,(4)}\Gamma^a_{bc} + {}^{(4)}G^{ab\,(4)}\Gamma^c_{bc}, \end{aligned}$$

we get

$$\partial_t {}^{(4)}G^{a0} = -\partial_i {}^{(4)}G^{ai} - {}^{(4)}G^{bc\,(4)}\Gamma^a_{bc} - {}^{(4)}G^{ab\,(4)}\Gamma^c_{bc}. \tag{1}$$

Since no term on the RHS of (1) contains third-time derivatives (or higher), this implies that there are no second-time derivatives contained in ${}^{(4)}G^{a0}$, and thus the four equations

$${}^{(4)}G^{a0} = 8\pi T^{a0} \tag{2}$$

do not yield any information whatsoever on how the fields evolve in time. Instead, they function as four *constraints* on the initial data, i.e., four relations that must be satisfied from the onset on the initial hypersurface at $x^0 = t$ if we are to have a physically meaningful system. Thus, we can see that the only true dynamical (*evolution*) equations are encoded in the remaining six field equations

---

[1] We adopt the usual convention in which $4D$ objects are distinguished from their $3D$ counterparts by using a superscript (4). There are exceptions to this rule, where $3D$ objects are denoted by different symbols from their $4D$ cousins (examples are T_{ab} and g_{ab} whose $3D$ counterparts are denoted by S_{ab} and γ_{ab}, respectively; these exceptions are merely a matter of convention, of course, but they are widely used in the numerical relativity literature).

[2] We will use geometric units ($G = c = 1$) throughout.

[3] We also use standard index convention, whereby the letters $a - h$ and $o - z$ are used for $4D$ spacetime indices that run from 0 to 3, whereas the letters $i - n$ are reserved for $3D$ spatial indices that run from 1 to 3. Lowercase Greek letters are reserved for components in a chosen basis (see [27] for reference).

$$^{(4)}G^{ij} = 8\pi T^{ij}. \tag{3}$$

We will see later on that certain projections of (2) and (3) onto the hypersurfaces will indeed yield the desired constraint and evolution equations of the system.

Thus, in essence, we are after a method to formulate the evolution of a gravitational field as a Cauchy problem; i.e., given adequate initial (and boundary) conditions, the fundamental equations must predict the future (or past) evolution of the system. However, an inconvenient hurdle that we immediately run into is that in GR space and time are treated on equal footing, thus making the notion of "time evolution" not as straightforward as our intuition dictates. Therefore, our first order of business is to somehow split the roles of space and time in a clear manner (and by this we do not mean "forget about GR and go back to Newtonian/Galilean gravity").

It turns out that there is a special class of spacetimes, known as *globally hyperbolic* spacetimes, that will allow us this sought-after time/space split. First, recall that a *Cauchy surface* is a spacelike hypersurface Σ embedded in an ambient manifold $\mathcal{M}$ such that each causal curve without endpoint in $\mathcal{M}$ intersects Σ once and only once. Equivalently, a Cauchy surface for a spacetime $\mathcal{M}$ is an *achronal* subspace $\Sigma \subset M$ (i.e., a subspace Σ in which no two points are timelike-related) which is met by every inextendible causal curve in $\mathcal{M}$. Now, we properly define the concept of global hyperbolicity:

Definition 1.1 A spacetime $\mathcal{M}$ is said to be *globally hyperbolic* if it admits a Cauchy surface. Equivalently, $\mathcal{M}$ is globally hyperbolic if it satisfies the *strongly causal condition* (i.e., if every $p \in \mathcal{M}$ has arbitrarily small neighbourhoods U in which every causal curve with endpoints in U is entirely contained in U) and if the "causal diamonds" $J^+(p) \cap J^-(q)$ are compact for all $p, q \in \mathcal{M}$.[4]

Global hyperbolicity is the standard condition in Lorentzian geometry that ensures the existence of maximal causal geodesic segments. Physically, this condition is closely connected to the issue of classical determinism and the strong cosmic censorship conjecture. Ringström [24] Even though this is by no means a condition satisfied a priori by all spacetimes, the $3 + 1$ formalism assumes that all physically reasonable spacetimes are of this type. This assumption is justified by the desire to have "nice" chronological/causal features in our spacetime (i.e., no *grandfather paradox* or any similar pathological behaviour). Moreover, the use of global hyperbolicity allows us to foliate our full $4D$ spacetime $\mathcal{M}$ in such a way that we can stack $3D$ spacelike Cauchy slices along a universal time axis, by virtue of $\mathcal{M}$ having topology $\Sigma \times \mathbb{R}$. This is certainly not the only way to foliate $\mathcal{M}$, but it is the most suitable one for the $3 + 1$ formalism.

[4] Here, we used the standard notation, where $J^+(p) = \{q \in \mathcal{M} \mid p \leq q\}$ and $J^-(p) = \{q \in \mathcal{M} \mid q \leq p\}$ are the *causal future* and *causal past*, respectively, of $p \in \mathcal{M}$.

1.1 ADM Variables and Adapted Coordinates

Given the foliation granted by the global hyperbolic condition described above, we can now determine the geometry of the region of spacetime contained between two adjacent hypersurfaces Σ_t and $\Sigma_{t+\mathrm{d}t}$ from just three basic ingredients:

(i) The $3D$ *metric* γ_{ij} (metric induced on Σ: $\gamma_{ab} \equiv \Phi^* g_{ab}$, where $\Phi\colon \Sigma \hookrightarrow \mathcal{M}$ is the embedding of Σ into $\mathcal{M}$) that measures proper distances within the hypersurface itself:

$$dl^2 = \gamma_{ij}\, \mathrm{d}x^i \mathrm{d}x^j .$$

The hypersurface is then said to be

$$\underbrace{- \; \textit{spacelike} \iff \gamma_{ab} \text{ is positive definite; i.e., it has signature } (+,+,+).}_{\text{our case}}$$

- *timelike* $\iff \gamma_{ab}$ is Lorentzian; i.e., it has signature $(-,+,+)$.
- *null* $\iff \gamma_{ab}$ is degenerate; i.e., it has signature $(0,+,+)$.

(We will shortly justify why we express the spatial metric both as $3D$ object (γ_{ij}) and as $4D$ object (γ_{ab}).)

(ii) The *lapse of proper time* between the hypersurfaces, as measured by observers whose worldlines extend along the direction normal to the hypersurfaces (these observers are usually referred to as *Eulerian observers*):

$$\mathrm{d}\tau = \alpha(t, x^i)\mathrm{d}t,$$

where α is known as the *lapse function* (this is denoted as N by some references, e.g., [15, 20]).

(iii) The relative velocity β^i between the Eulerian observers:

$$x^i_{t+\mathrm{d}t} = x^i_t - \beta^i(t, x^i)\mathrm{d}t, \qquad \text{for Eulerian observers.}$$

This 3-vector β^i measures how much the coordinates are shifted as we move from one slice to the next, and it is therefore conventionally named as the *shift vector*. (It is also denoted as N^i in the literature.)

Note that, as alluded to earlier, the foliation of $\mathcal{M}$ is not unique, and neither is the coordinates shift; α determines "how much slicing" is to be done, whilst β^i dictates how the spatial coordinates propagate from one hypersurface to the next. In fact, the latitude to choose a lapse function and shift vector demonstrates the gauge freedom that is inherent to the formulation of GR, a *covariant* theory.

From the *universal time function* t (given by the foliation), we can define the future-pointing timelike *unit normal* n^a to the slice Σ to be[5]

[5] The minus sign is chosen to ensure that n^a is always future-pointing.

$$n^a \equiv -\alpha \nabla^a t. \tag{4}$$

We think of n^a as the 4-velocity of an Eulerian observer, i.e., an observer whose worldline is always normal to the spatial slices. With this defined, we can see that the three scalar quantities that yield the spatial components of the shift vector, β^i, are given by

$$\beta^i = -\alpha \left(\vec{n} \cdot \vec{\nabla} x^i \right). \tag{5}$$

These three scalar quantities can then be used to form a full 4-vector β^a (orthogonal to n^a, by construction) which, in the adapted $3+1$ coordinates we are about to introduce, will have components $\beta^\mu = (0, \beta^i)$. Equipped with the unit normal and the shift vector, we can also define a *time vector* t^a given by

$$t^a \equiv \alpha n^a + \beta^a, \tag{6}$$

which is nothing but the vector tangent to the *time lines*, i.e., the congruence of lines of constant spatial coordinates x^i. The importance of this vector lies in the ability to *Lie drag* the hypersurfaces along it: the spatial basis vectors $e^a_{(i)}$ which are tangent to a particular slice Σ_t (i.e., that satisfy $\nabla_a t\, e^a_{(i)} = 0$) are *Lie dragged* along t^a,

$$\mathcal{L}_{\vec{t}}\, e^a_{(i)} = 0.$$

Remark 1.2 The *normal evolution vector* $m^a \equiv \alpha n^a$ can also be used to Lie drag the hypersurfaces (see, e.g., [15])

Now, since t^a is aligned with the basis vector $e^a_{(0)}$ whilst all remaining (spatial) coordinates remain constant along t^a, we get the basis components

$$t^\mu = e^\mu_{(0)} = \delta^\mu_0 = (1, 0, 0, 0). \tag{7}$$

This means that any Lie derivative along t^a will reduce to a partial derivative with respect to t: $\mathcal{L}_{\vec{t}} = \partial_t$ (we will use this later) Similarly, a straightforward derivation shows that in these adapted coordinates we have

$$\beta^\mu = (0, \beta^i) \tag{8a}$$

$$n^\mu = (\alpha^{-1}, -\alpha^{-1}\beta^i) \tag{8b}$$

$$n_\mu = (-\alpha, 0, 0, 0). \tag{8c}$$

These quantities will appear in the $3+1$ coordinates expression of the spatial metric γ_{ab}. To show this, we first need to introduce the *spatial projection operator*[6]

$$P_b^a \equiv \delta_b^a + n^a n_b, \tag{9}$$

which we then apply (twice; once per index) to the spacetime metric g_{ab} to get

$$P_a^c P_b^d g_{cd} = \left(\delta_b^a + n^a n_b\right)\left(\delta_b^a + n^a n_b\right) g_{cd} = g_{ab} + n_a n_b.$$

Since the induced metric is merely a projection of the spacetime metric onto the hypersurface, we have found an expression for our *spatial metric*:

$$\gamma_{ab} = g_{ab} + n_a n_b \tag{10}$$

and similarly the *inverse spatial metric*

$$\gamma^{ab} = g^{ac} g^{bd} \gamma_{cd} = g^{ab} + n^a n^b. \tag{11}$$

Hence, γ_{ab} is a projection tensor that discards components of $4D$ geometric objects that lie along n^a; it allows us to compute distances entirely within a slice Σ. Intuitively, γ_{ab} first calculates the spacetime distance with g_{ab} and then kills off the timelike contribution (normal to the spatial surface) with $n_a n_b$.

Remark 1.3 From (10), we see that, if we raise only one index of the spatial metric γ_{ab},

$$\gamma_b^a = g_b^a + n^a n_b = \delta_b^a + n^a n_b,$$

we find out that our projection operator is merely the spatial metric with one raised index

$$P_b^a = \gamma_b^a.$$

Therefore, from now on, we will exclusively use γ_b^a to denote the spatial projection operator.

Also, from (10) and from (8c), we can see that

[6] This operator projects a $4D$ tensor onto a spatial slice. For instance, if we take an arbitrary 4-vector v^a and hit it with the projection operator,

$$\underbrace{v^a}_{\text{arbitrary, } 4D} \overset{P_b^a}{\longmapsto} \underbrace{P_b^a v^b}_{\text{purely spatial}},$$

we get a purely spatial object that lies entirely on a hypersurface.

$$\gamma_{ij} = g_{ij} + \underbrace{n_i n_j}_{=0} = g_{ij}, \tag{12}$$

so that the spatial metric on Σ is just the spatial part of the spacetime 4-metric g_{ab}. Note also that, even though the covariant components do not necessarily vanish ($\gamma_{0\mu} = g_{0\mu} + n_0 n_\mu = g_{0\mu} + n_0 = g_{0\mu} + \alpha^2 \neq 0$, in general), any contribution to the timelike direction can be safely ignored since $n^a \gamma_{ab} = 0$. On the other hand, timelike components of spatial contravariant tensors do vanish, so we must have $\gamma^{a0} = 0$. Therefore, from (11), we get the components of the inverse spacetime metric in these adapted coordinates:

$$\begin{aligned}
g^{ab} &= \gamma^{ab} - n^a n^b \\
g^{0a} &= -n^0 n^a \implies g^{00} = -\alpha^{-2} \quad \& \quad g^{0i} = \alpha^{-2}\beta^i \\
g^{ij} &= \gamma^{ij} - n^i n^j = \gamma^{ij} - (-\alpha^{-1}\beta^i)(-\alpha^{-1}\beta^j) = \gamma^{ij} - \alpha^{-2}\beta^i\beta^j .
\end{aligned}$$

In matrix form,

$$g^{\mu\nu} = \begin{pmatrix} -1/\alpha^2 & \beta^i/\alpha^2 \\ \beta^j/\alpha^2 & \gamma^{ij} - \beta^i\beta^j/\alpha^2 \end{pmatrix}. \tag{13}$$

Now, by the condition $g^{ab} g_{bc} = \delta^a_c$, we can invert (13) to write the spacetime metric in $3+1$ coordinates:

$$g_{\mu\nu} = \begin{pmatrix} -\alpha^2 + \beta_k\beta^k & \beta_i \\ \beta_j & \gamma_{ij} \end{pmatrix}. \tag{14}$$

The covariant components β_i shown above come from lowering with the spatial metric, i.e., $\beta_i = \gamma_{ik}\beta^k$. We will always use the spatial metric to raise/lower indices of spatial objects because γ_{ij} and γ^{ij} are inverses of each other in the adapted coordinates:

$$\begin{aligned}
\gamma^{ik}\gamma_{kj} &= (g^{ik} + n^i n^k)(g_{kj} + n_k n_j) \\
&= g^{ik} g_{kj} + g^{ik} n_k n_j + n^i n^k g_{kj} + n^i n^k n_k n_j \\
&= \delta^i_j + n^i \underbrace{n_j}_{=0} + n^i \underbrace{n_j}_{=0} - n^i \underbrace{n_j}_{=0} = \delta^i_j .
\end{aligned}$$

Equation (14) shows that the line element of the full spacetime metric in $3+1$ coordinates is given by

$$ds^2 = \left(-\alpha^2 + \beta_i\beta^i\right) \mathrm{d}t^2 + 2\beta_i \,\mathrm{d}t\,\mathrm{d}x^i + \gamma_{ij}\,\mathrm{d}x^i \mathrm{d}x^j . \tag{15}$$

1.2 ADM Evolution and Constraints

Using the projection operator (9) (which we now know is just γ_b^a), we can define the *extrinsic curvature tensor* to be[7]

$$K_{ab} = -\gamma_a^c \gamma_b^d \nabla_c n_d. \tag{16}$$

This quantity measures how much n^a varies as we move from point to point on a particular slice Σ, and in doing so it describes how Σ is embedded in $\mathcal{M}$. Expanding (16), we get

$$K_{ab} = -\nabla_a n_b - n_a n^c \nabla_c n_b, \tag{17}$$

and, moreover, a straightforward calculation shows that

$$K_{ab} = -\frac{1}{2}\mathcal{L}_{\vec{n}}\gamma_{ab}. \tag{18}$$

From the latter and from the time vector (6), we get a natural time derivative of the metric:

$$\begin{aligned} K_{ab} = -\frac{1}{2}\mathcal{L}_{\vec{n}}\gamma_{ab} &= -\frac{1}{2}\mathcal{L}_{\frac{\vec{t}-\vec{\beta}}{\alpha}}\gamma_{ab} \\ &= -\frac{1}{2\alpha}\left(\mathcal{L}_{\vec{t}}\gamma_{ab} - \mathcal{L}_{\vec{\beta}}\gamma_{ab}\right) \\ &= -\frac{1}{2\alpha}\left(\partial_t \gamma_{ab} - \mathcal{L}_{\vec{\beta}}\gamma_{ab}\right), \end{aligned} \tag{19}$$

where on the last line we used the fact that, in the adapted coordinates, $\mathcal{L}_{\vec{t}}$ reduces to ∂_t. Thus, expanding the Lie derivative in (19) (and dropping timelike components), we have an *evolution equation of the spatial metric*:

$$\partial_t \gamma_{ij} = 2D_{(i}\beta_{j)} - 2\alpha K_{ij}. \tag{20}$$

Here, D is the affine connection compatible with the $3D$ metric (i.e., $D_c\gamma_{ab} = 0$), which is furnished by projecting *all* indices present in a $4D$ covariant derivative ∇ onto Σ; that is, for a $\binom{b}{c}$ tensor $\boldsymbol{T}$,

$$D_a T^{i_1 \ldots i_b}_{j_1 \ldots j_c} = \gamma_a^d \gamma^{i_1}_{k_1} \cdots \gamma^{i_b}_{k_b} \gamma^{\ell_1}_{j_1} \cdots \gamma^{\ell_c}_{j_c} \nabla_d T^{k_1 \ldots k_b}_{\ell_1 \ldots \ell_c}. \tag{21}$$

[7] The minus sign is merely a convention in the NR community. In the cosmology community, the sign is usually positive.

The evolution equation for the metric was not so difficult to derive; however, the remaining evolution equation (of K_{ij}) and the constraint equations need a bit more work. Given the brevity of this presentation, it would not be possible to show derivations in great detail, but nevertheless it should (in theory at least) entice the interested reader to derive all the results from scratch (this is the only true way to learn anyhow).

We need to find a way to formulate the EFEs in 3 + 1 form, which is a task we can accomplish by considering the following projections:

$$n^a n^b (* [^{(4)}]G_{ab} - 8\pi T_{ab}) = 0; \tag{22a}$$

$$\gamma_c^a \gamma_d^b (* [^{(4)}]G_{ab} - 8\pi T_{ab}) = 0; \tag{22b}$$

$$\gamma_c^b \left[n^a (* [^{(4)}]G_{ab} - 8\pi T_{ab}) \right] = 0. \tag{22c}$$

(Note that all other projections vanish identically thanks to the symmetries of the Riemann tensor.) These equations come about from projections of the $4D$ Riemann tensor,

$$\gamma_a^e \gamma_b^f \gamma_c^g \gamma_d^{h(4)} R_{efgh} = R_{abcd} + K_{ac}K_{bd} - K_{ad}K_{cb}; \tag{23a}$$

$$\gamma_a^e \gamma_b^f \gamma_c^g n^{h(4)} R_{efgh} = D_b K_{ac} - D_a K_{bc}; \tag{23b}$$

$$\gamma_a^q \gamma_b^r n^c n^{d(4)} R_{qcrd} = \mathcal{L}_{\vec{n}} K_{ab} + \frac{1}{\alpha} D_a D_b \alpha + K_b^c K_{ac}. \tag{23c}$$

These are the so-called *Gauss–Codazzi*, *Codazzi–Mainardi*, and *Ricci* equations, respectively. Note how (23a) and (23b) depend exclusively on the spatial metric, the extrinsic curvature, and their spatial derivatives; they will give rise to the constraint equations (they can be thought of as integrability conditions allowing the embedding of a $3D$ slice Σ with data (γ_{ab}, K_{ab}) inside the ambient spacetime manifold $\mathcal{M}$). On the other hand, the first term on the RHS of (23c) hints that we will get the evolution equation for the extrinsic curvature from this equation.

In fact, expanding (23a) and doing some algebra, we end up with

$$R + K^2 - K^{ab}K_{ab} = 16\pi\rho, \tag{24}$$

where ρ is the *total energy density as measured by a normal observer* n^a,

$$\rho \equiv n^a n^b T_{ab}. \tag{25}$$

Dropping timelike components, we have the *Hamiltonian constraint*, which must be satisfied on each slice of the foliation,

$$R + K^2 - K^{ij}K_{ij} = 16\pi\rho. \tag{26}$$

Similarly, from (23b), we get

$$D_b K^b_a - D_a K = 8\pi S_a, \tag{27}$$

where we used the *momentum density S_a as measured by a normal observer n^a*,

$$S_a \equiv -\gamma^b_a n^c T_{bc}. \tag{28}$$

Dropping timelike components and raising indices, we write (27) in its final form

$$D_j \left(K^{ij} - \gamma^{ij} K \right) = 8\pi S^i, \tag{29}$$

which is with the so-called *momentum constraints* that must also be satisfied on each hypersurface. Lastly, from (23c), some very messy algebra yields

$$\begin{aligned} \partial_t K_{ab} = \alpha(R_{ab} + K K_{ab} - 2K_{ac}K^c_b) - 8\pi\alpha \left(S_{ab} - \frac{1}{2}\gamma_{ab}(S - \rho) \right) \\ - D_a D_b \alpha + \mathcal{L}_{\vec{\beta}} K_{ab}, \end{aligned} \tag{30}$$

where S_{ab} is the *spatial stress*, given from a projection of the $4D$ energy–momentum tensor T_{ab},

$$S_{ab} \equiv \gamma^c_a \gamma^d_b T_{cd}, \tag{31}$$

and S is its *trace*,

$$S \equiv \gamma^{ab} S_{ab} = S^a_a. \tag{32}$$

Then, since the entire content of spatial tensors is available from their spatial components, we can write our results as

$$\begin{aligned} \partial_t K_{ij} = \alpha(R_{ij} + K K_{ij} - 2K_{ik}K^k_j) - 8\pi\alpha \left(S_{ij} - \frac{1}{2}\gamma_{ij}(S - \rho) \right) \\ - D_i D_j \alpha + \beta^k D_k K_{ij} + 2K_{k(j} D_{i)} \beta^k. \end{aligned} \tag{33}$$

And thus, we arrived at the evolution equation for the extrinsic curvature, our last piece of the puzzle.

2 BSSN Formalism of Numerical Relativity

The 3 + 1 ADM (à la York) decomposition of the EFEs presented in Sect. 1 is already, in theory at least, in a form suitable for evolution simulations on a computer. Unfortunately, however, a lot more work needs to be done before we take a crack at computing anything, since in practise one finds that this form of the 3 + 1 decomposition in fact results in large instabilities that develop during the simulation. This issue is known to be mainly due to the fact that the equations in this form are *weakly hyperbolic* rather than *strongly hyperbolic*, which means that they are not *well-posed*.[8] To get around this problem, many NR codes implement the so-called *BSSN* (a.k.a. *BSSNOK*) *formalism* which, together with the "1 + log" *slicing* and the "*gamma-driver*" gauge conditions (we will briefly discuss these in Sect. 3), does admit a strongly hyperbolic formulation of the EFEs.

Whereas the standard ADM equations involve evolution equations for the raw spatial metric and extrinsic curvature tensors, through the BSSN formalism we are going to modify the equations by factoring out a *conformal factor* and introducing three *conformal connection coefficients* $\bar{\Gamma}^i$, reducing in this manner the evolution equations to wave equations for the *conformal metric* components. In other words, instead of the ADM data $\{\gamma_{ij}, K_{ij}\}$, the BSSN formalism splits γ_{ij} into a conformal factor χ and a conformally related metric $\bar{\gamma}_{ij}$, and it also splits K_{ij} into its trace K and a traceless part A_{ij}. Moreover, three coefficients $\bar{\Gamma}^i$ of the conformal metric are introduced as well. Then, it is these quantities that are evolved instead of the original ADM ones ... long story short, the dynamical variables for the BSSN system will be given by

$$\{\chi, \bar{\gamma}_{ij}, \bar{A}_{ij}, K, \bar{\Gamma}^i\}. \tag{34}$$

We will present each of these quantities and derive their evolution equations in this section. Let us start by considering a conformal rescaling of the spatial metric of the form

$$\gamma_{ij} = \chi^{-1}\bar{\gamma}_{ij}, \tag{35}$$

where χ is some positive scaling factor called the *conformal factor*, and the background auxiliary metric $\bar{\gamma}_{ij}$ is known as the *conformally related metric* (or simply *conformal metric*). It may seem unclear why we scaled the spatial metric in this way, but let it suffice to say that this "trick" will actually yield a convenient and tractable system for the EFEs. Besides the mathematical convenience that such a conformal rescaling brings about, there is also the fact that equivalence classes of conformally related manifolds share some geometric properties. For example, it can be shown that two strongly causal Lorentzian metrics $g^{(1)}_{ab}$ and $g^{(2)}_{ab}$ for some

[8] For more on the concept of hyperbolicity, see the detailed analysis in [26].

manifold $\mathcal{M}$ determine the same future and past sets at all points (events) if and only if the two metrics are globally conformal, i.e., if $g_{ab}^{(1)} = \Psi g_{ab}^{(2)}$, for some smooth function $\Psi \in C^{\infty}(\mathcal{M})$. In this case, both spacetimes $(\mathcal{M}, g_{ab}^{(1)})$ and $(\mathcal{M}, g_{ab}^{(2)})$ belong to the same conformal class and share the same causal structure.

A somewhat natural representative of a conformal equivalence class is a metric $\bar{\gamma}_{ij}$ whose determinant is equal to that of the flat metric δ_{ij} in whatever coordinate system we are using, i.e., $\bar{\gamma} = \delta$. Thus, if we adopt a Cartesian coordinate system, we can always enforce that our conformal representative must have unit determinant, i.e., $\bar{\gamma} = 1$. Plugging this back into (35), we get

$$1 = \det \bar{\gamma}_{ij} = \det(\chi \gamma_{ij}) = \chi^3 \underbrace{\det \gamma_{ij}}_{\equiv \gamma} = \chi^3 \gamma.$$

This would correspond to the choice $\chi = \gamma^{-1/3}$, so that $\gamma_{ij} = \gamma^{1/3} \bar{\gamma}_{ij}$. Any spatial metric γ_{ij} in this conformal class yields the same value of $\bar{\gamma}_{ij}$. However, note that, since the determinant γ is coordinate-dependent, the conformal factor $\chi = \gamma^{-1/3}$ *is not* a scalar field. In fact, $\bar{\gamma}_{ij}$ is not a tensor field, but rather a *tensor density* of *weight* $-2/3$. To get around this issue of tensor densities, we could introduce a *background flat metric* δ_{ij} of Riemannian signature $(+, +, +)$ and set $\chi \equiv (\gamma/\delta)^{-1/3}$, so that χ becomes a scalar field in this manner, and we could then use non-Cartesian coordinates. However, for our purposes of implementing the standard BSSN formalism, it is convenient to stick to Cartesian coordinates; to see the implementation of this extended BSSN formalism (where non-Cartesian coordinates are used), the reader is referred to [15].

The conformal factor is one of the dynamical variables in the BSSN approach, and as such we need an evolution equation for it. A straightforward calculation shows that

$$\partial_t \chi = \frac{2}{3} \chi (\alpha K - \partial_i \beta^i) + \beta^i \partial_i \chi, \tag{36}$$

where K is the *trace of the extrinsic curvature*,

$$K \equiv g^{ab} K_{ab} = \gamma^{ij} K_{ij} = K^i_{\ i}. \tag{37}$$

(Note that the second equality holds because K_{ab} is a purely spatial object.) Now, before showing the evolution of the conformal metric (another BSSN dynamic variable; cf. (34)), we need to briefly discuss the split of the extrinsic curvature K_{ij} into its trace K and its traceless part A_{ij},

$$K_{ij} = A_{ij} + \frac{1}{3} \gamma_{ij} K. \tag{38}$$

Just as we rescaled the spatial metric, in the BSSN formalism we shall also rescale the traceless curvature A_{ij} as[9]

$$\bar{A}_{ij} = \chi A_{ij}, \tag{39}$$

which yields

$$\bar{A}_{ij} = \chi K_{ij} - \frac{1}{3}\bar{\gamma}_{ij} K. \tag{40}$$

In terms of these rescaled variables, it is straightforward to show that the *evolution of the conformal metric* is given by

$$\partial_t \bar{\gamma}_{ij} = -2\alpha \bar{A}_{ij} + \beta^k \partial_k \bar{\gamma}_{ij} + \bar{\gamma}_{ik} \partial_j \beta^k + \bar{\gamma}_{kj} \partial_i \beta^k - \frac{2}{3}\bar{\gamma}_{ij} \partial_k \beta^k, \tag{41}$$

where we had to use the fact that $\bar{\gamma}_{ij}$ is a tensor density of weight $-2/3$ when expanding the Lie derivative $\mathcal{L}_{\vec{\beta}} \bar{\gamma}_{ij}$. (The Lie derivative of a tensor density of weight ω is given by

$$\mathcal{L}_{\vec{x}} \boldsymbol{\tau} = [\mathcal{L}_{\vec{x}} \boldsymbol{\tau}]_{\omega=0} + \omega \boldsymbol{\tau}\, \partial_i x^i,$$

where the first term is the usual Lie derivative we would compute if $\boldsymbol{\tau}$ had zero weight (i.e., if $\boldsymbol{\tau}$ was a *tensor field* rather than a *tensor density*).) Furthermore, a somewhat involved calculation shows that the *evolution of the trace of the extrinsic curvature* is given by

$$\partial_t K = \alpha \left(\bar{A}_{ij} \bar{A}^{ij} + \frac{1}{3} K^2 \right) + 4\pi\alpha(\rho + S) - D^2 \alpha + \beta^i \partial_i K, \tag{42}$$

and another (yet even longer) computation yields the *evolution of the conformal traceless curvature*,[10]

$$\begin{aligned} \partial_t \bar{A}_{ij} = {} & \left[\chi(\alpha R_{ij} - 8\pi\alpha S_{ij} - D_i D_j \alpha)\right]^{\mathrm{TF}} - \alpha(2\bar{A}_{ik}\bar{A}^k_j + \bar{A}_{ij} K) \\ & + \beta^k \partial_k \bar{A}_{ij} + \bar{A}_{ik} \partial_j \beta^k + \bar{A}_{kj} \partial_i \beta^k - \frac{2}{3} \bar{A}_{ij} \partial_k \beta^k. \end{aligned} \tag{43}$$

You may notice, however, that there is something off-putting on both (42) and (43); we have covariant derivatives D of the lapse with respect to the physical metric

[9] Note, however, that a different scaling for A_{ij} is used when dealing with the initial data problem (we briefly discuss initial data in Sect. 3).

[10] Here, we use the notation $[\cdots]^{\mathrm{TF}}$ to denote the trace-free part of whatever object lies inside the brackets (e.g., $[K_{ij}]^{\mathrm{TF}} = A_{ij}$). In general, for a tensor $\boldsymbol{T}$ in a D-dimensional metric $\boldsymbol{g}$, we have $[\boldsymbol{T}]^{\mathrm{TF}} = \boldsymbol{T} - \boldsymbol{g}/D\, \mathrm{Tr}(\boldsymbol{T})$.

γ_{ij} as opposed to the BSSN conformal metric $\bar{\gamma}_{ij}$ (moreover, (43) also has the $3D$ Ricci tensor R_{ij} of γ_{ij} appearing in the expression). We correct these problems by introducing the *conformal connection* $\bar{D}$ of $\bar{\gamma}_{ij}$ and writing the Christoffel symbols of D in terms of those of $\bar{D}$,

$$\Gamma^i_{jk} = \bar{\Gamma}^i_{jk} - \frac{1}{2}\chi^{-1}\left(\delta^i_k\, \partial_j \chi + \delta^i_j\, \partial_k \chi - \bar{\gamma}_{jk}\bar{\gamma}^{i\ell}\, \partial_\ell \chi\right). \tag{44}$$

Using this relation, a quick calculation shows that

$$D_i D_j \alpha = \bar{D}_i \bar{D}_j \alpha + \frac{1}{2\chi}\left(2\bar{D}_{(i}\chi\, \bar{D}_{j)}\alpha - \bar{\gamma}_{ij}\, \bar{D}_k \chi\, \bar{D}^k \alpha\right), \tag{45}$$

which is to be inserted into both (42) and (43). Furthermore, we may show (after a very long calculation) that the Ricci tensor can be split as

$$R_{ij} = \bar{R}_{ij} + R^{\chi}_{ij}, \tag{46}$$

where

$$\begin{aligned}\bar{R}_{ij} &= -\frac{1}{2}\bar{\gamma}^{k\ell}\partial_k\partial_\ell\bar{\gamma}_{ij} + \bar{\gamma}_{k(i}\partial_{j)}\,\bar{\Gamma}^k + \bar{\Gamma}^k\,\bar{\Gamma}_{(ij)k}\\ &\quad + \bar{\gamma}^{k\ell}\left(2\,\bar{\Gamma}^m_{k(i}\,\bar{\Gamma}_{j)m\ell} + \bar{\Gamma}^m_{ik}\,\bar{\Gamma}_{j\ell m}\right)\end{aligned} \tag{47a}$$

$$\begin{aligned}R^{\chi}_{ij} &= \frac{1}{2}\left(\bar{D}_i\bar{D}_j(\log\chi) + \bar{\gamma}_{ij}\,\bar{D}_k\bar{D}^k(\log\chi)\right)\\ &\quad + \frac{1}{4}\left(\bar{D}_i(\log\chi)\bar{D}_j(\log\chi) - \bar{\gamma}_{ij}\,\bar{D}_k(\log\chi)\bar{D}^k(\log\chi)\right).\end{aligned} \tag{47b}$$

Remark 2.1 In (47a), we used the conformal coefficients $\bar{\Gamma}^i \equiv \bar{\gamma}^{jk}\,\bar{\Gamma}^i_{jk}$. The reason why we want to write $\bar{R}_{ij}$ in this form is because, with the exception of the Laplacian term $\bar{\gamma}^{k\ell}\partial_k\partial_\ell\bar{\gamma}_{ij}$, every other second derivative of the metric $\bar{\gamma}_{ij}$ is being absorbed into first derivatives of $\bar{\Gamma}^i$. This in turn makes the BSSN equations more *hyperbolic* (see, e.g., [26]).

Speaking of $\bar{\Gamma}^i$, this is the only remaining BSSN dynamic variable for which we need an evolution equation (cf. (34)). Here, we simply present it (this is another long derivation; try it):

$$\begin{aligned}\partial_t\bar{\Gamma}^i &= -2\alpha\left(\frac{3}{2\chi}\bar{A}^{ij}\,\bar{D}_j\chi + \frac{2}{3}\bar{D}^i K + 8\pi\,\bar{S}^i - \bar{\Gamma}^i_{jk}\bar{A}^{jk}\right) - 2\bar{A}^{ij}\,\bar{D}_j\alpha\\ &\quad + \beta^j\partial_j\bar{\Gamma}^i + \bar{\gamma}^{jk}\partial_j\partial_k\beta^i - \bar{\Gamma}^j\partial_j\beta^i + \frac{2}{3}\bar{\Gamma}^i\partial_j\beta^j + \frac{1}{3}\bar{\gamma}^{ij}\partial_j\partial_k\beta^k.\end{aligned} \tag{48}$$

Lastly, we close out this section by writing the constraints in BSSN variables:

$$\bar{R} + 2\,\bar{D}^2(\log\chi) - \frac{1}{2}\bar{D}_k(\log\chi)\bar{D}^k(\log\chi) + \frac{4}{3\chi}K^2 - \frac{1}{\chi}\bar{A}_{ij}\bar{A}^{ij} = 16\pi\,\bar{\rho} \tag{49}$$

$$\bar{D}_j\bar{A}^{ij} - \frac{3}{2\chi}\bar{A}^{ij}\,\bar{D}_j\chi - \frac{2}{3}\bar{D}^i K = 8\pi\,\bar{S}^i, \tag{50}$$

where we used the rescaling $\bar{\rho}^i \equiv \chi^{-1}\rho$ and $\bar{S}^i \equiv \chi^{-1}S^i$. Moreover, we used the transformation law for the spatial Ricci scalar R,

$$R = \chi\bar{R} + 2\,\chi\,\bar{D}^2(\log\chi) - \frac{1}{2}\,\chi\,\bar{D}_k(\log\chi)\bar{D}^k(\log\chi). \tag{51}$$

Equations (49) and (50) are the Hamiltonian and momentum constraints, respectively, in BSSN variables.

That was a very compact presentation of the BSSN formalism of numerical general relativity. Admittedly, this formalism is not nearly as intuitive and straightforward as the ADM alternative that we presented in Sect. 1, but it is nevertheless a much more robust formulation (numerically speaking). This is a running theme in physics (and science in general): analytical and numerical implementations rarely play fair ball with each other. The ADM formalism is important for historical (and pedagogical) reasons, but it is nearly useless for practical purposes. We remark, however, that BSSN is not by any means the only modern successful approach to numerical relativistic studies; other flourishing alternatives such as the *Generalised Harmonic Coordinates with Constraint Damping* (GHCD) [18, 22, 23] and Z4-like [4, 6, 7, 9, 25] formalisms are just as good (and in some cases even superior) as BSSN.

3 Further Considerations

Having presented two of the main formalisms of numerical relativity, we now turn to a brief discussion of some further considerations that must be taken into account: the initial data problem (Sect. 3.1) and gauge choice (Sect. 3.2). Yet another topic that we have made no mention of (nor will we get into, since it is way beyond our scope) is the actual numerical methods employed in the field of numerical relativity; suffice it to say that the two most widely used methods used by the NR community are the good old fashioned *finite difference methods* [21] and *spectral methods* [16]. The reader is encouraged to study those references to get up to speed on the numerical side of things. Last but not least, we close out this chapter by making brief mention of some potential applications of NR outside of the usual realm of black holes/neutron stars collisions (Sect. 3.3).

3.1 Initial Data

As we alluded to earlier, we are not free to impose whatever data we like on our initial time slice; the initial data has to be chosen in such a way that the Hamiltonian and momentum constraints are satisfied from the onset.[11] That being said, the constraints are just four equations that remove four degrees of freedom from the total twelve degrees of freedom of the system $\{\gamma_{ij}, K_{ij}\}$. Moreover, there is no a priori preference for which eight of the total data to use as free parameters and which remaining four to use in solving the constraint equations. This *initial data problem* is a difficult subject with a vast literature dedicated to it (see [13] or any of the standard textbooks, e.g., Chapter 9 of [15]); the two most popular approaches to tackle this problem are known as the *conformal transverse-traceless* (CTT) decomposition and the *conformal thin-sandwich* (CTS) decomposition. Both of these methods provide some guidance on how to choose which values will be free parameters and which will be constrained data, although in the absence of significant symmetries many of the choices are arbitrary.

3.2 Gauge Choice

Even though, in theory (i.e., analytically) all gauge choices should yield the same physical result, as it is often the case numerical simulations do not always play nice. Therefore, in order to achieve a long-term stable simulation, we need to specify the right gauge (choice for α and β^i) and determine how these quantities will evolve in coordinate time. Choosing static (i.e., time-independent) gauges is not a very good idea, since we have no *a priori* knowledge of which functions will serve us better; the best approach is to choose the lapse and shift dynamically as functions of the evolving geometry.

One may naively think, for instance, that setting $\alpha = 1$ would be an ideal choice (certainly, our calculations would simplify quite a bit).[12] Unfortunately, however, this turns out to be a terrible pick: the acceleration of a normal observer is given in terms of the lapse function as $n^b \nabla_b n_a = D_a \log \alpha$; thus, setting $\alpha = 1$ yields a vanishing acceleration of normal observers (hence, the choice $\alpha = 1$ is usually referred to as *geodesic slicing*, since Eulerian observers are in free fall). A detailed examination then shows that this almost always leads to a singularity; thus, singularity-avoiding techniques such as *maximal slicing* (computationally expensive) or $1 + \log$ *slicing* (lower computational cost) must be employed. The latter is the one that has been adopted by most modern NR codes; it is a generalised

[11] In addition, for non-vacuum spacetimes, the matter distribution (ρ, S^i) may have constraints of its own.

[12] This corresponds to an evenly spaced slicing, so that coordinate time coincides with proper time of Eulerian observers (recall that $\mathrm{d}\tau = \alpha \mathrm{d}t$).

hyperbolic slicing condition of Bona–Massó type [8] whose basic idea is to reduce the lapse in regions where the curvature is particularly strong. In general, the so-called *alpha-driver* condition is given by

$$\partial_t \alpha = -\zeta_1 \, \alpha^{\zeta_2} K + \zeta_3 \beta^i \partial_i \alpha, \tag{52}$$

with ζ_i being some positive scalar functions. From this equation, we get the $1 + \log$ slicing by fixing $\zeta_1 = 2$ and $\zeta_2 = \zeta_3 = 1$.

Similarly, we may also choose a vanishing shift vector ($\beta^i = 0$), so that the coordinates are not shifted as we move from slice to slice. This would also certainly simplify matters, and it is in fact a common gauge choice that works well in certain applications. However, in black hole spacetime simulations, if we use a vanishing β^i, the event horizon grows rapidly in coordinate space, due to the normal observers falling in, which causes the computational domain to end up eventually trapped inside the black hole (see [1]). Moreover, in order to counter the large field gradients (or "slice stretching")[13] incurred in the presence of a black hole, a nonvanishing β^i is required [2]. To deal with this slice stretching issue, gauge conditions were designed so that second-time derivatives of the shift are proportional to first-time derivatives of the coefficients $\bar{\Gamma}^i$ (i.e., $\partial_t^2 \beta^i \sim \partial_t \bar{\Gamma}^i$). In particular, a hyperbolic shift condition

$$\partial_t^2 \beta^i = \eta \partial_t \bar{\Gamma}^i - \xi \partial_t \beta^i, \tag{53}$$

where η and ξ are positive scalar fields, was introduced. This is the so-called *gamma-driver* shift condition. We may then use an auxiliary vector field $\mathcal{B}^i$ to perform the usual trick of rewriting a second-order derivative in first order,

$$\partial_t \mathcal{B}^i = \vartheta_1 \, \alpha^{\vartheta_2} \, \partial_t \bar{\Gamma}^i - \varrho_1 \, \mathcal{B}^i \tag{54a}$$

$$\partial_t \beta^i = \varrho_2 \, \mathcal{B}^i, \tag{54b}$$

where we also rewrote the scalar fields η and ξ in terms of four damping parameters $\vartheta_{1,2}$ and $\varrho_{1,2}$ that fine-tune the growth of the shift.

Once we have chosen our gauge, prescribed (and evolved, via the BSSN formalism presented above) the initial data, one must turn to a last (and key) step: interpret the obtained data in a gauge-independent way (e.g., one may wish to find event horizons or extract gravitational wave signals).

[13] $3 + 1$ simulations of black hole spacetimes without singularity excision and with singularity-avoiding lapse and vanishing shift fail after an evolution time of around 30–40M due to the so-called slice stretching (see [3]).

3.3 Potential Application to Cosmology

Thus far, we have only discussed purely geometric aspects of the 3 + 1 formulation of GR, without discussing constraints or evolution of any potential matter field that might be coupled to the EFEs; we now turn to this topic (for a thorough treatment, the reader may consult, e.g., Chapter 6 of [15]). Let us focus our succinct discussion in a scalar matter field ϕ minimally coupled to the EFEs. For such scalar field, the *Lagrangian* is given by

$$L_M = \frac{1}{2}\nabla_a\phi\nabla^a\phi + V(\phi), \tag{55}$$

where $V(\phi)$ is the *scalar potential*, which may be decomposed as

$$V(\phi) = \frac{1}{2}m^2\phi^2 + V_{\text{int}}(\phi), \tag{56}$$

m being the *mass* of the field and V_{int} the *interaction potential* (since we are interested in the minimally coupled case, the field is noninteracting; i.e., $V_{\text{int}} = 0$). If we then add the associated *scalar matter Lagrangian density* $\mathcal{L}_M = \sqrt{-g}L_M$ to the *gravitational Lagrangian density*, namely $\mathcal{L}_G = \sqrt{-g}R$, then minimisation of the (modified) *Einstein–Hilbert action*

$$S = \frac{1}{16\pi}\int (\mathcal{L}_G + \mathcal{L}_M)\, \mathrm{d}^4x \tag{57}$$

leads to the EFEs with stress–energy tensor defined by

$$T_{ab} = \nabla_a\phi\nabla_b\phi - \frac{1}{2}g_{ab}\left(\nabla_c\phi\nabla^c\phi + 2V(\phi)\right). \tag{58}$$

The addition of this scalar field to our 3 + 1 formulation of GR opens the doors to some interesting areas of study, for instance, *inhomogeneous cosmological inflation* [11, 12, 19] and *critical gravitational collapse* [10, 11, 17].

Moreover, from (55), the *Euler–Lagrange* equations yield our *equation of motion*, which coincides with the *Klein–Gordon* equation in curved spacetime:

$$\nabla^2\phi = \frac{\mathrm{d}V(\phi)}{\mathrm{d}\phi} = m^2\phi, \tag{59}$$

where $\nabla^2 = g^{ab}\nabla_a\nabla_b$. Since this equation of motion is of second order, it would be useful to cast it into first order for integration purposes; we accomplish this with the aid of new auxiliary variables Π and $\varkappa_i$ given by[14]

[14] Some references (e.g., [5]) define Π as the negative of ours; here, we follow the convention in [11].

$$\Pi \equiv \frac{1}{\alpha}\left(\partial_t\phi - \beta^i\partial_i\phi\right) \tag{60}$$

$$\varkappa_i \equiv \partial_i\phi. \tag{61}$$

Using these variables, (59) splits as

$$\partial_t\phi = \alpha\Pi + \beta^i\varkappa_i \tag{62a}$$

$$\partial_t\varkappa_i = \beta^j\partial_j\varkappa_i + \varkappa_j\partial_i\beta^j + \alpha\partial_i\Pi + \Pi\partial_i\alpha \tag{62b}$$

$$\partial_t\Pi = \beta^i\partial_i\Pi + g^{ij}\left(\alpha\partial_j\varkappa_i + \varkappa_j\partial_i\alpha - \Gamma^k_{ij}\varkappa_k\right) + \alpha\left(K\Pi + \frac{\mathrm{d}V(\phi)}{\mathrm{d}\phi}\right). \tag{62c}$$

These equations must be solved in conjunction with the gravitational field's $3+1$ equations in order to determine the complete evolution of a spacetime containing a scalar matter field. Note that (62a) is just the definition of Π (i.e., it is simply a rewriting of (60)) and (62b) follows directly from (61) and (62a) by commuting partial derivatives; the true equation of motion is in fact determined by (62c). The constraint given by (61), namely,

$$\mathcal{R}_i \equiv \varkappa_i - \partial_i\phi = 0, \tag{63}$$

must be preserved by the system (62). Of course, if it is solved *exactly*, (62) does guarantee the preservation of (63) throughout the evolution. The problem is at the *numerical* level, where truncation errors can give the residual $\mathcal{R}_i$ nonzero values; therefore, it is important to keep a close eye out on $\mathcal{R}_i$ (in addition to the Hamiltonian and momentum constraints) during the evolution to make sure that we are working with an accurate simulation.

Had we instead assumed homogeneity of the scalar field (i.e., $\nabla_i\phi = \partial_i\phi = 0$), then the equation of motion (59) for an FLRW metric would yield a relatively simple second-order ODE for the evolution of the "inflaton" scalar field $\phi(t)$,

$$\ddot{\phi} + 3H\dot{\phi} + m^2\phi = 0, \tag{64}$$

where, per usual notation, $H(t) = \dot{a}/a$ represents *Hubble's constant*, and $a(t)$ is the *expansion parameter* that appears in the FLRW metric. The much more complicated problem of dealing with an inhomogeneous inflaton requires the full power of numerical relativity, and it is currently a very active research area. For more on this topic, the reader is referred to references such as [11, 12, 14, 19].

Acknowledgments This work was made possible by the Domoschool organisers. Special thanks to them.

References

1. M. Alcubierre, *Introduction to 3+1 Numerical Relativity*. International Series of Monographs on Physics. (Oxford University Press, Oxford, 2008)
2. M. Alcubierre, B. Brügmann, Simple excision of a black hole in (3+1)-numerical relativity. Phys. Rev. **D63**, 104006 (2001)
3. M. Alcubierre, B. Brügmann, P. Diener, M. Koppitz, D. Pollney, E. Seidel, R. Takahashi, Gauge conditions for long term numerical black hole evolutions without excision. Phys. Rev. **D67**, 084023 (2003)
4. D. Alic, C. Bona-Casas, C. Bona, L. Rezzolla, C. Palenzuela, Conformal and covariant formulation of the Z4 system with constraint-violation damping. Phys. Rev. D **85**, 064040 (2012)
5. T.W. Baumgarte, S.L. Shapiro, *Numerical Relativity: Solving Einstein's Equations on the Computer*. (Cambridge University Press, Cambridge, 2010)
6. S. Bernuzzi, D. Hilditch, Constraint violation in free evolution schemes: comparing the BSSNOK formulation with a conformal decomposition of the Z4 formulation. Phys. Rev. D **81**, 084003 (2010)
7. C. Bona, C. Palenzuela, Dynamical shift conditions for the Z4 and BSSN formalisms. Phys. Rev. D **69**, 104003 (2004)
8. C. Bona, J. Masso, E. Seidel, J. Stela, A New formalism for numerical relativity. Phys. Rev. Lett. **75**, 600–603 (1995)
9. C. Bona, T. Ledvinka, C. Palenzuela, M. Žáček, General-covariant evolution formalism for numerical relativity. Phys. Rev. D **67**, 104005 (2003)
10. M.W. Choptuik, Universality and scaling in gravitational collapse of a massless scalar field. Phys. Rev. Lett. **70**, 9–12 (1993)
11. K. Clough, Scalar Fields in Numerical General Relativity: Inhomogeneous inflation and asymmetric bubble collapse. PhD thesis, King's College London, Cham (2017)
12. K. Clough, E.A. Lim, B.S. DiNunno, W. Fischler, R. Flauger, S. Paban, Robustness of inflation to inhomogeneous initial conditions. JCAP **1709**(09), 025 (2017)
13. G.B. Cook, Initial data for numerical relativity. Living Rev. Relativ. **3**(1), 5 (2000)
14. W.E. East, M. Kleban, A. Linde, L. Senatore, Beginning inflation in an inhomogeneous universe. JCAP **1609**(09), 010 (2016)
15. É. Gourgoulhon, $3 + 1$ *Formalism in General Relativity: Bases of Numerical Relativity* (Springer, Berlin, 2012)
16. P. Grandclément, J. Novak, Spectral methods for numerical relativity. Living Rev. Relativ. **12**(1), 1 (2009)
17. C. Gundlach, J.M. Martin-Garcia, Critical phenomena in gravitational collapse. Living Rev. Relativ. **10**, 5 (2007)
18. C. Gundlach, G. Calabrese, I. Hinder, J.M. Martín-García, Constraint damping in the Z4 formulation and harmonic gauge. Classical Quantum Gravity **22**(17), 3767–3773 (2005)
19. P. Laguna, H. Kurki-Suonio, R.A. Matzner, Inhomogeneous inflation: the initial-value problem. Phys. Rev. D **44**, 3077–3086 (1991)
20. C.W. Misner, K.S. Thorne, J.A. Wheeler, *Gravitation* (W.H. Freeman, New York, 1970)
21. *Numerical Analysis for Numerical Relativists*. Volume Lectures for VII Mexican School on Gravitation and Mathematical Physics (2006)
22. F. Pretorius, Evolution of binary black hole spacetimes. Phys. Rev. Lett. **95**, 121101 (2005)
23. F. Pretorius, Numerical relativity using a generalized harmonic decomposition. Classical Quantum Gravity **22**(2), 425–451 (2005)
24. H. Ringström, *The Cauchy Problem in General Relativity*. ESI Lectures in Mathematics and Physics. (European Mathematical Society, Zürich, 2009)

25. N. Sanchis-Gual, P.J. Montero, J.A. Font, E. Müller, T.W. Baumgarte, Fully covariant and conformal formulation of the Z4 system in a reference-metric approach: comparison with the BSSN formulation in spherical symmetry. Phys. Rev. D **89**, 104033 (2014)
26. O. Sarbach, G. Calabrese, J. Pullin, M. Tiglio, Hyperbolicity of the BSSN system of Einstein evolution equations. Phys. Rev. **D66**, 064002 (2002)
27. R. Wald, *General Relativity* (University of Chicago Press, Chicago, 1984)

Length-Contraction in Curved Spacetime

Colin MacLaurin

Contents

1 Introduction

Length-contraction is a staple of introductory special relativity, often encountered in high school physics. However, it receives no mention in a typical general relativity course nor in most curved spacetime literature. When first encountering the radial length $(1-2M/r)^{-1/2}dr$ in Schwarzschild spacetime, many students wonder what

C. MacLaurin (✉)
University of Queensland, Brisbane, QL, Australia
e-mail: colin.maclaurin@uqconnect.edu.au

S. L. Cacciatori, A. Kamenshchik (eds.), *Einstein Equations: Local Energy, Self-Force, and Fields in General Relativity*, Tutorials, Schools, and Workshops in the Mathematical Sciences, https://doi.org/10.1007/978-3-031-21845-3_9

happens inside the horizon, where the coefficient becomes imaginary. In one case, a lecturer explained that since there is no timelike Killing vector for $r \leq 2M$ there is no preferred length, which is a solid answer though an incomplete one. How does this relate to special relativity, where length depends on the relative motion between observers? Certain aspects of this foundational topic of length deserve careful reexamination.

Historically, length-contraction originated in Maxwell's electromagnetic theory and was developed by Heaviside, Fitzgerald, Lorentz, Larmor, Poincaré, Einstein, and others. The Michelson-Morley experiment influenced many of these. Various shape deformations were proposed, but the longitudinal contraction won out. This contraction was relative, not fixed by the frame of the luminiferous ether (as then assumed to exist). Yet its natural home, like the Lorentz transformation and electromagnetism generally, was in the new ontology of Einstein's relativity [1, §4].

In philosophy of relativity, the "ruler hypothesis" has received deserving attention. It states rulers achieve their intended purpose of measuring metric length. The hypothesis is not presented as being entirely true, but rather a statement for critical examination. Here, we will model a ruler by a spatial vector in the tangent space at a single point in curved spacetime. This may be considered part of an orthonormal frame. Equivalently, the ruler vector represents the orientation of a radar distance device. But while the motivation and intuition is literal rods, technically we analyse nothing more than the metric and several induced volume elements, including a certain transformation between frames. I hope this is reassuring to the many critics of the ruler concept in relativity; see also Sect. 6.

2 Proper Frame Measurement

This section relates proper distance within the local 3-dimensional space of a given observer, to a coordinate Φ. It summarises my earlier work, [6, §5] [7] with fresh explanation, plus careful justification which many readers should skip.

At a given point in spacetime, suppose $\mathbf{u}$ is a 4-velocity (a unit, timelike, future-pointing vector), $\boldsymbol{\xi}$ a unit vector orthogonal to $\mathbf{u}$, and Φ a scalar field differentiable at that point. In particular:

$$\langle \mathbf{u}, \boldsymbol{\xi} \rangle = 0, \qquad \langle \boldsymbol{\xi}, \boldsymbol{\xi} \rangle = 1.$$

The angle brackets denote the metric scalar product, where in general:

$$\langle \mathbf{a}, \mathbf{b} \rangle := g_{\mu\nu} a^\mu b^\nu = a_\mu b^\mu = g^{\mu\nu} a_\mu b_\nu,$$

for any pair of rank-1 tensors. We interpret $\mathbf{u}$ as an observer, $\boldsymbol{\xi}$ as an idealised "ruler", and Φ as a coordinate such as Schwarzschild r. We wish to compare the ruler to the coordinate gradient $d\Phi$, to determine the interval $\Delta\Phi$ spanned by the ruler with proper length $\Delta L = 1$ say. To be more precise, we seek the ratio at a point.

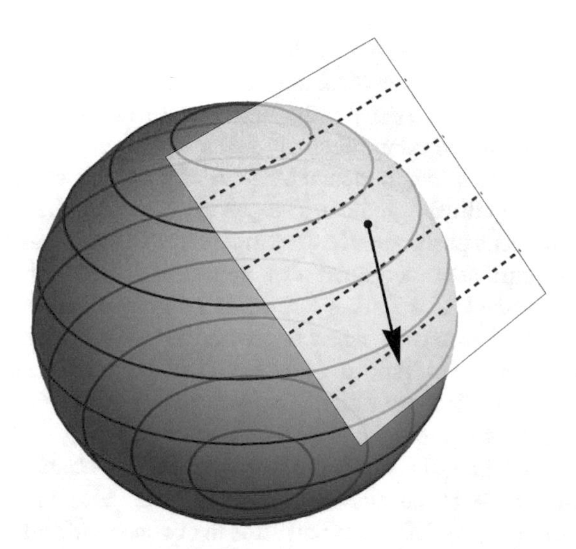

Fig. 1 A sphere with one tangent plane shown. For our later applications, think of this as a surface within Minkowski or Schwarzschild spacetime. The solid curves are level sets $\theta = \text{const}$. Within the tangent plane, $d\theta$ is visualised as hyperplanes (dotted lines). Our "ruler" $\boldsymbol{\xi}$ is a vector within the tangent space. Intuitively, it approximates a literal physical ruler's extent along the manifold itself (interpreting $\boldsymbol{\xi}$ as a line segment, then applying the exponential map), under minimal strain from tidal forces, acceleration, etc. But when taken as a pointwise 1-dimensional volume element on the manifold, it is precise

But the change in Φ by proper length, in the $\boldsymbol{\xi}$-direction, is $d\Phi/dL = \langle d\Phi, \boldsymbol{\xi} \rangle$. Rearranging:[1]

$$dL = \frac{1}{\langle d\Phi, \boldsymbol{\xi} \rangle} d\Phi. \tag{1}$$

Formally, we treat both sides as covectors or 1-forms. Recall the gradient $d\Phi$, equivalently written $\nabla\Phi$, has components $(d\Phi)_\mu = \partial\Phi/\partial x^\mu$ in a coordinate basis. There is an intuitive picture for the contraction $\langle d\Phi, \boldsymbol{\xi} \rangle$ of a 1-form and vector to return a scalar [12, §3.3] [8]. As in Fig. 1, $d\Phi$ may be visualised as a set of hyperplanes within a tangent space, or more roughly as hypersurfaces $\Phi = \text{const}$ within the manifold. Its action on a vector returns the number of hyperplanes spanned by that vector. The tangent space and quantities therein are essentially a linearisation or gradient at a point on the manifold, which is the motivation behind $\boldsymbol{\xi}$. Our approach is strictly local, which is more physically justified than claims of

[1] A negative coefficient is allowed. Also if $\langle d\Phi, \boldsymbol{\xi} \rangle = 0$, then Φ is constant along the $\boldsymbol{\xi}$ direction, so is of no use for demarcating this length. The analogy to Eq. (1) would be $d\Phi = 0$ but interpreted as 1-dimensional only.

"observer at infinity" measurement, as in Schwarzschild spacetime, for instance. The idea is to piece together such local measurements into an overall quantity.

dL is equivalent to infinitesimal radar distance. In fact it is simply the metric interval ds, restricted to the ruler direction! Just as the vector $\boldsymbol{\xi}$ represents a unit of length, the 1-form $\boldsymbol{\xi}^\flat$ is a length element.[2,3] So why bother, what is the use? The equation relates a coordinate gradient to a physical, measurable quantity. This is useful to find an object's position and extent in terms of the mathematical description of the theory—in particular, coordinates on spacetime.[4] Similarly an orthonormal tetrad, when interpreted as a rank-2 tensor, is nothing other than the metric in disguise! [3, §J]. Such a tetrad, when expressed in coordinates, relates physical quantities to coordinates. Our pair $(\mathbf{u}, \boldsymbol{\xi})$, or even just $\boldsymbol{\xi}$ alone, forms a sort of proto-tetrad. The orthonormal decomposition $\mathbf{g} = -\mathbf{u}^\flat \otimes \mathbf{u}^\flat + \boldsymbol{\xi}^\flat \otimes \boldsymbol{\xi}^\flat + \cdots$ exhibits the ruler's part within the full metric tensor.

Traditional tensor analysis adopts the coordinate basis vectors as measurement directions. Assuming ∂_Φ is spacelike, $\boldsymbol{\xi} := (g_{\Phi\Phi})^{-1/2}\partial_\Phi$ is a unit vector, for which Eq. (1) returns $dL = \sqrt{g_{\Phi\Phi}}\, d\Phi$. Similarly for the cobasis 1-form $d\Phi$, define $\boldsymbol{\xi} := (g^{\Phi\Phi})^{-1/2}(d\Phi)^\sharp$, then $dL = (g^{\Phi\Phi})^{-1/2} d\Phi$. The latter relies on Φ being spacelike, meaning $\langle d\Phi, d\Phi\rangle > 0$. The problem is, neither ruler is orthogonal to $\mathbf{u}$ in general, meaning these are not measurements in a given observer's frame! To correct this, use the spatially projected vector $(d\Phi)^\sharp + \langle d\Phi, \mathbf{u}\rangle\mathbf{u}$. This is orthogonal to $\mathbf{u}$, hence spatial or zero. If nonzero, normalise it and apply Eq. (1) to get:

$$dL_{\Phi,3\text{-gradient}} = \frac{1}{\sqrt{\langle d\Phi, d\Phi\rangle + \langle d\Phi, \mathbf{u}\rangle^2}} d\Phi. \tag{2}$$

If Φ is taken as a coordinate, we may rewrite as components $\langle d\Phi, d\Phi\rangle = g^{\Phi\Phi}$ and $\langle d\Phi, \mathbf{u}\rangle = u^\Phi$. This ruler direction is a natural definition of "the Φ-direction" relative to $\mathbf{u}$. Recall the normal vector $(d\Phi)^\sharp$ is orthogonal to the level sets $\Phi =$

[2] Technically, we define dL only on the 1-dimensional subspace of the cotangent space spanned by $\boldsymbol{\xi}^\flat$. This is because the rate $d\Phi/dL$ is a derivative along the ruler direction specifically. With this restriction understood, Eq. (1) is generally covariant. In fact dL is merely $\boldsymbol{\xi}^\flat$, as seen from contracting both with the single basis vector $\boldsymbol{\xi}$. If using the notation $\boldsymbol{\xi}^\flat$, it seems fine to treat it as 4-dimensional.

[3] Compare H. Brown that "[rods and clocks] are not the analogue of thermometers of the spacetime metric. They measure in the weaker sense that their behaviour correlates with aspects of spacetime structure..." [2]. Exceptions include Weyl's theory and some recent alternate gravity theories [1].

[4] Coordinates are useful for bookkeeping, to "chart" positions and build a global picture, an extrinsic standard. In other cases they are an interim computational tool, such as for volume integrals. It is often stated coordinates have no physical meaning, with Einstein quotes for support; however, in the present context I prefer to nuance: no *direct* physical meaning, which is also backed up by Einstein quotes such as "...immediate metrical significance". Norton puts it well: "In Einstein's words the '[coordinate] system... has no physical reality.' We might phrase this more cautiously by saying that the coordinate system has no reality independent of the metric, for the combination of coordinate system and metric certainly do represent aspects of physical reality" [11, §5.5.3].

const. If Φ is spacelike, $(d\Phi)^\sharp$ is the direction of greatest increase of Φ, per length of vector. Similarly the projected gradient gives the greatest increase of Φ, amongst any vectors orthogonal to $\mathbf{u}$. It maximises $d\Phi/dL$ and minimises $dL/d\Phi$, for any vector in $\mathbf{u}$'s 3-space. If $\langle d\Phi, \mathbf{u}\rangle = 0$, the measurement reduces to the traditional quantity.

3 Length-Contraction and Measurement

This section considers another "observer" $\mathbf{n}$, in whose frame the local 3-space of $\mathbf{u}$ is length-contracted. We derive a 4-vector formalism for length-contraction, introduce "de-contraction", and apply these to length measurements. From now on, we mostly term $\mathbf{u}$ "fluid" or "matter", to distinguish from the frame doing the observing. Shortly, $\mathbf{n}$ will be generalised to be timelike, null, or spacelike. It stands for "normal" to a hypersurface on the manifold, or at least to a hyperplane in a tangent space.

3.1 Relative 3-Velocity

Recall the prototypical Lorentz boost $t' = \gamma(t - \beta x)$ and $x' = \gamma(x - \beta t)$ between global inertial frames in Minkowski spacetime, where $\gamma = (1 - \beta^2)^{-1/2}$ is the Lorentz factor. There is an analogous transformation on 4-vectors rather than coordinates, which easily accommodates arbitrary boost directions:

$$\mathbf{n} = \gamma(\mathbf{u} + \beta\hat{\vec{\mathbf{n}}}), \qquad -\hat{\vec{\mathbf{u}}} = \gamma(\hat{\vec{\mathbf{n}}} + \beta\mathbf{u}).$$

In the following, $\mathbf{n}$ is timelike. In our notation, which is motivated by the usage below, $\vec{\mathbf{n}} \equiv \beta\hat{\vec{\mathbf{n}}}$ is the relative velocity of $\mathbf{n}$ according to $\mathbf{u}$'s frame, and conversely $\vec{\mathbf{u}} \equiv \beta\hat{\vec{\mathbf{u}}}$ is the velocity of $\mathbf{u}$ in $\mathbf{n}$'s frame, where β is the relative speed. The hats signify unit vectors, so these are the boost directions. The boost sends $\mathbf{u}$ to $\mathbf{n}$, and $\vec{\mathbf{n}}$ to $-\vec{\mathbf{u}}$ [5, §4]. Note it is not an inverse boost; the plus signs above are because vectors transform oppositely to coordinates. In curved spacetime, one applies a separate *local Lorentz boost* at every point, which is a transformation within each tangent space.

Supposing $\mathbf{u}$ and $\mathbf{n}$ are provided, we may recover:

$$\gamma = -\langle \mathbf{u}, \mathbf{n}\rangle,$$

and hence:

$$\vec{\mathbf{n}} = \gamma^{-1}\mathbf{n} - \mathbf{u}.$$

One can also show:

$$\vec{\mathbf{u}} = \gamma^{-1}\mathbf{u} - \mathbf{n}.$$

The relative velocity vectors are purely spatial, according to the observer doing the measuring:

$$\langle \mathbf{u}, \vec{\mathbf{n}} \rangle = 0 = \langle \mathbf{n}, \vec{\mathbf{u}} \rangle.$$

The relative speed is recovered from $\beta = \sqrt{1 - \gamma^{-2}}$, or:

$$\langle \vec{\mathbf{n}}, \vec{\mathbf{n}} \rangle = \beta^2 = \langle \vec{\mathbf{u}}, \vec{\mathbf{u}} \rangle.$$

Our arrow notation imitates 3-vectors; however, we use 4-vector formalism.[5] We stress that $\vec{\mathbf{n}}$ depends on both $\mathbf{n}$ and $\mathbf{u}$, and likewise for $\vec{\mathbf{u}}$.

3.2 Length-Contraction Vector

We derive a 4-vector formalism for length-contraction. This algebraic approach is useful for comparing measurements between frames.

In relativity, length-contraction results from relative motion between frames. In Fig. 2, the observer $\mathbf{n}$ determines $\mathbf{u}$'s 3-space to be contracted along the direction of relative motion (the horizontal axis, which is aligned with $\vec{\mathbf{u}}$). This is due to the relativity of simultaneity. This diagram is in principle standard, apart from the vector

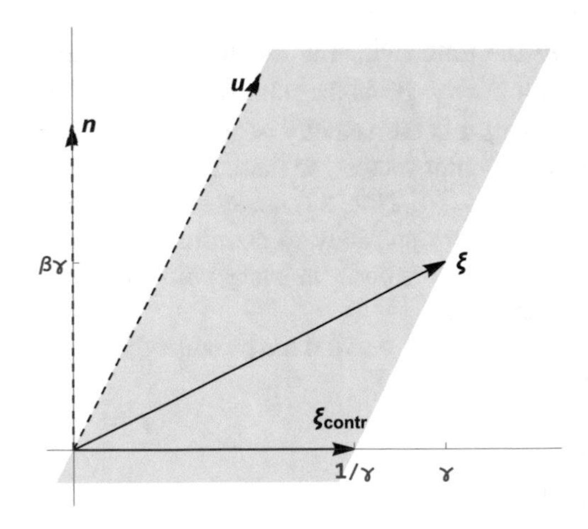

Fig. 2 Spacetime diagram showing length-contraction, in the case of $\mathbf{n}$ timelike. The observer $\mathbf{u}$ is holding a unit ruler $\boldsymbol{\xi}$, which sweeps out the grey worldsheet over time. The intersection with $\mathbf{n}$'s 3-space is the vector $\boldsymbol{\xi}_{\text{contr}}$

[5] 3-vector treatments may equate spatial vectors with their boosted image, because the components are the same in the boosted Minkowski coordinates.

notation. It is found in Schutz, for example, [12, §1.8] but appears absent from most recent textbooks.

The figure applies to both inertial macroscopic objects in flat spacetime, or locally in curved spacetime within a single tangent space. Either way, in Minkowski coordinates aligned with the boost direction, the ruler is $\xi^\mu = (\beta\gamma, \gamma, 0, 0)$, whereas the contracted ruler is $\xi^\mu_{\text{contr}} = (0, 1/\gamma, 0, 0)$. Note $\boldsymbol{\xi}_{\text{contr}}$ is not the boost of $\boldsymbol{\xi}$, nor is it orthogonal projection onto $\mathbf{n}$'s 3-space. These are all parallel—at least for purely spatial vectors aligned with the boost direction—but have lengths $1/\gamma$, 1, and γ, respectively, compare Jantzen, Carini, and Bini [5, Figure 1].

To obtain $\boldsymbol{\xi}_{\text{contr}}$ as an abstract vector (not just components in an adapted coordinate system), observe from the diagram we must add to $\boldsymbol{\xi}$ a multiple of $\mathbf{u}$ so that the sum is orthogonal to $\mathbf{n}$. Require $\langle \boldsymbol{\xi} + \lambda \mathbf{u}, \mathbf{n} \rangle = 0$ say, with unique solution $\lambda = -\langle \boldsymbol{\xi}, \mathbf{n} \rangle / \langle \mathbf{u}, \mathbf{n} \rangle$ aside from some exceptional cases.[6] Hence:

$$\boldsymbol{\xi}_{\text{contr}} = \boldsymbol{\xi} - \frac{\langle \boldsymbol{\xi}, \mathbf{n} \rangle}{\langle \mathbf{u}, \mathbf{n} \rangle} \mathbf{u}. \tag{3}$$

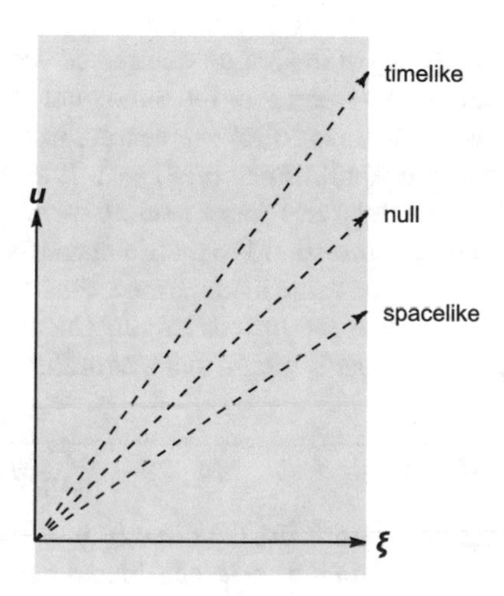

Fig. 3 Length-contraction, where the observer or normal $\mathbf{n}$ has arbitrary nature. The vertical axis is $\mathbf{u}$'s time, in contrast to Fig. 2, and $\mathbf{n}$ is a different vector in this case. For rulers aligned with the boost or "tilt" direction, if the hyperplane normal $\mathbf{n}$ is the timelike vector shown, then $\boldsymbol{\xi}_{\text{contr}}$ is the spacelike vector, and vice versa. For an aligned ruler and $\mathbf{n}$ null, $\boldsymbol{\xi}_{\text{contr}} = \mathbf{u} + \boldsymbol{\xi}$, which is proportional to $\mathbf{n}$. For unaligned rulers, the length-contracted ruler may remain spacelike, irrespective of the sign of $\langle \mathbf{n}, \mathbf{n} \rangle$

[6] If $\mathbf{n}$ is orthogonal to both $\mathbf{u}$ and $\boldsymbol{\xi}$, then the entire worldsheet lies in the hyperplane. If $\mathbf{n}$ is orthogonal to $\mathbf{u}$ only, the intersection is the infinite line spanned by $\mathbf{u}$.

This holds for any $\boldsymbol{\xi}$ (orthogonal to $\mathbf{u}$), not just those aligned with the boost direction.[7] It also holds for hypersurfaces of any character (hyperplanes, within a tangent space): timelike, null, or spacelike. See Fig. 3. Equation (3) is unchanged by rescaling $\mathbf{n}$, so $\mathbf{n}$ need not be normalised, which is reassuring for the null case especially. Hence $\mathbf{n}$ may be any nonzero vector. Note that while $\boldsymbol{\xi}$ is associated with $\mathbf{u}$, since $\langle \mathbf{u}, \boldsymbol{\xi} \rangle = 0$, the contracted vector $\boldsymbol{\xi}_{\text{contr}}$ is orthogonal to $\mathbf{n}$, hence may be spacelike, null, or even timelike. This generalises length-contraction to 4-vectors and arbitrary "observer" hyperplanes.

Two sources deserve special commendation. Møller presents the only textbook vector treatment of length-contraction I am aware of, though he uses 3-vectors [9, §19]. Jantzen, Carini, and Bini define various projection and boost maps, including a certain $P(\mathbf{u}, \mathbf{n})^{-1}$ which acts as: $P(\mathbf{u}, \mathbf{n})^{-1}\boldsymbol{\xi} = P_{\mathbf{u}}\boldsymbol{\xi} + \langle \vec{\mathbf{n}}, \boldsymbol{\xi} \rangle \mathbf{u}$. This is equivalent to Eq. (3). They also describe a scaled boost $\gamma^{-1} B(\mathbf{n}, \mathbf{u})\boldsymbol{\xi}$, which is equivalent only for $\boldsymbol{\xi}$ aligned with the boost direction [5, §4]. Both sources assume $\mathbf{n}$ is timelike.

3.3 Measurement

The squared-norm $\langle \boldsymbol{\xi}_{\text{contr}}, \boldsymbol{\xi}_{\text{contr}} \rangle$ of the length-contracted vector may take any value in $(-\infty, 1]$. But physically we interpret it as the same ruler, just in a different frame. It retains "unit" length in the various following senses. In the ruler's *proper* frame (that is, the frame $\mathbf{u}$) it has unit length, since $\langle \boldsymbol{\xi}, \boldsymbol{\xi} \rangle = 1$. If the ruler has tick marks or labels like "1/10", "2/10", etc., then any frame reading them agrees the stated length is 1, even though this does not correspond to metric distance in that frame. Further, the mass or number of atoms is the same in any frame. Finally, any frame can correct for the relative motion, to recover the proper length. Under this interpretation, the ratio between tick mark (or "proper") length and a coordinate gradient is:

$$dL_{\text{ticks}} = \frac{1}{\langle d\Phi, \boldsymbol{\xi}_{\text{contr}} \rangle} d\Phi = \frac{1}{\langle d\Phi, \boldsymbol{\xi} \rangle - \frac{\langle \boldsymbol{\xi}, \mathbf{n} \rangle}{\langle \mathbf{u}, \mathbf{n} \rangle} \langle d\Phi, \mathbf{u} \rangle} d\Phi. \tag{4}$$

This is from substituting $\boldsymbol{\xi}_{\text{contr}}$ from Eq. (3) directly into Eq. (1). That formula is valid because the vector is "unit" in the sense considered here (tick mark length, not metric length). Note the coefficient is not simply $\gamma / \langle d\Phi, \boldsymbol{\xi} \rangle$, in general.

One can also compare two rulers directly, not against a coordinate. Suppose $\boldsymbol{\rho}$ is a ruler in some third frame. Then:

$$\langle \boldsymbol{\xi}_{\text{contr}}, \boldsymbol{\rho}_{\text{contr}} \rangle$$

expresses their overlap in $\mathbf{n}$'s frame. It accounts for both length-contraction, and their co-alignment in $\mathbf{n}$'s space.

[7] We give an alternate derivation for the case of $\mathbf{n}$ timelike, for reassurance. Decompose $\boldsymbol{\xi}$ into orthogonal components in the $\mathbf{n}$ and $\hat{\vec{\mathbf{u}}}$ directions, plus a remainder. Remove the $\mathbf{n}$-component, divide the boost direction component by γ^2, and preserve the remainder. Note that dividing the $\hat{\vec{\mathbf{u}}}$ component by γ^2 is equivalent to subtracting $1 - \gamma^{-2} = \beta^2$ times $\hat{\vec{\mathbf{u}}}$, from $\boldsymbol{\xi}$. This leaves $\boldsymbol{\xi} + \langle \boldsymbol{\xi}, \mathbf{n} \rangle \mathbf{n} - \langle \boldsymbol{\xi}, \vec{\mathbf{u}} \rangle \vec{\mathbf{u}}$ after absorbing the β's, which simplifies to Eq. (3).

4 Volume Forms

There are at least three different volume elements an observer **n** might attribute to a fluid **u**, or vice versa. These include the usual metric volume, a length-contracted volume, and a "de-contracted" volume which corrects for length-contraction. As previously we allow **n** to be timelike, null, or spacelike.

4.1 Metric Volume

Recall the volume form on some oriented region of a p-dimensional submanifold is:

$$\sqrt{|g_{(p)}|}\, dx^1 \wedge \cdots \wedge dx^p, \tag{5}$$

where the x^i are coordinates on the submanifold, $g_{(p)}$ is the determinant of the induced metric, and $|\cdot|$ its absolute value. Geometrically, at each point the scalar $\sqrt{|g_{(p)}|}$ is the volume of the p-dimensional parallelepiped spanned by the coordinate basis vectors ∂_i. To see this, first recall this parallelepiped has oriented volume:

$$\mathbf{a} \wedge \mathbf{b} \wedge \cdots ,$$

where we relabel the ∂_i as **a**, **b**, etc. See the leftmost illustration in Fig. 4. This is a p-vector, with "$\wedge$" the wedge product, so the expression is the sum of all signed permutations of its elements. It seems convenient to omit the $1/p!$ normalisation convention for our application. For a bivector, $\mathbf{a} \wedge \mathbf{b} = \mathbf{a} \otimes \mathbf{b} - \mathbf{b} \otimes \mathbf{a}$. Now the scalar volume is formed from contracting the (dual of the) p-vector with itself, which is given by the determinant of the metric products between the spanning sides:

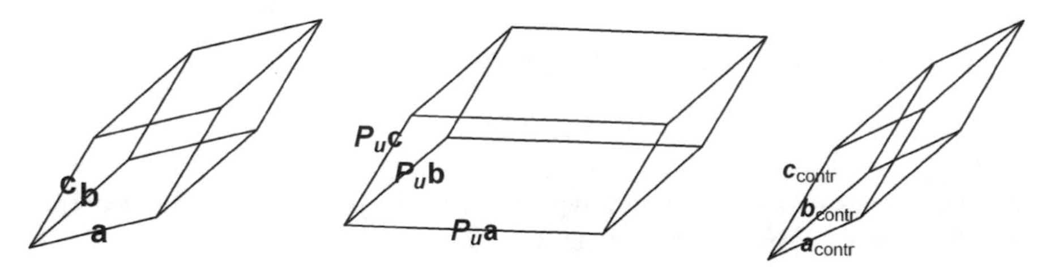

Fig. 4 Parallelepipeds representing the metric volume, length de-contracted volume, and length-contracted volume elements, respectively. All reside at a single point in curved spacetime. In the diagrams $p = 3$, and the boost or tilt direction aligns roughly with "**a**". In general, for each case the wedge product of the p side vectors is a p-vector representing the oriented volume. Its magnitude is the scalar volume. $\mathbf{P_u}$ is the spatial projector. This diagram is a more literal visualisation, in contrast to the spacetime diagrams (Figs. 2 and 3) which are more abstract. Both pictures are insightful, and complementary

$$\left|\langle \mathbf{a} \wedge \mathbf{b} \wedge \cdots, \mathbf{a} \wedge \mathbf{b} \wedge \cdots \rangle\right|^{1/2} = \begin{Vmatrix} \langle \mathbf{a}, \mathbf{a} \rangle & \langle \mathbf{a}, \mathbf{b} \rangle & \cdots \\ \langle \mathbf{b}, \mathbf{a} \rangle & \langle \mathbf{b}, \mathbf{b} \rangle & \cdots \\ \vdots & \vdots & \ddots \end{Vmatrix}^{1/2} . \tag{6}$$

On the RHS, the inner $|\cdot|$ is called a Gram determinant, and the outer $|\cdot|$ is an absolute value required by the Lorentzian spacetime signature. Since $\langle \partial_i, \partial_j \rangle = g_{ij}$, the matrix entries are simply metric components, so this is indeed $\sqrt{|g_{(p)}|}$ as claimed. We interpret the parallelepiped as lying within the tangent space at a single point in curved spacetime, which leads to an exact (as in precise) "infinitesimal" volume.

4.2 De-Contracted Volume

To the observer (or hyperplane normal) $\mathbf{n}$, the fluid $\mathbf{u}$ is length-contracted, so the usual metric volume element on the hyperplane does *not* coincide with the proper volume of fluid. A "de-contracted" volume element compensates for this.

Consider an arbitrary vector "**a**", interpreted as a line segment. This vector intersects part of the fluid's worldline(s); however, in general the fluid will be length-contracted along this direction through space and time. We wish to transform the vector to one representing the proper length of fluid intersected. This is analogous to recovering $\boldsymbol{\xi}$ from $\boldsymbol{\xi}_{\text{contr}}$, which is achieved by:

$$\boldsymbol{\xi} = \boldsymbol{\xi}_{\text{contr}} + \langle \boldsymbol{\xi}_{\text{contr}}, \mathbf{u} \rangle \mathbf{u}. \tag{7}$$

To see this, evaluate the RHS, or inspect the previous length-contraction diagrams. It is simply the usual spatial projection using $P^a{}_b := g^a{}_b + u^a u_b$. Call the map $\boldsymbol{\xi}_{\text{contr}} \mapsto \boldsymbol{\xi}$ "de-contraction". While length-contraction depends on both observers $\mathbf{n}$ and $\mathbf{u}$, the reverse process requires only $\mathbf{u}$. (And whereas $\boldsymbol{\xi}_{\text{contr}}$ is obtained by a certain projection orthogonal to $\mathbf{n}$, the recovery of $\boldsymbol{\xi}$ is an *orthogonal* projection orthogonal to $\mathbf{u}$.)

In the same way, the vector "**a**" is de-contracted as:

$$\mathbf{a} \mapsto \mathbf{a} + \langle \mathbf{a}, \mathbf{u} \rangle \mathbf{u}. \tag{8}$$

We place no restriction on the resulting length, unlike for $\boldsymbol{\xi}$. This vector recovers a fluid frame quantity, locally.[8] Its squared-length is:

[8] Over a macroscopic region—a submanifold—each pointwise de-contraction is into **u**'s frame. However, the overall result still depends on the choice of simultaneity, that is the "time" slice or tilt of the submanifold. This matters if the fluid has intrinsic expansion or contraction over time. The example in Sect. 5.4 considers this carefully.

$$\big\langle \mathbf{a} + \langle \mathbf{a}, \mathbf{u}\rangle \mathbf{u}, \mathbf{a} + \langle \mathbf{a}, \mathbf{u}\rangle \mathbf{u}\big\rangle = \langle \mathbf{a}, \mathbf{a}\rangle + \langle \mathbf{a}, \mathbf{u}\rangle^2, \tag{9}$$

which gives the proper length of material intersected by **a**. Note the vector "**a**" may be spacelike, null, or timelike. But its de-contraction is orthogonal to the timelike vector **u**, hence is spacelike or the zero vector.

Now consider a curve parametrised by some λ, which need not be affine. Label the tangent vector field $a^\mu := dx^\mu/d\lambda$. The usual metric length is $\int \sqrt{|\langle \mathbf{a}, \mathbf{a}\rangle|}\, d\lambda$. However, at each point, "**a**" may have arbitrary spacetime direction, so the local fluid element is length-contracted along this **a**-direction. (Here we need not specify the "observer" **n**, as the curve itself sets the tilt. We know only that $\langle \mathbf{n}, \mathbf{a}\rangle = 0$, in principle.) Hence the proper length of fluid along the curve is:

$$L = \int \sqrt{\langle \mathbf{a}, \mathbf{a}\rangle + \langle \mathbf{a}, \mathbf{u}\rangle^2}\, d\lambda.$$

This extends to higher dimensions. Consider a volume element $\mathbf{a} \wedge \mathbf{b} \wedge \cdots$ (typically, formed from coordinate basis vectors at a single point of a submanifold). De-contract each of the sides, then recombine them using wedge products:

$$\begin{aligned}&\big(\mathbf{a} + \langle \mathbf{a}, \mathbf{u}\rangle \mathbf{u}\big) \wedge \big(\mathbf{b} + \langle \mathbf{b}, \mathbf{u}\rangle \mathbf{u}\big) \wedge \cdots \\ &\qquad = \mathbf{a} \wedge \mathbf{b} \wedge \cdots + \mathbf{u} \wedge \big(\mathbf{u}^\flat \lrcorner (\mathbf{a} \wedge \mathbf{b} \wedge \cdots)\big),\end{aligned} \tag{10}$$

after simplifying. The "$\lrcorner$" symbol means $\mathbf{u}^\flat$ is substituted "into" the leftmost argument of: $\mathbf{a}\wedge\mathbf{b}\wedge\cdots$; for background see Vaz and da Rocha [13, §2.6]. The effect is a spatial projection, returning a new parallelepiped within **u**'s space. Compare Eq. (8), which is the 1-dimensional case. In particular, the expression vanishes upon substituting $\mathbf{u}^\flat$. The projected parallelepiped's volume scalar represents the proper volume of fluid contained in the original (coordinate basis) parallelepiped. It follows by combining results similar to Eqs. (6) and (9):

$$\begin{vmatrix} \langle \mathbf{a}, \mathbf{a}\rangle + \langle \mathbf{a}, \mathbf{u}\rangle^2 & \langle \mathbf{a}, \mathbf{b}\rangle + \langle \mathbf{a}, \mathbf{u}\rangle \langle \mathbf{b}, \mathbf{u}\rangle & \cdots \\ \langle \mathbf{b}, \mathbf{a}\rangle + \langle \mathbf{b}, \mathbf{u}\rangle \langle \mathbf{a}, \mathbf{u}\rangle & \langle \mathbf{b}, \mathbf{b}\rangle + \langle \mathbf{b}, \mathbf{u}\rangle^2 & \cdots \\ \vdots & \vdots & \ddots \end{vmatrix}^{1/2}.$$

No absolute value sign is required. This is the coefficient in the de-contracted volume p-form: $|\cdots|^{1/2} dx^1 \wedge \cdots \wedge dx^p$, in contrast to Eq. (5). If **u** lies within the original volume element $\mathbf{a} \wedge \mathbf{b} \wedge \cdots$, the de-contracted element vanishes. Geometrically, this is because there is zero spatial length along the time direction, and the p-volume of the remaining orthogonal $(p-1)$-parallelepiped vanishes. Another important special case is when either volume element includes the boost direction. Then the fluid volume is simply the Lorentz factor times the metric volume. But the power of our formalism is for volumes aligned with neither **u** nor the boost (generally, "tilt") direction.

One could easily assume that since the spatial projection $P_{\mathbf{u}}\mathbf{a}$ is onto $\mathbf{u}$'s space, it is a measurement by the observer $\mathbf{u}$, in its own frame $\mathbf{u}$. It is not, at least in the following senses. It is $\mathbf{a}_{\text{contr}}$ which represents the length determined in $\mathbf{u}$'s frame (at least for '$\mathbf{a}$' spacelike). Rather $P_{\mathbf{u}}\mathbf{a}$ is used by $\mathbf{n}$, it is the fluid length in $\mathbf{n}$'s frame, which corrects $\mathbf{n}$'s volume element for length-contraction.

4.3 Contracted Volume

A third volume measure is a *contraction* of space, in contrast to the de-contraction examined above. The fluid frame $\mathbf{u}$ might wonder, "To the observer $\mathbf{n}$, my 3-space is length-contracted. Which volume element would $\mathbf{n}$ attribute to my space, or a subspace thereof?" In this interpretative choice, $\mathbf{n}$ does not adjust for length-contraction. (The result is not just the metric volume in $\mathbf{n}$'s space, at least not directly, since the volume element does not lie orthogonal to $\mathbf{n}$. It is only after length-contraction that the volume element is orthogonal to $\mathbf{n}$.) The square in Sect. 5.2 is an intentionally simple example, to clarify the motivating idea.

Consider a parallelepiped $\mathbf{a} \wedge \mathbf{b} \wedge \cdots$ lying entirely in $\mathbf{u}$'s space, so all sides are orthogonal to $\mathbf{u}$. To $\mathbf{n}$, each side is length-contracted, for example:

$$\mathbf{a}_{\text{contr}} = \mathbf{a} - \frac{\langle \mathbf{a}, \mathbf{n} \rangle}{\langle \mathbf{u}, \mathbf{n} \rangle}\mathbf{u}.$$

This is equivalent to Eq. (3), except we do not require "$\mathbf{a}$" to have unit length. The vector $\mathbf{a}_{\text{contr}}$ may be timelike, null, or spacelike, since $\mathbf{n}$ may have any nature. The p-vector obtained by wedging together the contracted sides represents the length-contracted parallelepiped:

$$\begin{aligned} &\left(\mathbf{a} - \frac{\langle \mathbf{a}, \mathbf{n} \rangle}{\langle \mathbf{u}, \mathbf{n} \rangle}\mathbf{u}\right) \wedge \left(\mathbf{b} - \frac{\langle \mathbf{b}, \mathbf{n} \rangle}{\langle \mathbf{u}, \mathbf{n} \rangle}\mathbf{u}\right) \wedge \cdots \\ &\qquad = \mathbf{a} \wedge \mathbf{b} \wedge \cdots - \frac{1}{\langle \mathbf{u}, \mathbf{n} \rangle}\mathbf{u} \wedge \left(\mathbf{n}^\flat \lrcorner (\mathbf{a} \wedge \mathbf{b} \wedge \cdots)\right), \end{aligned} \tag{11}$$

after simplifying. See Fig. 4. This lies within the hyperplane orthogonal to $\mathbf{n}$, as seen by substituting $\mathbf{n}^\flat$. Note it depends only on the combination $\mathbf{a} \wedge \mathbf{b} \wedge \cdots$, and not the individual sides; the same is true of the de-contracted volume in Eq. (10). The scalar product of two length-contracted sides is, for example:

$$\left\langle \mathbf{a} - \frac{\langle \mathbf{a}, \mathbf{n} \rangle}{\langle \mathbf{u}, \mathbf{n} \rangle}\mathbf{u}, \mathbf{b} - \frac{\langle \mathbf{b}, \mathbf{n} \rangle}{\langle \mathbf{u}, \mathbf{n} \rangle}\mathbf{u} \right\rangle = \langle \mathbf{a}, \mathbf{b} \rangle - \frac{\langle \mathbf{a}, \mathbf{n} \rangle \langle \mathbf{b}, \mathbf{n} \rangle}{\langle \mathbf{u}, \mathbf{n} \rangle^2}, \tag{12}$$

using $\langle \mathbf{a}, \mathbf{u} \rangle = 0 = \langle \mathbf{b}, \mathbf{u} \rangle$. Hence the length-contracted volume scalar is, by analogy with Eq. (6):

$$\left\| \begin{matrix} \langle \mathbf{a}, \mathbf{a} \rangle - \frac{\langle \mathbf{a}, \mathbf{n} \rangle^2}{\langle \mathbf{u}, \mathbf{n} \rangle^2} & \langle \mathbf{a}, \mathbf{b} \rangle - \frac{\langle \mathbf{a}, \mathbf{n} \rangle \langle \mathbf{b}, \mathbf{n} \rangle}{\langle \mathbf{u}, \mathbf{n} \rangle^2} & \cdots \\ \langle \mathbf{b}, \mathbf{a} \rangle - \frac{\langle \mathbf{b}, \mathbf{n} \rangle \langle \mathbf{a}, \mathbf{n} \rangle}{\langle \mathbf{u}, \mathbf{n} \rangle^2} & \langle \mathbf{b}, \mathbf{b} \rangle - \frac{\langle \mathbf{b}, \mathbf{n} \rangle^2}{\langle \mathbf{u}, \mathbf{n} \rangle^2} & \cdots \\ \vdots & \vdots & \ddots \end{matrix} \right\|^{1/2} .$$

Again, the double lines $\| \cdot \|$ mean absolute value of the determinant. The contracted volume p-form is then $\| \cdots \|^{1/2} dx^1 \wedge \cdots \wedge dx^p$. A special case is when the boost or tilt direction lies within the parallelepiped, then the volume scalar is just γ^{-1} times the metric volume.

5 Examples

The Lorentz boost and square examples below are simple rectilinear motions used to illustrate the concepts. The rotating disc seems paradoxical but is largely understood in the literature. The Schwarzschild spacetime example is highly original.

5.1 Lorentz Boost

Consider again the prototypical Lorentz boost $t' = \gamma(t - \beta x)$ and $x' = \gamma(x - \beta t)$ in Minkowski spacetime. Define two timelike observers, a "stationary" one with $\mathbf{u} = \partial_t$ and ruler $\boldsymbol{\xi} = \partial_x$, and a "moving" observer $\mathbf{n} = \partial_{t'}$ with ruler $\boldsymbol{\rho} = \partial_{x'}$. All components will be expressed in the unprimed coordinates $(t, x, \cdots)$. Then $u^\mu = (1, 0, \cdots)$, $\xi^\mu = (0, 1, \cdots)$, $n^\mu = (\gamma, \beta\gamma, \cdots)$, $\rho^\mu = (\beta\gamma, \gamma, \cdots)$, $(dx)_\mu = (0, 1, \cdots)$, and $(dx')_\mu = (-\beta\gamma, \gamma, \cdots)$. From Eq. (1), the stationary observer measures $dL = dx$, while the moving observer measures unit length $dL = dx'$, each using their own ruler. (Do not read anything into the repeated "dL" notation, these quantities are distinct.) If the stationary observer compares their ruler with the moving observer's coordinate x', they find:

$$dL = \gamma^{-1} dx'$$

in the proper frame, which is occasionally termed "length-expansion" because the ruler takes up an interval $\Delta x' = \gamma$, though of course we interpret this philosophically as "really" the ruler having fixed unit length and the coordinate level sets $x' =$ const being length-contracted. Likewise, the moving observer determines $dL = \gamma^{-1} dx$. To the stationary frame, the moving ruler is $\rho^\mu_{\text{contr}} = (0, \gamma^{-1}, 0, 0)$, for which its tick mark readings claim:

$$dL_{\text{ticks}} = \gamma \, dx,$$

so it occupies a shortened interval $\Delta x = \gamma^{-1}$. Yet also $dL_{\text{ticks}} = dx'$ since both ruler and x'-coordinate are length-contracted.

5.2 *Square Flux*

Continuing from the previous example, now consider a square spanned by the x and z-axes, with oriented area the 2-vector:

$$\partial_x \wedge \partial_z,$$

which has proper area scalar 1. Consider a fluid with 4-velocity $\tilde{u}^\mu = (\gamma, \beta\gamma, 0, 0)$. The square is like a window onto the fluid. In the "square frame", the fluid is length-contracted across this window, so it has intrinsic area greater than 1. The boost direction is aligned with the square. Hence the fluid proper flux is γ, which is straightforward once the scenario is understood. In terms of the formalism in Sect. 4.2, the side ∂_x has de-contraction vector:

$$(\beta\gamma^2, \gamma^2, 0, 0),$$

and ∂_z needs no de-contraction. The above vector has length γ, and since the de-contracted sides remain orthogonal, the area is simply their product γ. Note we did not need to specify the "motion" of the square—that is, the 4-velocity or normal $\mathbf{n}$ of an associated observer or frame—nor whether the square persists through time (or space) at all! One only knows that in principle, such an $\mathbf{n}$ must be orthogonal, hence be a linear combination of ∂_t and ∂_y.

But now assume the square is attached to the timelike observer $\tilde{\mathbf{n}} = \partial_t$. According to the $\tilde{\mathbf{u}}$ frame, the square is length-contracted, so its area is reduced to γ^{-1}. (For this contracted volume measure, we do not compensate for length-contraction, which is a choice and definition.) Using the Sect. 4.3 formalism, the side ∂_x effectively becomes $(\beta, 1, 0, 0)$, which has length γ^{-1} as expected. The contracted sides are orthogonal, hence the contracted area scalar is simply their product γ^{-1}. This square example helps to illustrate the conceptual motivation behind the three volume elements.

5.3 *Rotating Disc*

Take Minkowski spacetime in polar coordinates (t, r, ϕ). In the coordinate or "laboratory" frame $\mathbf{n} := \partial_t$, the disc rotates at $d\phi/dt = \Omega$, a constant for all particles on the disc. These particles $\mathbf{u}$ have:

$$u^\mu = \gamma(1, 0, \Omega),$$

where $\gamma = -\langle \mathbf{n}, \mathbf{u} \rangle = (1 - \Omega^2 r^2)^{-1/2}$. The disc is bounded by $r < \Omega^{-1}$, to remain subluminal. The natural ruler in the tangential direction, for a disc particle, is:

$$\xi^\mu = \gamma\big(\Omega r, 0, \frac{1}{r}\big),$$

which is orthogonal to $\mathbf{u}$ and ∂_r, and concurs with the maximal direction given in Sect. 2. What is the disc circumference at given r? In one sense it is the Euclidean value, according to the lab observer if they ignore length-contraction, in which case the disk may as well be absent entirely.

A better answer is the disc's *proper* circumference. In the *lab* frame, disc rulers are length-contracted to:

$$\xi^\mu_{\text{contr}} = \big(1, 0, \frac{1}{\gamma r}\big).$$

Hence the unit rulers are shortened to γ^{-1} (recall this is a statement about vectors in local tangent spaces; any ratios apply to "infinitesimal" measurements on the actual disc). We have:

$$dL_{\text{ticks}} = \gamma r \, d\phi,$$

meaning a greater length fits into a given $\Delta\phi$ coordinate interval than the Euclidean value. Hence the proper circumference is $L = 2\pi r\gamma$! This strange result is the consensus view, which we thoroughly affirm.

Now consider the frame of a disc particle. In this frame, a disc ruler measures:

$$dL = \gamma^{-1} r \, d\phi. \tag{13}$$

Paradoxically, there is *less* ruler length in a given coordinate interval $\Delta\phi$ than the Euclidean value, in this frame. Likely few sources state this result clearly. The result holds locally but paradoxically does not lead to a total circumference $2\pi r\gamma^{-1}$. It may seem the problem is the vorticity of the disc particles, so no simultaneity surface exists which is orthogonal to the worldlines, on any open neighbourhood of the surface. See Fig. 5. (However, even for fluids without vorticity, the simultaneity hypersurfaces do not align with the desired measurement direction, in general.) For an integral curve of the field $\boldsymbol{\xi}$ at fixed r, it turns out a coordinate interval $\Delta\phi = 2\pi\gamma^2$ is needed to return to the starting particle on the disc, because the particle has moved in the meantime. This recovers the $L = 2\pi r\gamma$ global result.

Complications of this nature occur whenever the observer field $\mathbf{n}$ has vorticity. While our formalism can handle this, a simpler and safer approach is to pick a hypersurface or other submanifold, and "de-contract" it. That the *fluid* $\mathbf{u}$ has vorticity is inconsequential, in this approach. For the circumference, an obvious choice is the circle at fixed t and r. After de-contraction, the result is again $L = 2\pi r\gamma$. Another safe approach is to define a "rotating" coordinate $\tilde{\phi} := \phi - \Omega t$.

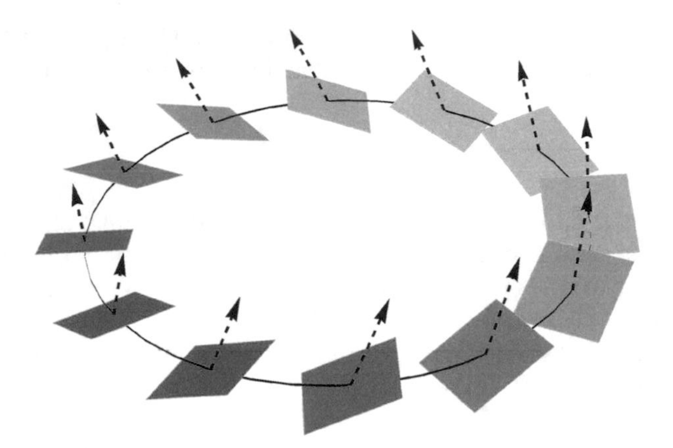

Fig. 5 The failure of synchrony on the rotating disc, due to its rotation. The vertical axis is Minkowski time t, and the loop is a circle at fixed r. Shown are the 4-velocity vectors of selected disc particles, with each square representing the orthogonal subspace within each local tangent space. A "circumference" curve tangent to all such planes does not close up but forms a helix in spacetime. Hence the length measurement requires careful conceptual reasoning

It is comoving with the disc particle worldlines, since $\langle d\tilde{\phi}, \mathbf{u}\rangle = 0$. Hence in *any* observer frame $\mathbf{n}$, the contracted ruler gives the same reading $dL_{\text{ticks}} = \gamma r\, d\tilde{\phi}$, using Eq. (4). Similarly one revolution is $\Delta\tilde{\phi} = 2\pi$, along *any* reasonable hypersurface. Again, the proper circumference is $2\pi r\gamma$.

5.4 Schwarzschild Spacetime

It is often claimed the properties of a hypersurface $t = \text{const}$ of Schwarzschild time are measurement by an "observer at infinity". However, interpretations using local observers are more physically justified. The normal vector field is $(dt)^\sharp$, which is timelike for $r > 2M$. After normalising, these are the static observers $(1-2M/r)^{-1/2}\partial_t$, hence in the local observer field approach, $t = \text{const}$ is physically interpreted as the space of static observers.

But how do non-static observers measure Schwarzschild spacetime? For now, assume the motion is restricted to the equatorial plane $\theta = \pi/2$. Parametrise observers $\mathbf{u}$ by their energy per mass e and angular momentum per mass ℓ, defined from the usual Killing vector fields:

$$e := -\langle \mathbf{u}, \partial_t\rangle, \qquad \ell := \langle \mathbf{u}, \partial_\phi\rangle.$$

These are constant along geodesics, but either way:

$$u^\mu = \left(e\left(1-\frac{2M}{r}\right)^{-1}, \pm\sqrt{e^2-\left(1-\frac{2M}{r}\right)\left(1+\frac{\ell^2}{r^2}\right)}, 0, \frac{\ell}{r^2}\right). \tag{14}$$

The "$\pm$" is an extra degree of freedom when outside the horizon but must be a minus for $r \le 2M$. It is not obvious which choice of ruler orientation is best. The coordinate gradient rulers yield measurements: [7]

$$dL_{r,3\text{–gradient}} = \frac{1}{\sqrt{e^2-(1-2M/r)\ell^2/r^2}}dr, \qquad dL_{(\phi)} = \frac{r}{\sqrt{1+\ell^2/r^2}}d\phi, \tag{15}$$

from Eq. (2). The corresponding ruler vectors (omitted) are orthogonal for $\ell = 0$, but apparently not otherwise. The "θ" ruler is trivial, with measurement $dL = r\,d\theta$. For $\ell = 0$ the motion is radial, and the above proper frame measurements reduce to $|e|^{-1}dr$ and $r\,d\phi$. In fact these can be seen by inspection of the line element expressed in generalised Gullstrand-Painlevé coordinates, as examined in my previous DomoSchool contribution [6]. For a static observer, additionally $e = \sqrt{1-2M/r}$, so the familiar radial length element $(1-2M/r)^{-1/2}dr$ is recovered. r is often called a "curvature coordinate", which is accurate; however, a nuanced description would mention spherically symmetric submanifolds, or observers with zero angular momentum.

What is the 3-volume in Schwarzschild spacetime? If choosing an $r = $ const slice, the metric volume inside the horizon is infinite, unless the black hole formed by collapse and later evaporates. One author obtains a radial distance $\approx 10^{77}$ lightyears for a solar mass black hole, despite a coordinate interval of just $\Delta r = 2M \approx 3\,\text{km}$. In our local observers approach, one must choose an observer field. One reasonably natural choice is to fix e and ℓ as constants across all observers, after checking the range of r is consistent. The presence of vorticity in the observer field adds complication, as the rotating disc scenario showed. Hence we cannot naively extrapolate the proper frame measurements in Eq. (15) to a total volume. Instead, it is simpler to choose submanifolds and "de-contract" them.

Consider the circular curve parametrised by ϕ, at fixed t and r, where we allow any $r \in (0,\infty)$. The tangent vector is just ∂_ϕ, which de-contracts (that is, spatially projects orthogonal to the 4-velocity in Eq. (14)) to a vector of length $\sqrt{1+\ell^2/r^2}\,r$, yielding a total circumference:

$$\sqrt{1+\ell^2/r^2}\,2\pi r.$$

On the other hand, a great circle in the ∂_θ direction has Euclidean circumference:

$$2\pi r,$$

if misaligned with the fluid. (Imagine extending the field $\mathbf{u}$ by revolving the equatorial plane about an axis passing through $\phi = 0, \pi$. This also requires a certain redefinition of ℓ for consistency. Choose the great circle at $\phi = \pi/2, 3\pi/2$.)

Now consider a 2-sphere at fixed t and r, parametrised by θ and ϕ. The tangent vector ∂_θ is already de-contracted, has length r, and is orthogonal to the de-contraction of ∂_ϕ. Hence the 2-sphere area determined by $\mathbf{u}$ follows simply:

$$\sqrt{1 + \ell^2/r^2}\, 4\pi r^2,$$

by invoking symmetry. Now the radial direction is less obvious. We seek the most natural length corresponding to a given coordinate interval, say $r \in (0, 2M]$. However, the length depends on the curve chosen, even for our de-contracted length element, intuitively because the tilt affects how many worldlines of $\mathbf{u}$ are intersected. The natural choice is a radial curve orthogonal to $\mathbf{u}$. We do not need to solve for it explicitly, just its tangent:

$$a^\mu := \Big(-\frac{\sqrt{e^2 - (1 + \ell^2/r^2)(1 - 2M/r)}}{e(1 - 2M/r)}, 1, 0, 0 \Big),$$

which is the unique vector orthogonal to $\mathbf{u}$, $d\theta$ and $d\phi$, up to scalar multiple. We chose $a^r = 1$, which corresponds to a curve parametrised by r. This tangent vector has norm $|e|^{-1}\sqrt{1 + \ell^2/r^2}$ and is already de-contracted since it is orthogonal to $\mathbf{u}$. It follows the radial length element is:

$$dL_{\text{radial}} = |e|^{-1}\sqrt{1 + \ell^2/r^2}\, dr. \tag{16}$$

This is valid both locally in the proper frame $\mathbf{u}$, and for our sought macroscopic length (contrast Eq. (13)). For $\ell \neq 0$ the coefficient diverges as $r \to 0^+$, but this is not the case for the "r-gradient" ruler (Eq. (15)), so not a physical disaster. An antiderivative is:

$$L_{\text{radial}} = \frac{1}{e}\Big(\sqrt{1 + \ell^2/r^2}\, r - \ell \tanh^{-1} \frac{\ell}{\sqrt{1 + \ell^2/r^2}\, r} \Big).$$

For $\ell \neq 0$ this also diverges as $r \to 0$. But at least the 3-dimensional volume remains finite, it turns out. The idea is to find a hypersurface whose tilt agrees as closely with the fluid frames as possible. Because of the vorticity, it cannot be orthogonal to the worldlines. We choose a spherically symmetric hypersurface, to avoid the rotation, but one which aligns with the fluid's space in the radial direction. But this is just the one spanned by the vectors ∂_θ, ∂_ϕ, and "$\mathbf{a}$" above, hence the de-contracted 3-volume element is:

$$\frac{1 + \ell^2/r^2}{e} r^2\, dr \wedge d\theta \wedge d\phi,$$

at the equator. Invoking symmetry, the integration is trivial, and the total scalar volume between $r \to 0$ and $r = R \in (0, \infty)$ is:

$$\frac{1}{e}\Big(\frac{4}{3}\pi R^3 + 4\pi \ell^2 R\Big). \tag{17}$$

The effect of angular momentum is to increase the proper volume, as for the rotating disc. For $\ell = 0$ and $e = \pm 1$, the 3-volume is Euclidean. This case was noticed a century ago, from the line element in Gullstrand-Painlevé coordinates, but not understood. The $e = 0$ case is distinct, with natural hypersurfaces at constant $r < 2M$, being topologically $\mathbb{R} \times S^2$. Compare Fig. 5 in Ref. [6].

6 Discussion

In summary, we have introduced two new volume measures: a length-contracted volume and a "de-contracted" volume, on submanifolds of spacetime. They are derived from the usual metric volume, plus physical consideration of an (active) transformation between two frames: a timelike "fluid" $\mathbf{u}$, and an "observer" normal $\mathbf{n}$ which may be timelike, null, or spacelike. The wedge product tool handles this in a beautiful and elegant way. The results hold in arbitrary dimension, with the expression "3-space" really meaning space in contrast to spacetime. The algebraic formulation is simpler than reasoning from first principles alone. It applies to the rotating disc, conveniently handling the vorticity which makes this scenario so paradoxical. This is extended to Schwarzschild spacetime, returning the spatial volume relative to a two-parameter family of fluid fields (Eq. (17)). In particular, the radial distance is relative to the observer. Primarily our results are geometrical, for which they are exact, and only secondarily a *model* of an extended physical ruler.[9]

The very mention of length and volume measurement brings to mind introductory textbook material, such as the proper length $\int ds$ of a curve, or generally the metric volume $\int \sqrt{|g_{(p)}|}\, d^p x$ of a submanifold. Indeed the topic is foundational, but the results herein result from years of careful thinking and refinement. We speculate development was hindered by the overreach of various correct results, as follows.

[9] The vectors like $\boldsymbol{\xi}$ and "$\mathbf{a}$" are crude models, from a materials science or engineering perspective. But a "constructive" approach incorporating all known physics would be extremely complicated. It would include the stress-strain response of an extended material ruler to the undeformed configuration (the metric length, as analysed in this work), using relativistic elasticity theory or even our best quantum description of matter. The distortions from curvature, tidal forces, and any external acceleration or forces on the ruler, may induce sound modes which depend on the past history of the worldsheet. Its stress-energy has back-action on spacetime. Also a true ruler is not 1-dimensional. Einstein was right, both to treat rods and clocks as fundamental, and to later acknowledge the shortcomings [2]. Perhaps quantum fields would be effective rulers, using the spreading of a wavepacket or the size of a hydrogen orbital.

General covariance was a tremendous advance, but it does not imply coordinates have no relation to physical meaning, at least when observers are specified.[10,11] Born-rigid motion is very limited, and intrinsically rigid objects are unphysical, but approximate rulers including radar devices work well in reasonable environments, and length does not reduce to time measurement. The rotating disc is perplexing but does not imply relativity can only accommodate pointlike—not extended—objects. There are other contributing factors besides these "overreaches". Alternate voices including Gullstrand, Painlevé and Sagnac raised important questions, but their (gravely mistaken) conclusion that relativity is self-contradictory surely did not motivate others to sift out the gold. Yet perhaps the main reason is, in relativity the topic of observers and physical measurement is under-promoted, despite some excellent technical treatments [4, 5].

In future work we offer more diagrams and explanation, including separate papers on black hole spacetimes, and a rotating disc in an axially symmetric spacetime. The formalism applies to any abstract "fluid" where its proper flux is significant (the spacelike hypersurface cases may be largely known already, but presumably not lower-dimensional submanifolds with arbitrary tilt). Example flows are the vector potential in Maxwell's electromagnetism, or the Noether current in various quantum theories. A signed volume scalar would accommodate multiple crossings of the same fluid worldline over a chosen submanifold. There will be connections with Stokes' theorem. Another application is averaging or coarse-graining of fluid quantities, within a different frame. A straightforward extension would be length-contraction using simultaneity other than the usual Poincaré-Einstein convention, or within theories without Lorentz invariance. We assumed a field of observers, but many applications require extension of a single frame, such as satellite radar distance in Solar System astrometry. Our results would form a first-order approximation for an extended frame, but the error should be quantified.

7 DomoSchool Memories

Domodossola is stunning, including from Cima Lariè (2144m). :) *Grazie* to Sergio Cacciatori and Andrea Cottini for warm hospitality. It was a big privilege to meet Donato Bini, world expert on observers and spacetime splitting: thank-you for

[10] In Norton's view, [10, §2] "That coordinate systems can be used to represent significant physical content is not the modern view and it is tempting to think that no other view is possible. But that narrowmindedness is quite incorrect". Equation (1) is also relational, in that it depends on the ruler orientation, which is in turn constrained by the observer 4-velocity $\mathbf{u}$.

[11] On the other hand, Schwarzschild t has been afforded too much meaning, as the (only) "time at infinity", the historical rejection of black hole collapse, and the repeated assertion by textbooks that r is not metric distance but $(1 - 2M/r)^{-1/2}dr$ is.

discussion and gelati. It was fun to use our coffee and meal tickets, exploring various cafes and restaurants. Thanks to Orville Damaschke, Jiří Ryzner, and Bini for tolerating my order of tea with pizza.

References

1. H. Brown, *Physical Relativity: Space-Time Structure from a Dynamical Perspective* (Oxford, England, 2005)
2. H. Brown, The behaviour of rods and clocks in general relativity and the meaning of the metric field, in *Beyond Einstein*, ed. by D. Rowe, T. Sauer, S. Walter (Springer, Berlin, 2018)
3. S. Carroll, *Spacetime and Geometry: An Introduction to General Relativity* (Addison Wesley, New York, 2004)
4. F. de Felice, D. Bini, *Classical Measurements in Curved Space-Times* (Cambridge Uninversity, Cambridge, 2010)
5. R. Jantzen, P. Carini, D. Bini, The many faces of gravitoelectromagnetism. Ann. Phys. **215**(1), 1–50 (1992)
6. C. MacLaurin, *Schwarzschild Spacetime Under Generalised Gullstrand-Painlevé Slicing* (Birkhäuser, Cham, 2019)
7. C. MacLaurin, *Clarifying Spatial Distance Measurement* (World Scientific, Singapore, 2021). Submitted, and book forthcoming
8. C. Misner, K. Thorne, J. Wheeler, *Gravitation* (W.H. Freeman and Co., New York, 1973)
9. C Møller, *The Theory of Relativity* (1952)
10. J. Norton, Einstein's triumph over the spacetime coordinate system: A paper presented in honor of roberto torretti, in *Diálogos: Revista de filosofía de la Universidad de Puerto Rico* (2002)
11. J. Norton, *A Conjecture on Einstein, the Independent Reality of Spacetime Coordinate Systems and the Disaster of 1913* (2005)
12. B. Schutz, *A First Course in General Relativity* (Cambridge University, Cambridge, 2009)
13. J. Vaz, R. da Rocha, *An Introduction to Clifford Algebras and Spinors* (Oxford, England, 2016)

Exact Solutions of Einstein–Maxwell(-Dilaton) Equations with Discrete Translational Symmetry

Jiří Ryzner and Martin Žofka

Contents

Mathematics Subject Classification (2000) 83C15, 83C22, 83C57

1 Introduction

Einstein equations are non-linear set of differential equations, yet there exist exact solutions, whose geometry and matter content lead to linearity. This is also the case of the multiple extremal[1] black hole solutions, which first appeared in the 1940s due to the works of Majumdar [1] and Papapetrou [2], yet their interpretation had to

[1] The term extremal refers to the fact that the black hole horizons are degenerate, which in this case translates to their charges equal to their masses.

J. Ryzner (✉) · M. Žofka
Institute of Theoretical Physics, Charles University, Prague, Czech Republic
e-mail: j8.ryzner@gmail.com; zofka@mbox.troja.mff.cuni.cz

S. L. Cacciatori, A. Kamenshchik (eds.), *Einstein Equations: Local Energy, Self-Force, and Fields in General Relativity*, Tutorials, Schools, and Workshops in the Mathematical Sciences, https://doi.org/10.1007/978-3-031-21845-3_10

wait until the 1970s and the classical paper by Hartle and Hawking [3]. The family of Majumdar–Papapetrou (MP) spacetimes is a manifestation of the curious fact that the balance between electrostatic and gravitational forces in classical physics is preserved even in general relativity. This also extends to the linearity of the field equations and thus the principle of superposition holds in GR as well. In our previous works on the subject, [4] and [5], we were interested in solutions exhibiting axial and cylindrical symmetry, respectively. The cylindrically symmetric solution was the field of an infinitely extended extremally charged string (ECS) located along the axis of symmetry. One of the questions arising from the paper was whether it is possible to produce the same field asymptotically far from the axis and due to an infinite number of isolated, equidistantly distributed, identical point sources of a mass equal to their charge. This was motivation for our next work [6], where we investigated a few ways to construct such solutions. The interesting feature of the resulting spacetimes is that it has a discrete translational symmetry everywhere while asymptotically, it would be fully cylindrically symmetric. We now build on this work, review some facts and show newly discovered features.

The chapter is organized as follows: in the first section, we review the general MP solution in an arbitrary dimension with an arbitrary source. In Sect. 2, we look at the first solution which is an infinite alternating "crystal" constructed of alternating positive and negative charges. The solution is in the form of an infinite series of functions: we review its convergence, asymptotics and derivatives to be able to infer the symmetries of the spacetime and its interpretation. In Sect. 3, we study the case of a uniform crystal composed of identical charges. In Sect. 4, we show the solution of a smooth "crystal", which is a solution of MP and charged dust. Section 5 presents a way of constructing the 4D infinite crystal out of a closed-form 5D solution via dimensional reduction. We conclude with some final remarks and open questions.

1.1 Majumdar–Papapetrou Solution

The metric ${}^D g$ of the Majumdar–Papapetrou solution in arbitrary dimension[2] $D = n + 1, n \geq 3$, reads [7]

$$ {}^D g = -U^{-2}\mathrm{d}t^2 + {}^n h_{ij}\mathrm{d}x^i\mathrm{d}x^j, \tag{1} $$

where t is a time-like Killing coordinate, so that the metric is static with the function $U = U(x^i)$ only depending on Cartesian-like spatial coordinates[3] x^i. The spatial metric ${}^n h_{ij}$ is conformally flat

[2] The number of spacelike dimensions is denoted as n.

[3] Latin indices range over $1, \ldots, n$ and label only spatial components, and Greek indices are $0, \ldots, n$.

$$ {}^{n}h = U^{\frac{2}{n-2}} \cdot {}^{n}\delta_{ij}\mathrm{d}x^{i}\mathrm{d}x^{j}. \tag{2} $$

These coordinates describe well the region above the horizons. The electromagnetic potential A and the electromagnetic field tensor F read

$$ A = c_n \frac{\mathrm{d}t}{U}, \quad F = \mathrm{d}A = -c_n \sum_{i=1}^{n} \frac{U_{,i}}{U^2}\mathrm{d}x^i \wedge \mathrm{d}t \tag{3} $$

with $c_n = \sqrt{\frac{n-1}{2(n-2)}}$. The corresponding stress–energy tensor, T, is

$$ T^{\mu\nu} = \frac{1}{4\pi}\left(F^{\mu}_{\ \beta}F^{\nu\beta} - \frac{\mathcal{F}}{4}g_{\mu\nu}\right), \tag{4} $$

where

$$ \mathcal{F} = F_{\mu\nu}F^{\mu\nu} = c_n^2\,(n-2)^2 \sum_{i=1}^{n}\left[\frac{\partial\left(U^{\frac{-1}{(n-2)}}\right)}{\partial x^i}\right]^2 \tag{5} $$

is the Maxwell scalar. The non-vanishing components of the stress–energy tensor are

$$ 16\pi T_0^{\ 0} = \mathcal{F}, \quad 4\pi T_i^{\ j} = -c_n^2 U^{\frac{2-2n}{n-2}} U_{,i}U_{,j} - \frac{\mathcal{F}}{4}\delta_i^{\ j}. \tag{6} $$

Einstein and Maxwell equations then have the form

$$ \mathrm{Ric}_{\mu\nu} - \frac{R}{2}g_{\mu\nu} = 8\pi T_{\mu\nu}, \quad \nabla_\nu F^{\mu\nu} = 4\pi J^{\mu}, \quad \nabla_\mu J^{\mu} = 0. \tag{7} $$

The 4-current J^{μ} due to the charge density $\varrho(x^j)$ leads to a single Einstein–Maxwell equation

$$ J^{\mu} = -c_n \frac{\varrho(x^j)}{\sqrt{-\mathfrak{g}}}\delta_0^{\mu} \Rightarrow \Delta_{\delta}U = \sum_{i=1}^{n} U_{,ii} = -4\pi\varrho(x^j). \tag{8} $$

Here, $\mathfrak{g}$ is the determinant of the metric $g_{\mu\nu}$ and Δ_δ denotes the flat-space Laplacian.[4] In case of Majumdar–Papapetrou, $\varrho(x)$ is assumed to be a distribution of point charges, which means that $J^{\mu} = 0$ away from the sources. One particular solution, in which we are interested, is a multi-black hole spacetime of the form

[4] The Laplacian for the spatial metric h is defined as $\Delta_h f \equiv h^{ij}\nabla_i\nabla_j f = \frac{1}{\sqrt{\mathfrak{h}}}\left(\sqrt{\mathfrak{h}}h^{ij}f_i\right)_{,j}$.

$$U(x) = 1 + \sum_{i=1}^{N} \frac{M_i}{r_i^{n-2}}, \; r_i^2 = \sum_{a=1}^{n} (x^a - x_i^a)^2, \tag{9}$$

with the corresponding charge current [7]

$$\sqrt{-\mathrm{g}} J^0 = -\frac{c_n}{4\pi} \Delta_\delta U = \frac{c_n \pi^{\frac{n}{2}-1}}{\Gamma\left(\frac{n}{2}-1\right)} \sum_{i=1}^{N} M_i \cdot {}^n\delta\left(x - x_i\right). \tag{10}$$

Here, M_i are constants, Γ is the Gamma function and ${}^n\delta$ is the n-dimensional Dirac delta function. It can be shown that M_i is the mass and also charge of each black hole, and for $M_i > 0$ the puncture located at $r_i = 0$ looks like a point, but in fact it represents the surface of a sphere $\mathbb{S}^{n-1}$ of dimension $n - 1$ (and for $M_i < 0$ the surface $r_i = 0$ corresponds to the location of a naked singularity). In $D = 4$, there exists a coordinate transformation, which regularizes the metric at a (arbitrarily chosen) horizon $r_i = 0$ and the horizon is smooth [3]. However, in $D > 4$, this holds only for a single black hole ($N = 1$). For $N = 2, 3$, it was shown that the horizon is not smooth [8], while for a higher number of black holes the situation is still unclear.

2 Alternating Crystal

We aim to construct a solution, which would exhibit a discrete translational symmetry along an axis due to which we shall call it a "crystal". To achieve this, we use linearity of the field equations in the MP spacetime. First, we construct a spacetime with an anti-periodic potential, since it is easier to investigate the convergence of the sums than in the case of a symmetric potential.

2.1 Constructing the Solution

We seek the function U in the form

$$U = 1 + \lambda\chi, \lambda = \frac{Q}{k}, \tag{11}$$

where k is the crystal lattice constant, Q is the charge (and mass) of each black hole and χ corresponds to the classical potential. From (8), we know that χ has to satisfy Laplace's equation away from the sources

$$\chi_{,\rho\rho} + \frac{\chi_{,\rho}}{\rho} + \chi_{,zz} = 0. \tag{12}$$

Let us review the main results from [6]. We construct χ by the following ansatz:

$$\chi = \chi_0 + \sum_{n=1}^{\infty} (-1)^n \chi_n, \; \chi_0 = \frac{1}{r}, \; \chi_{n \neq 0} = \frac{1}{r_n} + \frac{1}{r_{-n}}, \; r_n = \sqrt{\rho^2 + (z-n)^2}. \quad (13)$$

The idea is to prove uniform convergence for $0 \leq z \leq 1/2$ and then to extend the definition of χ to other regions using symmetries of χ. One finds that

$$|\chi_n| \sim \frac{1}{n}, \left|\frac{\partial \chi_n}{\partial \rho}\right| \sim \left|\frac{\partial \chi_n}{\partial z}\right| \sim \frac{1}{n^2}, \left|\frac{\partial^2 \chi_n}{\partial \rho^2}\right| \sim \left|\frac{\partial^2 \chi_n}{\partial z^2}\right| \sim \frac{1}{n^3}. \quad (14)$$

This behaviour is independent of ρ which grants us uniform convergence of the corresponding infinite sums, and we can also exchange derivatives with the infinite summation. This ensures that χ constructed in this manner is a valid solution of Laplace's equation. We also get its symmetries:

$$\chi(\rho, z) = \chi(-\rho, z) = -\chi(\rho, z+1). \quad (15)$$

The charge density can be expressed via the Dirac comb[5] Ш distribution

$$\varrho = -\frac{1}{4\pi} \Delta_\delta \chi = \frac{Ш_2(z) - Ш_2(z-1)}{2\pi\rho} \delta(\rho). \quad (16)$$

The function χ is periodic, and it is thus natural to ask for Fourier coefficients for a fixed ρ. Since the potential diverges at $r = 0$ as $1/r$, which is not integrable near the origin, the coefficients will exist only for $\rho > 0$. Finding the coefficients directly is hard, and we thus proceed in another way by assuming

$$\chi(\rho, z) = \sum_{n=1}^{\infty} A_n(z) B_n(\rho), \; \rho > 0. \quad (17)$$

We plug this into (12) and get

$$-\alpha_n^2 = \frac{A_n''(z)}{A_n(z)} = -\frac{1}{B_n}\left(B_n''(\rho) + \frac{B_n'(\rho)}{\rho}\right). \quad (18)$$

[5] Dirac comb is a periodic tempered distribution defined as

$$Ш_T(t) \equiv \sum_{n \in \mathbb{Z}} \delta(t - nT) = \frac{1}{T} \sum_{n \in \mathbb{Z}} e^{2\pi i n t/T} = \frac{1}{T} + \frac{2}{T} \sum_{n=1}^{\infty} \cos\left(\frac{2\pi n t}{T}\right).$$

Here, α_n is a separation constant. We see that A_n is solved by sines and cosines and B_n is solved by Bessel functions. We use the symmetries of (15) and asymptotic flatness in the ρ direction (e.g., from the Moore-Osgood theorem) and get

$$\chi = \sum_{l=1}^{\infty} f_l \cos\left[\alpha_l z\right] K_0\left[\alpha_l \rho\right], \alpha_l = \pi(2l-1), \rho > 0. \tag{19}$$

Here, K_0 is the modified Bessel function,[6] which diverges at $\rho = 0$ due to the non-integrability of χ at the origin. We determine the unknown coefficients f_l formally from the charge enclosed in a cylinder of radius R and height $2h$, which is aligned with the z axis and centred at the origin:

$$4\pi Q(R,h) = -2\pi \int_0^R \chi_{,z}\Big|_{z=-h}^{z=+h} \rho \, \mathrm{d}\rho - 2\pi \int_{-h}^{h} \left(\chi_{,\rho}\rho\right)\Big|_{\rho=R} \mathrm{d}z. \tag{20}$$

We denote the first integral by q_1 and the second one by q_2. The first term yields

$$q_1 = 2\sum_{l=1}^{\infty} f_l \sin\left(\alpha_l h\right)\left[R K_1\left(\alpha_l R\right) - \frac{1}{\alpha_l}\right]. \tag{21}$$

However, in the limit $R \to 0$, this term vanishes. The second term gives

$$q_2 = \sum_{l=1}^{\infty} \int_{-h}^{h} f_l \cos\left[\alpha_l z\right] \alpha_l R K_1\left[\alpha_l R\right] \mathrm{d}z. \tag{22}$$

We now write the charge using charge density (16) and determine the coefficients

$$\lim_{R\to 0} 4\pi Q(R,h) = \int_{-h}^{h} \left[Ш_2(z) - Ш_2(z-1)\right] \mathrm{d}z \Rightarrow f_l = 4. \tag{23}$$

We see that the potential decays exponentially, and we determine the first leading term by using the sum and integral inequality,[7] which is independent of z, as $\chi_{,z} \to$

[6] Modified Bessel function of the second kind K_ν is defined as

$$K_\nu(x) \equiv \int_0^{\infty} \exp\left(-x\cosh t\right)\cosh\left(\nu t\right) \mathrm{d}t, x > 0.$$

[7] From the Newton integral test, we get the following inequality for integrable, non-increasing and non-negative C_l:

$$C_l \geq 0, \frac{\partial C_l}{\partial l} \leq 0 \,\forall l \geq 1, \sum_{l=1}^{\infty} C_l < \infty \Rightarrow \int_1^{\infty} C_l \mathrm{d}l \leq \sum_{l=1}^{\infty} C_l \leq C_1 + \int_1^{\infty} C_l \mathrm{d}l.$$

0 for large ρ. We thus set $z = 0$ and get

$$\chi(\rho, z = 0) = \sum_{l=1}^{\infty} C_l \left[1 + O\left(\frac{1}{\rho}\right)\right], C_l = 4\frac{\sqrt{\pi} e^{-\alpha_l \rho}}{\sqrt{2\alpha_l \rho}}. \tag{24}$$

We use the integral inequality and get the following expression:

$$\frac{\sqrt{2}e^{-\pi\rho}}{\sqrt{\rho}}\left[\frac{1}{\pi\rho} + O\left(\frac{1}{\rho^2}\right)\right] \leq \sum_{l=1}^{\infty} C_l \leq \frac{\sqrt{2}e^{-\pi\rho}}{\sqrt{\rho}}\left[2 + \frac{3}{4\pi\rho} + O\left(\frac{1}{\rho^2}\right)\right]. \tag{25}$$

This gives us strong bounds on the potential, although we do not get the leading term precisely. From numerical computation, we see that the first leading term is

$$\chi(\rho, z = 0) = c\frac{e^{-\pi\rho}}{\sqrt{\rho}} + O\left(\frac{e^{-\pi\rho}}{\rho^{3/2}}\right), c \approx 2.74. \tag{26}$$

We see that the potential decays faster than any multipole of an isolated system.

2.2 *Geometry*

In the cylindrical coordinates, the fields read

$$g = -U^{-2}\mathrm{d}t^2 + U^2\left(\mathrm{d}\rho^2 + \rho^2\mathrm{d}\phi^2 + \mathrm{d}z^2\right), A = \frac{\mathrm{d}t}{U}, U = 1 + \lambda\chi. \tag{27}$$

Note that χ is anti-periodic, but U is not. Near the origin, the potential behaves as a single extremal Reissner–Nordström solution:

$$\chi(r, \theta) = \frac{1}{r} - 2\ln 2 - \frac{\bar{\zeta}(3)}{8}\left[1 - 3\cos(2\theta)\right]r^2 + O(r^4), r \ll 1, \tag{28}$$

where we use spherical coordinates r and θ. Therefore, it is possible to find a suitable coordinate transformation, which regularizes the metric at $r = 0$. We apply the transformation [3]

$$v = t + W^2(r), W(r) = 1 + \frac{\lambda}{r} - 2\lambda\ln 2. \tag{29}$$

The metric transforms to

$$g = -\frac{\mathrm{d}v^2}{U^2} + 2\frac{W^2}{U^2}\mathrm{d}v\mathrm{d}r + \left(U^2 - \frac{W^4}{U^2}\right)\mathrm{d}r^2 + U^2r^2\mathrm{d}\Omega_2^2. \tag{30}$$

The coefficients near origin behave as

$$\frac{1}{U^2} = \frac{r^2}{\lambda^2} + \frac{r^3}{\lambda^3}(\lambda \ln 16 - 2) + O\left(r^4\right), \tag{31}$$

$$\frac{W^2}{U^2} = 1 + \frac{3}{4}\zeta(3)\,(3\cos(2\theta) + 1)\, r^3 + O\left(r^4\right), \tag{32}$$

$$U^2 - \frac{W^4}{U^2} = -\frac{3}{2}\lambda^2\zeta(3)\,(3\cos(2\theta) + 1)\, r + O\left(r^2\right). \tag{33}$$

Near the origin $r \approx 0$, we have for the metric and its determinant

$$g = -2\mathrm{d}v\mathrm{d}r + \lambda^2\mathrm{d}\Omega_2^2 + O(r),\, \mathfrak{g} = -\lambda^4\sin^2\theta + O(r). \tag{34}$$

We see that the metric is regular on the surface $r = 0$, which is a two-dimensional sphere of radius λ, and it is possible to extend it to $r < 0$, where the metric takes the form

$$\mathrm{d}s^2 = -\tilde{U}^{-2}\mathrm{d}t^2 + \tilde{U}^2\left(\mathrm{d}\tilde{r}^2 + \tilde{r}^2\mathrm{d}\Omega^2\right), \tag{35}$$

where $\tilde{r} \geq 0$ is the new radial coordinate and the functions read

$$\tilde{U} = 1 + \lambda\tilde{\chi},\, \tilde{\chi}(\tilde{r}, \theta) = \chi(\tilde{r}, \theta) - \frac{2}{\tilde{r}}. \tag{36}$$

We can also see that $\lambda \to -\lambda$ corresponds to the $r \to \tilde{r}$ region. The Maxwell invariant expanded as a series for $r \ll 1$ reads

$$\mathcal{F} = \frac{2}{\lambda^2} - \frac{8r(1 - \lambda\ln 4)}{\lambda^3} + \frac{20r^2\left(\lambda^2\ln^2 8 - 2\lambda\ln 4 + 1\right)}{\lambda^4} + O(r^3),\, r \ll 1. \tag{37}$$

For Kretschmann invariant, we have

$$\mathcal{K} = \frac{8}{\lambda^4} - \frac{64r(1 - \lambda\ln 4)}{\lambda^5} + \frac{336r^2\left(\lambda^2\ln^2 8 - 2\lambda\ln 4 + 1\right)}{\lambda^6} + O(r^3),\, r \ll 1. \tag{38}$$

We can thus see that the point $r = 0$ is in fact a non-singular surface of non-zero area.

3 Uniform Crystal

In the previous section, we constructed a solution with alternating charges. The alternating sign granted us uniform convergence and asymptotic flatness for large ρ; however, the resulting potential χ is anti-periodic, and we always have singularities.

It is natural to ask about the existence of a superposition of individual charges with potential $1/r$ for each of the charges distributed equidistantly along the z-axis.

3.1 Constructing the Solution

We seek the function U in the form

$$U = 1 + \lambda\varphi, \lambda = \frac{Q}{k}, \tag{39}$$

where φ is a symmetric periodic potential. We will show that λ is the linear charge (and mass) density to the first leading order of the weak-field limit and the spacetime becomes the ECS spacetime for large ρ. We construct φ for $0 \le z \le 1/2$ as

$$\varphi = \varphi_0 + f(\rho)\sum_{n=1}^{\infty}\frac{\varphi_n}{f(\rho)}, \varphi_0 = \chi_0, \varphi_{n\neq 0} = \chi_n - \frac{2}{n}. \tag{40}$$

Because φ_n goes as $1/n$ for large n, we need to regularize the terms to achieve point-wise convergence. For finite ρ, the sum converges uniformly, but for large ρ we need to introduce a suitable regulator function—we choose $f = e^{2\rho}$. Then, the individual terms and their derivatives read

$$\left|e^{-2\rho}\varphi_n\right| \le \frac{1}{2n^3} + \frac{4}{n^2}, \nabla\varphi_n = \nabla\chi_n. \tag{41}$$

For the derivatives, we are allowed to take f out of the sum as we can use our results from the alternating crystal (14), and we see that the potential φ and its first and second derivatives converge uniformly. Then, we extend the potential for any z via its symmetries

$$\varphi(\rho, z) = \varphi(-\rho, z) = \varphi(\rho, z+1). \tag{42}$$

The charge density can be expressed as

$$\varrho = -\frac{1}{4\pi}\Delta_\delta\varphi = \frac{1}{2\pi\rho}Ш_1(z)\,\delta(\rho). \tag{43}$$

The function χ is periodic, and it is thus natural to ask for the Fourier coefficients for a fixed ρ. Since the potential diverges at $r = 0$ as $1/r$ and also diverges for large ρ (as can be seen, e.g., from the sum estimates for the first ρ derivative), the coefficients will exist only for a positive finite ρ. We again search for the coefficients via the ansatz

$$\varphi(\rho, z) = f_0 \ln \rho + \sum_{n=1}^{\infty} A_n(z) B_n(\rho),\ \rho > 0. \tag{44}$$

The first term corresponds to the $n = 0$ case. We apply (42) and get

$$\varphi = f_0 \ln \rho + \sum_{n=1}^{\infty} f_n \cos\left[\alpha_n z\right] K_0\left[\alpha_n \rho\right]. \tag{45}$$

We determine f_l formally from the electric charge as in (20)

$$4\pi Q(R, h) = -2\pi \int_0^R \varphi_{,z}\Big|_{z=-h}^{z=+h} \rho\, \mathrm{d}\rho - 2\pi \int_{-h}^{h} (\varphi_{,\rho}\rho)\big|_{\rho=R} \mathrm{d}z = 2\pi(q_1 + q_2). \tag{46}$$

This time we get

$$q_1 = \sum_{l=1}^{\infty} f_l 2\alpha_l \sin\left[\alpha_l h\right] \left[\frac{R K_1\left[\alpha_l R\right]}{\alpha_l} - \frac{1}{\alpha_l^2}\right]. \tag{47}$$

We see that q_1 vanishes in the limit $R \to 0$. The second term gives

$$q_2 = \int_{-h}^{h} \left[-f_0 + \sum_{l=1}^{\infty} f_l \cos\left(\alpha_l z\right) \alpha_l R K_1\left(\alpha_l R\right)\right] \mathrm{d}z. \tag{48}$$

We apply the limit $R \to 0$, compare with charge density (43) and get

$$\lim_{R\to 0} 4\pi Q(R, h) = \int_{-h}^{h} \text{Ш}_1(z) \mathrm{d}z \Rightarrow f_0 = -2,\ f_l = 4. \tag{49}$$

We explicitly identify the ECS contribution here, see [5].

3.2 Geometry

In cylindrical coordinates, the fields read

$$g = -U^{-2}\mathrm{d}t^2 + U^2\left(\mathrm{d}\rho^2 + \rho^2 \mathrm{d}\phi^2 + \mathrm{d}z^2\right),\ A = \frac{\mathrm{d}t}{U},\ U = 1 + \lambda\varphi. \tag{50}$$

Note that φ is periodic and so is U. Near the origin, the potential behaves as single extremal Reissner–Nordström solution:

$$\varphi(r,\theta) = \frac{1}{r} + \frac{\bar{\zeta}(3)}{2}\left[1 - 3\cos(2\theta)\right] r^2 + O(r^4),\, r \ll 1. \tag{51}$$

Therefore, we apply a similar transformation as in [3]

$$v = t + W^2(r),\, W(r) = 1 + \frac{\lambda}{r}. \tag{52}$$

The metric in the new coordinates has the same form as in (30). The coefficients near the origin behave as

$$\frac{1}{U^2} = \frac{r^2}{\lambda^2} + \frac{2r^3}{\lambda^3} + O\left(r^4\right), \tag{53}$$

$$\frac{W^2}{U^2} = 1 + \zeta(3)(3\cos(2\theta) + 1)r^3 + O\left(r^4\right), \tag{54}$$

$$U^2 - \frac{W^4}{U^2} = -2\lambda^2\zeta(3)\,(3\cos(2\theta) + 1)\, r + O\left(r^2\right). \tag{55}$$

We see that the metric is regular at $r = 0$ and that it is possible to extend it to $r < 0$ as in (35), and we just formally replace $(r, U) \rightarrow (\tilde{r}, \tilde{U})$, where $\tilde{r} \geq 0$ is the new radial coordinate and the new function $\tilde{U}$ reads

$$\tilde{U} = 1 + \lambda\tilde{\varphi},\, \tilde{\varphi}(\tilde{r},\theta) = \varphi(\tilde{r},\theta) - \frac{2}{\tilde{r}}. \tag{56}$$

4 Smooth Crystal

In the previous chapters, we constructed two unique solutions. The first one consisted of alternating charges, which was useful for proving uniform convergence. However, the range of the potential χ is $\mathbb{R}$, which inevitably results in naked singularities. At least we get an asymptotically flat spacetime at cylindrical infinity. In the second solution, we summed only positive charges, but we had to regularize the sum twice, and in the end, we also ended up with naked singularities. Therefore, it is natural to ask—is there any solution, where we would get an extremal RN in the vicinity of the sources which would not have any singularities? What is its behaviour at infinity? Our idea is to construct a superposition of individual charges generating the Yukawa potential, i.e., $e^{-\alpha r}/r$ for each of the charges distributed equidistantly along the z-axis. The constant α appearing in the potential determines its effective range $1/\alpha$. Thanks to the exponential suppression, the uniform convergence will be ensured. However, this potential is not a vacuum solution and involves charged dust. We can easily modify the MP solution by adding dust. The stress–energy tensor is then decomposed as

$$T^{\mu\nu} = E^{\mu\nu} + M^{\mu\nu}, \tag{57}$$

where $E^{\mu\nu}$ is the electromagnetic part

$$E^{\mu\nu} = \frac{1}{4\pi}\left(F^{\mu}_{\ \beta}F^{\nu\beta} - \frac{\mathcal{F}}{4}g_{\mu\nu}\right), \tag{58}$$

while $M^{\mu\nu}$ corresponds to the dust

$$M_{\mu\nu} = \frac{\varrho(x^j)}{U^3}u_\mu u_\nu, u^\mu = U\delta^\mu_0. \tag{59}$$

Field equations for the metric, electromagnetic field and dust yield only a single equation

$$\Delta_\delta U = -4\pi\varrho. \tag{60}$$

4.1 Constructing the Solution

We seek U in the form

$$U = 1 + \lambda\sigma, \tag{61}$$

where the potential σ satisfies (60) in cylindrical coordinates

$$\sigma_{,\rho\rho} + \frac{\sigma_{,\rho}}{\rho} + \sigma_{,zz} = -4\pi\varrho. \tag{62}$$

We assume the potential σ to be of the following form:

$$\sigma = \sigma_0 + \sum_{n=1}^{\infty}\sigma_n, \sigma_0 = e^{-\alpha r}\chi_0, \sigma_{n\neq 0} = \frac{e^{-\alpha r_n}}{r_n} + \frac{e^{-\alpha r_{-n}}}{r_{-n}}, \alpha > 0. \tag{63}$$

We use bounds for χ_n (14) and get

$$|\sigma_n| \sim \frac{e^{-\alpha n}}{n}, \left|\frac{\partial\sigma_n}{\partial z}\right| \sim \frac{\rho}{n}\left|\frac{\partial\sigma_n}{\partial\rho}\right| \sim \frac{\alpha}{n}e^{-\alpha n}\cosh(\alpha z). \tag{64}$$

For the second derivatives, we obtain

$$\left|\frac{\partial^2\sigma_n}{\partial z^2}\right| \sim \alpha\left|\frac{\partial\sigma_n}{\partial z}\right|, \left|\frac{\partial^2\sigma_n}{\partial\rho^2}\right| \sim \alpha\left|\frac{\partial\sigma_n}{\partial\rho}\right|. \tag{65}$$

We see that we achieved uniform convergence of the potential up to its second derivatives. The symmetries are the same as for the uniform crystal (42)

$$\sigma(\rho, z) = \sigma(-\rho, z) = \sigma(\rho, z+1). \tag{66}$$

The charge density can be expressed as

$$\varrho = -\frac{1}{4\pi}\Delta_\delta \sigma = \frac{1}{2\pi\rho}\text{Ш}_1(z)\,\delta(\rho) - \frac{\alpha^2}{4\pi}\sigma, \tag{67}$$

where we recognize the distributional and functional parts of the density. Let us now seek the Fourier coefficients using the ansatz

$$\sigma(\rho, z) = f_0 K_0(\alpha\rho) + \sum_{n=1}^{\infty} A_n(z) B_n(\rho),\ \rho > 0. \tag{68}$$

The first term corresponds to $n = 0$. We get a separable set of equations for A_n and B_n

$$-\beta_n^2 = \frac{A_n''(z)}{A_n(z)} = \alpha^2 - \frac{\rho B_n''(\rho) + B_n'(\rho)}{\rho B_n(\rho)}. \tag{69}$$

Here, β_n is the separation constant. Using the symmetries (66), we have

$$\sigma = f_0 K_0(\alpha\rho) + \sum_{n=1}^{\infty} f_n \cos\left[\beta_n z\right] K_0\left[\gamma_n \rho\right],\ \gamma_n = \sqrt{\alpha^2 + \beta_n^2}. \tag{70}$$

The coefficients are determined from the charge formally

$$4\pi\, Q(R, h) = -2\pi \int_0^R \sigma_{,z}\big|_{z=-h}^{z=+h} \rho\, \mathrm{d}\rho - 2\pi \int_{-h}^{h} (\sigma_{,\rho}\rho)\big|_{\rho=R}\, \mathrm{d}z = 2\pi(q_1 + q_2). \tag{71}$$

The first term yields

$$q_1 = 2\sum_{l=1}^{\infty} f_l \beta_l \sin\left[\beta_l h\right]\left[\frac{R K_1(\gamma_l R)}{\gamma_l} - \frac{1}{\gamma_l^2}\right]. \tag{72}$$

Again it vanishes for $R \to 0$. The second term gives

$$q_2 = \int_{-h}^{h}\left[f_0 \alpha R K_1(\alpha R) + \sum_{l=1}^{\infty} f_l \cos(\beta_l z)\, \gamma_l R K_1(\gamma_l R)\right] \mathrm{d}z. \tag{73}$$

The linear density here is more tricky, as we have two terms:

$$-\lambda\!\!\!\!-(z) = \lim_{R\to 0}\int_0^{2\pi}\int_0^R \varrho(\rho,z)\rho\,\mathrm{d}\phi\,\mathrm{d}\rho = \mathrm{III}_1(z) - \lim_{R\to 0}\frac{\alpha^2}{2}\int_0^R \rho\varrho(\rho,z)\mathrm{d}\rho. \tag{74}$$

The second part vanishes, and we get $-\lambda\!\!\!\!-(z) = \mathrm{III}_1(z)$. We plug it in and determine the coefficients

$$\lim_{R\to 0} 4\pi Q(R,h) = \int_{-h}^{h}\mathrm{III}_1(z)\mathrm{d}z \Rightarrow f_0 = 2,\ f_l = 4. \tag{75}$$

4.2 Geometry

The metric and electromagnetic potential in cylindrical coordinates read

$$g = -U^{-2}\mathrm{d}t^2 + U^2\left(\mathrm{d}\rho^2 + \rho^2\mathrm{d}\phi^2 + \mathrm{d}z^2\right),\ A = \frac{\mathrm{d}t}{U},\ U = 1 + \lambda\sigma. \tag{76}$$

Both σ and U are periodic functions. It is no surprise that near the origin we retain a single extremal Reissner–Nordström solution:

$$\sigma(r,\theta) = \frac{1}{r} - \alpha - 2\ln\left(1 - e^{-\alpha}\right) + \frac{\alpha^2}{2}r + O(r^2). \tag{77}$$

We can regularize the metric at $r = 0$ by the transformation

$$v = t + W^2(r),\ W(r) = 1 - \lambda\alpha - 2\lambda\ln\left(1 - e^{-\alpha}\right) + \frac{\lambda}{r}. \tag{78}$$

The metric transforms to

$$g = -\frac{\mathrm{d}v^2}{U^2} + 2\frac{W^2}{U^2}\mathrm{d}v\mathrm{d}r + \left(U^2 - \frac{W^4}{U^2}\right)\mathrm{d}r^2 + U^2r^2\mathrm{d}\Omega_2^2. \tag{79}$$

The coefficients near the origin behave as

$$\frac{1}{U^2} = \frac{r^2}{\lambda^2} + \frac{2r^3\left(\lambda\alpha + 2\lambda\ln\left(1 - e^{-\alpha}\right) - 1\right)}{\lambda^3} + O\left(r^4\right), \tag{80}$$

$$\frac{W^2}{U^2} = 1 - \alpha^2r^2 + O\left(r^3\right), \tag{81}$$

$$U^2 - \frac{W^4}{U^2} = 2\lambda^2\alpha^2 + O\left(r\right). \tag{82}$$

We see that the metric is regular here. Contrary to the uniform and alternating crystals, we see that g_{rr} is not zero on the horizon. Near the origin $r \approx 0$, the metric and its determinant can be written as

$$g = -2\mathrm{d}v\mathrm{d}r + 2\lambda^2\alpha^2\mathrm{d}r^2 + \lambda^2\mathrm{d}\Omega_2^2 + O(r),\, \mathfrak{g} = -\lambda^4\sin^2\theta + O(r). \tag{83}$$

Under the horizon, the function U is replaced by $\tilde{U}$, which takes the form

$$\tilde{U} = 1 + \lambda\tilde{\sigma},\, \tilde{\sigma}(\tilde{r},\theta) = \sigma(\tilde{r},\theta) - \frac{2}{\tilde{r}}, \tag{84}$$

with $\tilde{r} \geq 0$ being the new radial coordinate.

5 Uniform Reduced Crystal

In the previous sections, we have constructed three solutions possessing a discrete translational symmetry. However, the master function U was always expressed as an infinite sum with no closed formula. In this section, we use a different approach to get a solution with a closed formula.

5.1 *Constructing the Solution*

We take a 5D crystal solution in the form

$$^5g = -U^{-2}\mathrm{d}t^2 + U\left(\mathrm{d}\rho^2 + \rho^2\mathrm{d}\phi^2 + \rho^2\sin^2\phi\,\mathrm{d}\xi^2 + \mathrm{d}z^2\right),\, ^5A = \frac{\sqrt{3}}{2}\frac{\mathrm{d}t}{U}. \tag{85}$$

Here, U satisfies 4D Laplace's equation and reads [9]

$$U = 1 + \mu\eta,\, \eta = \sum_{n=-\infty}^{\infty}\frac{1}{\rho^2 + (z-n)^2} = \frac{\pi}{\rho}\frac{\sinh(2\pi\rho)}{\cosh(2\pi\rho) - \cos(2\pi z)},\, \mu = \frac{M}{L^2}, \tag{86}$$

where $M > 0$ is mass of the individual black holes and $L > 0$ is spacing of the grid. The metric is independent of the coordinate ξ, which we use for dimensional reduction (for details, see [6]). We end up with 4D fields

$$^4\bar{g} = -U^{-2}\mathrm{d}t^2 + U\left(\mathrm{d}\rho^2 + \rho^2\mathrm{d}\phi^2 + \mathrm{d}z^2\right),\, \Phi = \rho\sqrt{U\sin^2\phi} \tag{87}$$

$$^4A = \frac{\sqrt{3}}{2}\frac{\mathrm{d}t}{U},\, ^4F = \frac{\sqrt{3}}{2}\frac{\mathrm{d}t}{U^2}\wedge\left(U_{,\rho}\mathrm{d}\rho + U_{,z}\mathrm{d}z\right). \tag{88}$$

Here, Φ is an additional scalar field, which is produced as a result of the dimensional reduction. From now on, we work only in 4D, so we drop the dimensional index 4. The reduced quantities satisfy different equations than in the MP solutions:

$$U_{,\rho\rho} + U_{,zz} + 2\frac{U_{,\rho}}{\rho} = 0, 3\Box_g \Phi = \Phi R. \tag{89}$$

The Ricci scalar is no longer zero and equals the negative trace of the total stress–energy tensor T

$$\mathcal{F} = R = -3\frac{U_{,\rho}^2 + U_{,z}^2}{2U^3} = -T. \tag{90}$$

The Fourier series of η reads

$$\eta = \frac{\pi}{\rho} + \sum_{n=1}^{\infty} \frac{e^{-2\pi n\rho}}{\rho} \cos(2\pi nz). \tag{91}$$

From the series expansion for large ρ, we see that the spacetime is cylinder-asymptotically flat:

$$\eta = \frac{\pi}{\rho} + O(\rho^{-2}), \rho \gg 1. \tag{92}$$

5.2 Geometry

Near the origin, the potential behaves as

$$\eta(r, \theta) = \frac{1}{r^2} + \frac{\pi^2}{3} - \frac{\pi^4}{45}\left[2\cos(2\theta) - 1\right] r^2 + O(r^4), r \ll 1. \tag{93}$$

The metric is apparently singular at $r = 0$, but this can be removed by the following procedure. We follow up [10] and start with a transformation of the radial coordinate:

$$r = \sqrt{\sigma}, \mathrm{d}r = \frac{\mathrm{d}\sigma}{2\sigma^{1/2}}, \sigma > 0. \tag{94}$$

The metric transforms to

$$\mathrm{d}s^2 = -\frac{\mathrm{d}t^2}{U^2} + \frac{U}{4\sigma}\mathrm{d}\sigma^2 + \sigma U \mathrm{d}\theta^2 + \sigma U \sin^2\theta \, \mathrm{d}\phi^2. \tag{95}$$

Now, we need to introduce another coordinate

$$\mathrm{d}v_{\pm} = \mathrm{d}t \pm [V(\sigma,\theta)\mathrm{d}\sigma + W(\sigma,\theta)\mathrm{d}\theta], \tag{96}$$

where functions V and W read

$$V(\sigma,\theta) = \frac{1}{2\sqrt{\sigma}} U^{3/2}(\sigma,\theta),\ W(\sigma,\theta) = \int_0^{\sigma} \frac{\partial V(\sigma',\theta)}{\partial\theta} \mathrm{d}\sigma'. \tag{97}$$

This brings the metric to the form

$$\begin{aligned} \mathrm{d}s^2\Big|_{\sigma>0} &= -\frac{\mathrm{d}v^2}{U^2} - \frac{2\mathrm{d}\theta\mathrm{d}\sigma\, V W}{U^2} + \sigma U \sin^2\theta \mathrm{d}\phi^2 \mp \\ &\mp 2\mathrm{d}v \left(\frac{W\mathrm{d}\theta + V\mathrm{d}\sigma}{U^2}\right) + \mathrm{d}\theta^2 \left(\sigma U - \frac{W^2}{U^2}\right). \end{aligned} \tag{98}$$

For $\sigma \ll 1$, the metric coefficients behave as

$$V = \frac{\mu^{3/2}}{2\sigma^2} + \frac{(\pi^2\mu + 3)\sqrt{\mu}}{4\sigma} + O\left(\sigma^0\right),\ W = -\frac{1}{15}\pi^4\mu^{3/2}\sigma\sin(2\theta) + O\left(\sigma^2\right). \tag{99}$$

Therefore, in the near-horizon limit, the metric becomes

$$\mathrm{d}s^2\Big|_{|\sigma|\ll 1,\sigma>0} = \mp\frac{1}{\sqrt{\mu}}\mathrm{d}v\mathrm{d}\sigma + \mu\mathrm{d}\Omega^2 + O\left(\sqrt{\sigma}\right). \tag{100}$$

This is clearly non-singular at $\sigma = 0$. However, one needs to check whether the horizon is smooth,[8] which is the case here as proven in [10]. In the region $\sigma < 0$, one repeats the process with functions

$$\tilde{U} = 1 + \mu\tilde{\eta},\ \tilde{V} = \frac{1}{2\sqrt{|\sigma|}}\left|\tilde{U}\right|^{3/2},\ \tilde{W} = \int_0^{\sigma}\frac{\partial\tilde{V}}{\partial\theta}\mathrm{d}\sigma', \tag{101}$$

where the new potential reads

$$\tilde{\eta}(\sigma,\theta) = \sum_{m\in\mathbb{Z}\setminus\{0\}} \frac{1}{|\sigma| + m^2 - 2m\sqrt{|\sigma|}\cos\theta} - \frac{1}{|\sigma|}. \tag{102}$$

Then, the metric reads

$$\mathrm{d}s^2\Big|_{\sigma<0} = -\frac{\mathrm{d}v^2}{\tilde{U}^2} - \frac{2\mathrm{d}\theta\mathrm{d}\sigma\,\tilde{V}\tilde{W}}{\tilde{U}^2} + \sigma\tilde{U}\sin^2\theta\mathrm{d}\phi^2 \mp \tag{103}$$

[8] For simple binary MP black holes in 5D, the horizon is not smooth. Thanks to the alignment of all black holes in the crystal, the horizons of individual black holes are smooth.

$$\mp 2\mathrm{d}v\left(\frac{\tilde{W}\mathrm{d}\theta+\tilde{V}\mathrm{d}\sigma}{\tilde{U}^2}\right)+\mathrm{d}\theta^2\left(\sigma\tilde{U}-\frac{\tilde{W}^2}{\tilde{U}^2}\right).$$

One can check that the metric in region $\sigma > 0$ matches the one in region $\sigma < 0$:

$$\mathrm{d}s^2\Big|_{|\sigma|\ll 1,\sigma<0}=\mp\frac{1}{\sqrt{\mu}}\mathrm{d}v\mathrm{d}\sigma+\mu\mathrm{d}\Omega^2+O\left(\sqrt{-\sigma}\right). \tag{104}$$

6 Conclusions and Summary

In this proceedings contribution, we have studied the properties of spacetimes exhibiting a discrete translational symmetry. The first solution consisted of alternating positive and negative charges located on a straight line. This had an impact on the asymptotics—the potential decreased exponentially for large ρ. We thus obtained a cylindrically symmetric asymptotic solution. However, the alternating signs of the charges caused the potential to have both positive and negative values. This results in naked singularities.

The second solution consisted only of positive charges. In the cylinder-asymptotic region we obtained the ECS solution representing an extremally charged string, as expected. However, due to the regularization, the range of the potential contained both positive and negative values, and we ended up with singularities.

The third solution consisted of positive Yukawa-like charges forming thus a "smooth crystal", as the charge is distributed throughout the whole spacetime. Due to that, we needed to include dust in the Einstein equations. Summing Yukawa potentials resulted in an improved convergence of the sum, cylinder-asymptotic flatness and no singularities, as the potential is always positive. Comparing the three solutions, the smooth crystal is the best one of them regarding these criteria.

The fourth solution was constructed via dimensional reduction to obtain a closed expression for the corresponding infinite sum. This resulted in an additional scalar field, and the geometry was different from the corresponding 4D MP spacetime. The resulting spacetime is cylinder-asymptotically flat, horizons are smooth and we have no singularities.

We thus compared all four solutions and pointed out their similarities and differences. Plots of potentials defining the solutions are shown in Figs. 1 and 2.

In our future work, we plan to try to include fluid with pressure. This modifies Einstein equations as well as the geometry. The spatial 3D section is then conformal to 3D spaces of constant curvature depending on the value of the pressure. It is then natural to ask about the convergence of the sum appearing in the solution, its asymptotics and the existence of naked singularities and horizons.

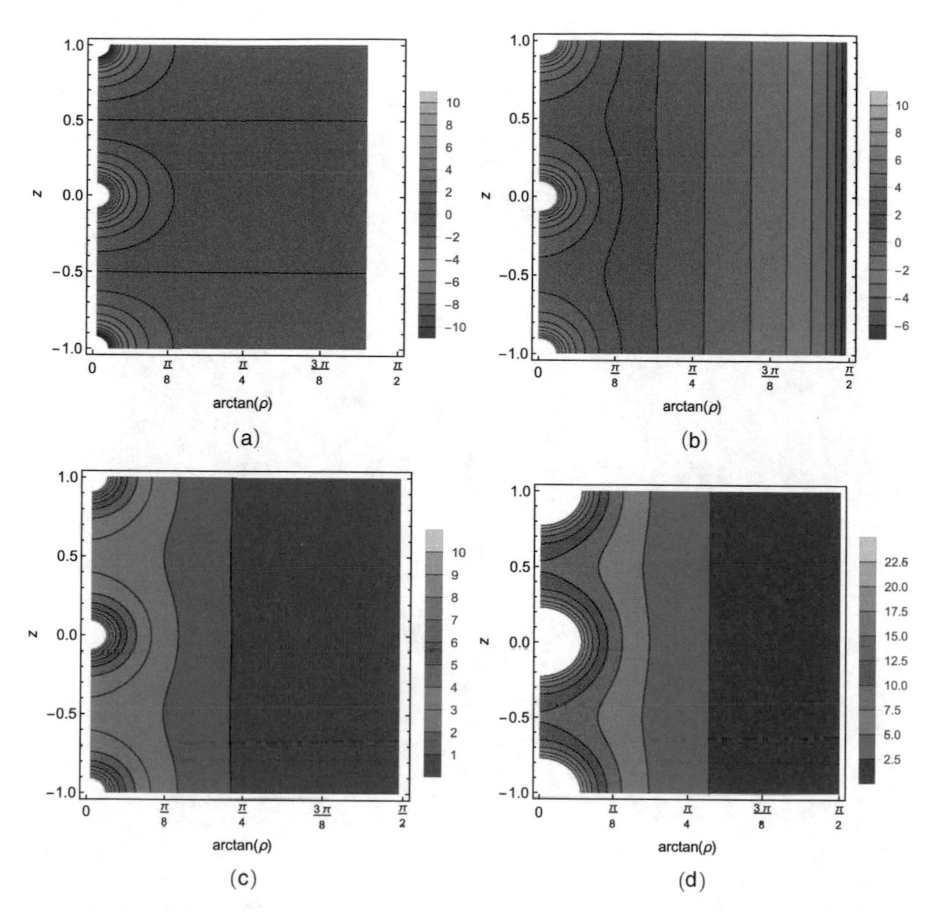

Fig. 1 Conformal contour plot of **(a)** χ (alternating crystal), **(b)** φ (uniform crystal), **(c)** σ (smooth crystal) **(d)** and η (uniform reduced crystal)

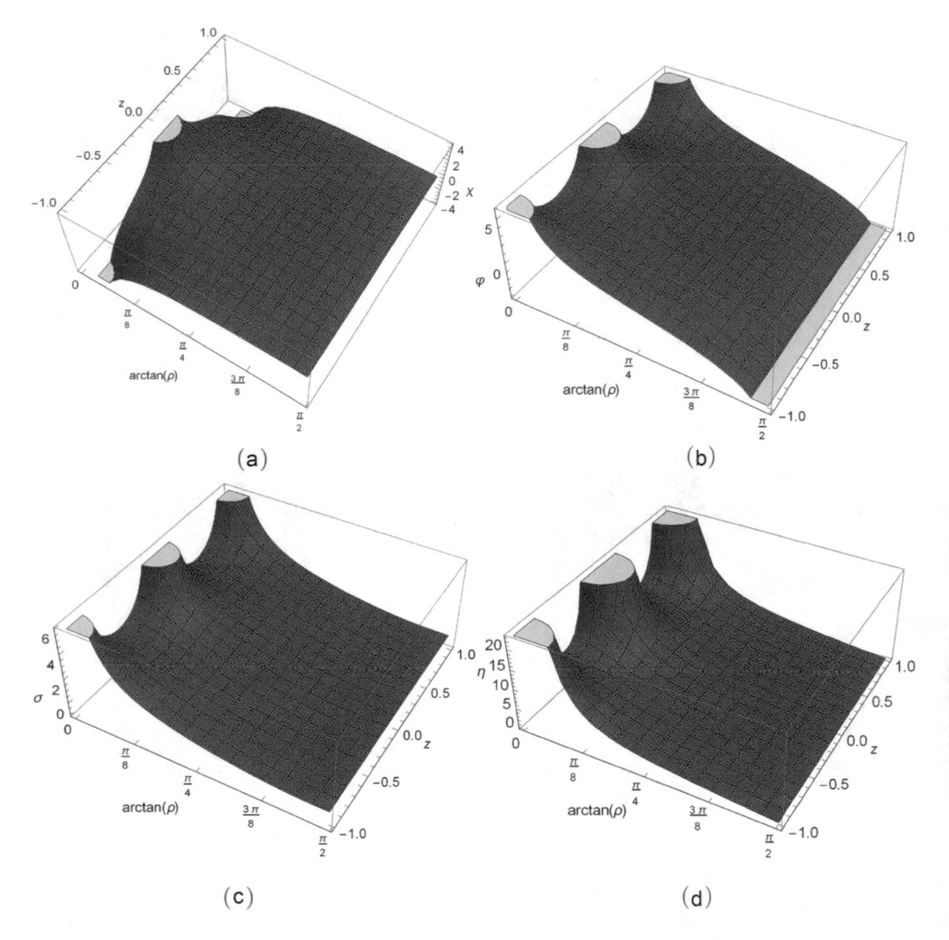

Fig. 2 Conformal plot of (**a**) χ (alternating crystal), (**b**) φ (uniform crystal), (**c**) σ (smooth crystal) and (**d**) η (uniform reduced crystal)

Acknowledgments J.R. was supported by grant GAUK 80918. M.Ž. acknowledges support by GACR 17-13525S.

Grants GAUK 80918, GACR 17-13525S.

References

1. S.D. Majumdar, A class of exact solutions of Einstein's field equations. Phys. Rev. **72**, 390–398 (1947)
2. A. Papaetrou, A static solution of the equations of the gravitational field for an arbitrary charge distribution. Proc. Roy. Irish Acad. (Sect. A) **A51**, 191–204 (1947)
3. J.B. Hartle, S.W. Hawking, Solutions of the Einstein-Maxwell equations with many black holes. Commun. Math. Phys. **26**(2), 87–101 (1972)

4. J. Ryzner, M. Žofka, Electrogeodesics in the di-hole Majumdar-Papapetrou spacetime. Classical Quantum Gravity **32**(20), 205010 (2015)
5. J. Ryzner, M. Žofka, Extremally charged line. Classical Quantum Gravity **33**(24), 245005 (2016)
6. J. Ryzner, M. Žofka, *Einstein Equations: Physical and Mathematical Aspects of General Relativity* (Springer-Verlag GmbH, Berlin, 2019)
7. J.P.S. Lemos, V.T. Zanchin, Class of exact solutions of Einstein's field equations in higher dimensional spacetimes, $d \geq 4$: Majumdar-Papapetrou solutions. Phys. Rev. D **71**, 124021 (2005)
8. D.L. Welch, On the smoothness of the horizons of multi-black-hole solutions. Phys. Rev. D **52**, 985–991 (1995)
9. V.P. Frolov, A. Zelnikov, Scalar and electromagnetic fields of static sources in higher dimensional Majumdar-Papapetrou spacetimes. Phys. Rev. D **85**, 064032 (2012)
10. R.C. Myers, Higher-dimensional black holes in compactified space-times. Phys. Rev. D **35**, 455–466 (1987)

Exact Solutions of the Einstein Equations for an Infinite Slab with Constant Energy Density

Tereza Vardanyan and Alexander Yu. Kamenshchik

Contents

1 Introduction

It is known that even in the absence of matter sources the Einstein equations of General relativity can have very nontrivial solutions. Historically, the first such solution was the external Schwarzschild solution for a static spherically symmetric geometry [1]. It was extremely useful for the study of general relativistic corrections to the Newtonian gravity and for the description of such effects as the precession of the Mercury perihelion and the light deflection in the gravitational field. This solution also opened a fruitful field of black hole physics. The Schwarzschild

T. Vardanyan (✉)
Dipartimento di Fisica e Astronomia, Università di Bologna and INFN, Bologna, Italy
e-mail: tereza.vardanyan@bo.infn.it

A. Yu. Kamenshchik
Dipartimento di Fisica e Astronomia, Università di Bologna and INFN, Bologna, Italy

L.D. Landau Institute for Theoretical Physics, Russian Academy of Sciences, Moscow, Russia
e-mail: kamenshchik@bo.infn.it

S. L. Cacciatori, A. Kamenshchik (eds.), *Einstein Equations: Local Energy, Self-Force, and Fields in General Relativity*, Tutorials, Schools, and Workshops in the Mathematical Sciences, https://doi.org/10.1007/978-3-031-21845-3_11

solution contains a genuine singularity in the centre of the spherical symmetry. To avoid it and to describe real spherically symmetric objects like stars, Schwarzschild also invented an internal solution [2] generated by a ball with constant energy density and with isotropic pressure. At the boundary of the ball, the pressure disappears and the external and internal solutions are matched. In this case, there is no singularity in the centre of the ball. Later, more general spherically symmetric geometries were studied in the papers by Tolman [3], Oppenheimer–Volkoff [4], Buchdahl [5] and many others. Similar problems with cylindrical axial symmetry were also studied (see e.g. [6] and the references therein). In paper [7], the question of existence of solutions of the Einstein equations in the presence of concentrated matter sources, described by the generalised functions (distributions), was studied. It was shown that in contrast to the case of electrodynamics, where the charged ball can be contracted to the point and the charge density becomes proportional to the Dirac delta function while the Poisson equation is still valid, we cannot do it in the General Relativity. The reason lies in the non-linearity of the Einstein equations. It was shown in [7] that the solutions with distributional sources cannot exist for zero-dimensional (point-like particles) and one-dimensional (strings) objects but can exist for two-dimensional (shells) objects. This fact makes the study of geometries possessing plane symmetries particularly interesting. Indeed, the plane-symmetric solutions of the Einstein equations were also studied in the literature (see e.g. [8, 9] and the references therein).

However, to our knowledge, exact static solutions of the Einstein equations, in the spacetimes with plane symmetry in the presence of an infinite slab with a finite thickness, were not studied. Thus, our objective in the present work was to find such solutions with the matching between the geometry inside the slab and that outside of it. Here, we would like to say that the first static solution in an empty spacetime possessing plane symmetry is almost as old as the Schwarzschild solution. This is the spatial Kasner solution [10] found in 1921 and its particular case—the Weyl–Levi-Civita solution [11, 12], found even earlier. Hence, we wanted to find for the case of plane symmetry some analog of matching between Schwarzschild external and internal solutions. We have considered an infinite slab with a finite thickness and a constant mass (energy) density and have found two particular solutions: one with isotropic pressure and one for tangential pressure equal to zero. In both cases, we require that all components of pressure vanish at the boundary of the slab, just like in the case of the Schwarzschild internal solution. The structure of the chapter is the following: in Sect. 2, we write down some general formulae for the spacetimes with the spatial geometry possessing plane symmetry, in Sect. 3, we describe the solution with isotropic pressure, while Sect. 4 is devoted to the solutions with vanishing tangential pressure. Section 5 contains some concluding remarks.

The results presented here were published in [13].

2 Einstein Equations for Spacetimes with Spatial Geometry Possessing Plane Symmetry

Let us consider the metric with plane symmetry, where the metric coefficients depend on one spatial coordinate x:

$$ds^2 = a^2(x)dt^2 - dx^2 - b^2(x)dy^2 - c^2(x)dz^2. \tag{1}$$

Before plunging into technical details connected with the search for the solutions for the thick slab, let us recall briefly what is known about the empty spacetime solutions and the solutions in the presence of thin shells.

For the metric (1) in the empty spacetime, we have two general solutions. One of them is the Minkowski metric, where $a = b = c = 1$ and another one is the Kasner solution [10] with

$$a(x) = a_0(x - x_1)^{p_1}, \; b(x) = b_0(x - x_1)^{p_2}, \; c(x) = c_0(x - x_1)^{p_3}, \tag{2}$$

where the Kasner indices p_1, p_2 and p_3 satisfy the equations

$$p_1 + p_2 + p_3 = p_1^2 + p_2^2 + p_3^2 = 1. \tag{3}$$

The Kasner solution is more often used in a "cosmological form":

$$ds^2 = dt^2 - a_0^2 t^{2p_1} dx^2 - b_0^2 t^{2p_2} dy^2 - c_0^2 t^{2p_3}. \tag{4}$$

This form of the Kasner metric was rediscovered in papers [14–16] and has played an important role in cosmology. The study of Kasner dynamics in paper [16] has led to the discovery of the oscillatory approach to the cosmological singularity [17], known also as the Mixmaster universe [18]. The further development of this line of research has brought the establishment of the connection between the chaotic behaviour of the universe in superstring models and the infinite-dimensional Lie algebras [19].

Coming back to the spatial form of the Kasner metric (2)–(3), one sees that the requirement of symmetry in the plane between the y and z directions implies the condition

$$p_2 = p_3. \tag{5}$$

It is easy to see that there are two solutions of Eqs. (3) satisfying the condition (5). One of them is the Rindler spacetime [20] with

$$p_1 = 1, \; p_2 = p_3 = 0. \tag{6}$$

It is well known that the Rindler spacetime represents a part of the Minkowski spacetime rewritten in the coordinates connected with an accelerated observer. There is a coordinate singularity (horizon) at $x = x_1$. Another solution is

$$p_1 = -\frac{1}{3}, \; p_2 = p_3 = \frac{2}{3}. \tag{7}$$

This particular solution was found by Weyl [11] and Levi-Civita [12] before the work of Kasner.[1] This solution describes a universe, where a real curvature singularity is present at $x = x_1$.

The detailed account of the solutions in the presence of a thin plate with constant energy density was given in paper [9]. These solutions have some distinguishing features. First of all, the energy density of the plate and its tangential pressure should both be proportional to the delta function, while the component of the pressure perpendicular to the plate is equal to zero. Furthermore, the metric is continuous everywhere, but its derivative has a finite jump at the location of the plate. The spacetimes on the right and on the left from the plane are of the type described above: Minkowski, Rindler or Weyl–Levi-Civita. The reflection symmetry is present, i.e. the spacetimes on both sides are the same, if and only if the energy density and the pressure are connected by the relation $p = -\frac{1}{4}\rho$ or $\rho = 0$. Otherwise, this symmetry is lost.

In the paper [9], the solutions in the presence of a finite-thickness slab were also discussed. Some features of such solutions were analysed qualitatively or numerically, but exact solutions were not found. One of these interesting features is the absence of the reflection symmetry. Here, we obtain some exact solutions manifesting this feature. Concerning the properties of the matter constituting the slab, being inspired by the internal Schwarzschild solution [2], we assume that the energy density is constant, while the pressure should disappear at the boundaries of the slab.

Now, we write down some general formulae necessary for the metric with the plane symmetry (1). The non-vanishing Christoffel symbols are

$$\begin{aligned} &\Gamma^x_{tt} = a'a, \; \Gamma^x_{yy} = -b'b, \; \Gamma^x_{zz} = -c'c, \\ &\Gamma^t_{tx} = \frac{a'}{a}, \; \Gamma^y_{yx} = \frac{b'}{b}, \; \Gamma^z_{zx} = \frac{c'}{c}. \end{aligned} \tag{8}$$

The components of the Ricci tensor are

---

[1] In paper [16], a convenient parametrization of the Kasner indices was presented:

$$p_1 = -\frac{u}{1+u+u^2}, \; p_2 = \frac{1+u}{1+u+u^2}, \; p_3 = \frac{u(1+u)}{1+u+u^2}.$$

In terms of this parametrization, the Rindler solution corresponds to $u = 0$, while the Weyl–Levi-Civita solution is given by $u = 1$.

$$
\begin{aligned}
R_{tt} &= a''a + \frac{a'b'a}{b} + \frac{a'c'a}{c}, \\
R_t^t &= \frac{a''}{a} + \frac{a'b'}{ab} + \frac{a'c'}{ac}, \\
R_{xx} &= -\frac{a''}{a} - \frac{b''}{b} - \frac{c''}{c}, \\
R_x^x &= \frac{a''}{a} + \frac{b''}{b} + \frac{c''}{c}, \\
R_{yy} &= -b''b - \frac{a'b'b}{a} - \frac{b'c'b}{c}, \\
R_y^y &= +\frac{b''}{b} + \frac{a'b'}{ab} + \frac{b'c'}{bc}, \\
R_{zz} &= -c''c - \frac{a'c'c}{a} - \frac{b'c'c}{b}, \\
R_z^z &= \frac{c''}{c} + \frac{a'c'}{ac} + \frac{b'c'}{bc}.
\end{aligned} \tag{9}
$$

The Ricci scalar is

$$
R = 2\left(\frac{a''}{a} + \frac{b''}{b} + \frac{c''}{c} + \frac{a'b'}{ab} + \frac{a'c'}{ac} + \frac{b'c'}{bc}\right). \tag{10}
$$

The energy–momentum tensor for a fluid with isotropic pressure is

$$
T_{\mu\nu} = (\rho + p(x))u_\mu u_\nu - p(x)g_{\mu\nu}, \tag{11}
$$

where we shall write

$$
\rho = \frac{4k^2}{3} = \text{constant} \tag{12}
$$

for convenience. Then,

$$
u_t = a,\ u_x = u_y = u_z = 0. \tag{13}
$$

The equation

$$
T^\nu_{\mu;\nu} = 0 \tag{14}
$$

for $\mu = x$ gives

$$
p' = -\frac{a'}{a}(\rho + p), \tag{15}
$$

where "prime" means the derivative with respect to x. The integration of Eq. (15) gives

$$p = -\frac{4k^2}{3} + \frac{p_0}{a}, \tag{16}$$

where p_0 is an arbitrary constant. The Einstein equations are

$$-\frac{b''}{b} - \frac{c''}{c} - \frac{b'c'}{bc} = \frac{4k^2}{3}, \tag{17}$$

$$\frac{a'b'}{ab} + \frac{a'c'}{ac} + \frac{b'c'}{bc} = p, \tag{18}$$

$$+\frac{a''}{a} + \frac{c''}{c} + \frac{a'c'}{ac} = p, \tag{19}$$

$$+\frac{a''}{a} + \frac{b''}{b} + \frac{a'b'}{ab} = p. \tag{20}$$

Introducing new functions

$$A = \frac{a'}{a},\ B = \frac{b'}{b},\ C = \frac{c'}{c}, \tag{21}$$

we can rewrite the Einstein equations (17)–(20) as follows:

$$-B' - B^2 - C' - C^2 - BC = \frac{4k^2}{3}, \tag{22}$$

$$AB + AC + BC = p, \tag{23}$$

$$A' + A^2 + C' + C^2 + AC = p, \tag{24}$$

$$A' + A^2 + B' + B^2 + AB = p. \tag{25}$$

3 Solution with Isotropic Pressure

In what follows, we shall consider only the solutions where the symmetry between the directions along the coordinate axes y and z is present. Then,

$$B = C, \tag{26}$$

and we obtain from Eq. (22)

$$-2B' - 3B^2 = \frac{4k^2}{3}. \tag{27}$$

Integrating this equation, we obtain

$$B = C = -\frac{2}{3}k \tan k(x + x_0). \tag{28}$$

Using the definitions (21), we obtain

$$b = b_0(\cos k(x + x_0))^{\frac{2}{3}}, \tag{29}$$

$$c = c_0(\cos k(x + x_0))^{\frac{2}{3}}. \tag{30}$$

Let us note that in order to not have singularities in the metric, we need to require that

$$[-L + x_0, L + x_0] \subset (-\pi/2, \pi/2), \tag{31}$$

where $x = -L$ and $x = L$ are the locations of the boundary of the slab. Substituting Eqs. (28) and (16) into Eq. (23), we obtain

$$-\frac{a'}{a}\frac{4k}{3}\tan k(x + x_0) + \frac{4k^2}{9}\tan^2 k(x + x_0) = -\frac{4k^2}{3} + \frac{p_0}{a}. \tag{32}$$

This equation can be rewritten as

$$a' - \frac{k}{3}\tan k(x + x_0)a - k\cot k(x + x_0)a + \frac{3p_0}{4k}\cot k(x + x_0) = 0. \tag{33}$$

The general solution of the corresponding homogeneous equation is

$$a(x) = a_1 \sin k(x + x_0)(\cos k(x + x_0))^{-\frac{1}{3}}, \tag{34}$$

where a_1 is an integration constant. We shall look for the solution of the inhomogeneous equation (33) in the following form:

$$a(x) = \tilde{a}(x) \sin k(x + x_0)(\cos k(x + x_0))^{-\frac{1}{3}}. \tag{35}$$

Substituting the expression (35) into Eq. (33), we have

$$\tilde{a}' = -\frac{3p_0}{4k}\frac{(\cos k(x + x_0))^{\frac{4}{3}}}{\sin^2 k(x + x_0)}. \tag{36}$$

Integrating by parts, we obtain

$$\tilde{a}(x) = \frac{3p_0}{4k^2}\cot k(x+x_0)(\cos k(x+x_0))^{\frac{4}{3}} + \frac{p_0}{k}\int dx\,(\cos k(x+x_0))^{\frac{4}{3}} + a_2, \tag{37}$$

where a_2 is an integration constant. Introducing a variable

$$u \equiv \sin^2 k(x+x_0),$$

one can find that

$$\frac{p_0}{k}\int_{-x_0}^{x} dy(\cos k(y+x_0))^{\frac{4}{3}} = \frac{p_0}{2k^2}\mathcal{B}\left(\sin^2 k(x+x_0); \frac{1}{2}, \frac{7}{6}\right)\text{Sign}[\sin k(x+x_0)], \tag{38}$$

where the incomplete Euler function is defined as

$$\mathcal{B}(x, r, s) \equiv \int_0^x du u^{r-1}(1-u)^{s-1}. \tag{39}$$

Thus, the general solution of Eq. (33) is

$$a(x) = \frac{3p_0}{4k^2}\cos^2 k(x+x_0) + \frac{p_0}{2k^2}(\cos k(x+x_0))^{\frac{1}{3}}|\sin k(x+x_0)|\mathcal{B}\left(\sin^2 k(x+x_0); \frac{1}{2}, \frac{7}{6}\right) + a_3 \sin k(x+x_0)(\cos k(x+x_0))^{-\frac{1}{3}}. \tag{40}$$

Looking at the expression (16), we see that the disappearance of the pressure on the boundary of the slab is equivalent to the requirement that

$$a(-L) = a(L) = \frac{3p_0}{4k^2}. \tag{41}$$

On using Eq. (40), this condition can be rewritten as

$$-\frac{3p_0}{4k^2}\sin^2 k(\pm L+x_0) + \frac{p_0}{2k^2}(\cos k(\pm L+x_0))^{\frac{1}{3}}|\sin k(\pm L+x_0)|\mathcal{B}\left(\sin^2 k(\pm L+x_0); \frac{1}{2}, \frac{7}{6}\right) + a_3 k \sin(\pm L+x_0)(\cos k(\pm L+x_0))^{-\frac{1}{3}} = 0. \tag{42}$$

Now, we have two free parameters x_0 and a_3, which we can fix in such a way to provide the disappearance of the pressure on the border of the slab. Let us first choose

$$x_0 = L. \tag{43}$$

It guarantees that

$$a(-L) = \frac{3p_0}{4k^2} \tag{44}$$

and, hence,

$$p(-L) = 0. \tag{45}$$

With this choice of x_0, the requirement (31) becomes

$$2kL < \frac{\pi}{2}. \tag{46}$$

It is easy to see that if the inequality (46) is broken, the cosine is equal to zero at some value of the coordinate x inside the slab and one encounters a singularity.

Now, substituting the value (43) into Eq. (42), we can choose the constant a_3 requiring the disappearance of the pressure on the other border of the slab $x = L$. This constant is

$$a_3 = \frac{p_0}{4k^2}\left(3\sin 2kL\cos^{1/3} 2kL - 2\cos^{2/3} 2kL\ \mathcal{B}(\sin^2 2kL; 1/2, 7/6)\right). \tag{47}$$

Finally, we can write

$$\begin{aligned} a(x) &= \frac{3p_0}{4k^2}\cos^2 k(x+L) \\ &+\frac{p_0}{2k^2}(\cos k(x+L))^{\frac{1}{3}}\sin k(x+L)\mathcal{B}\left(\sin^2 k(x+L); \frac{1}{2}, \frac{7}{6}\right) \\ &+\frac{p_0}{4k^2}\left(3\sin 2kL\cos^{1/3} 2kL - 2\cos^{2/3} 2kL\ \mathcal{B}(\sin^2 2kL; 1/2, 7/6)\right) \\ &\times \sin k(x+L)(\cos k(x+L))^{-\frac{1}{3}}. \end{aligned} \tag{48}$$

Thus, we have obtained a complete solution of the Einstein equations in the slab, where the energy density is constant and the pressure disappears on the boundary between the slab and an empty space. Let us make some comments here. First, the scale factors a, b and c and hence the metric coefficients are not even, and the solution is not invariant with respect to the inversion

$$x \to -x.$$

However, making the change $x \to -x$, we obtain another solution of our equations. It can be obtained also by choosing $x_0 = -L$ instead of $x_0 = L$ and by the corresponding change of the expression for the coefficient a_3, which is reduced to the change of the sign of the argument of the trigonometrical functions. There is no qualitative difference between these two solutions. Thus, we shall study the first one. Let us emphasise that the choice $x_0 = \pm L$ is obligatory in order for the pressure to vanish on both boundaries of the slab, and, hence, the asymmetry of these two solutions is an essential feature of the problem. It arises in spite of the initial symmetry of the Einstein equations and of the position of the slab. Thus, one can speak about some kind of symmetry breaking phenomenon.

Let us consider the question of matching of the solutions in the slab with the vacuum solutions outside the slab. Our solution inside the slab possesses symmetry in the plane (y, z). Thus, we shall try to match it at $x < -L$ and at $x > L$ with one of these three solutions: Minkowski, Rindler or Weyl–Levi-Civita (7).

Consider the plane $x = -L$. We shall require that

$$a_{\text{ext}}(-L) = a(-L),\ b_{\text{ext}}(-L) = b(-L),\ c_{\text{ext}}(-L) = c(-L),$$
$$a'_{\text{ext}}(-L) = a'(-L),\ b'_{\text{ext}}(-L) = b'(-L),\ c'_{\text{ext}}(-L) = c'(-L). \tag{49}$$

Looking at the expressions (48), (29) and (30), we see that at $x = -L$ the derivatives of b and c disappear (provided $x_0 = L$), while the derivative of a at this point is different from zero. Thus, we should choose the Rindler geometry for $x < -L$

$$ds^2 = a_R^2(x - x_R)^2 dt^2 - dx^2 - b_R^2(dy^2 + dz^2). \tag{50}$$

We can consider the analogous matching conditions at $x = L$. Here, the derivatives of all three scale factors are non-vanishing. Thus, for $x > L$, we have a Weyl–Levi-Civita solution

$$ds^2 = a_{WLC}^2(x - x_{WLC})^{-2/3} dt^2 - dx^2 - b_{WLC}^2(x - x_{WLC})^{4/3}(dy^2 + dz^2). \tag{51}$$

Let us now discuss these matching conditions in more detail. On the plane $x = -L$, we have

$$\frac{3p_0}{4k^2} = a_R(-L - x_R), \tag{52}$$

to match the scale factors (the subscript "R" means "Rindler") and

$$a_3 k = a_R, \tag{53}$$

where a_3 is given by Eq. (47) to match their derivatives. It follows from Eqs. (52) and (53) that

$$x_R = -L - \frac{3p_0}{4a_3k^3}. \tag{54}$$

Plotting (47) as a function of $2kL$, we can see that for values smaller than $2kL \approx 1.05$, $a_3 < 0$ and thus

$$x_R > -L. \tag{55}$$

Therefore, there is no horizon for these values of kL.

At the boundary $x = L$, it is more convenient to write down the conditions of matching of the tangential scale factors b:

$$b_0(\cos 2kL)^{2/3} = b_{WLC}(L - x_{WLC})^{2/3}, \tag{56}$$

$$-\frac{2}{3}b_0 k(\cos 2kL)^{-1/3}\sin 2kL = \frac{2}{3}b_{WLC}(L - x_{WLC})^{-1/3}. \tag{57}$$

From these two equations, we easily find that

$$x_{WLC} = L + \frac{1}{k}\cot 2kL. \tag{58}$$

Provided the condition (46), we see that x_{WLC} is necessarily bigger than L, and we cannot avoid having a singularity in the space on the right side of the slab, at least not if the energy density ρ of the slab is positive. To obtain the solution for the case $\rho < 0$, we can replace k by ik in the solution that we already have. Then, trigonometric functions turn into hyperbolic ones, and the expression (58) is replaced by

$$x_{WLC} = L - \frac{1}{k}\coth 2kL. \tag{59}$$

The above expression is smaller than L; therefore, there is no singularity. In the case of an infinitely thin slab, the conclusion that the singularity is unavoidable for $\rho > 0$ was obtained in [9].

The expression for x_{WLC} given by Eq. (58) guarantees the satisfaction of the matching conditions also for the scale factor a and its derivative. It follows from the fact that for both the Weyl–Levi-Civita solution and our internal solution,

$$\frac{a'}{a}(L) = -\frac{1}{2}\frac{b'}{b}(L), \tag{60}$$

which in turn follows from Eq. (23) and from the disappearance of the pressure on the border of the slab.

As we mentioned earlier, the solution (48) is not invariant with respect to the inversion of the coordinate x. However, for a particular value of kL, one can have an even solution, invariant with respect to this inversion. Indeed, we can transform the general solution for the scale factor a (40) into an even function of x by putting $a_3 = 0$ and $x_0 = 0$. Then, also $b(x)$ and $c(x)$ become even. One can check numerically that at $kL \approx 1.05$ the expression for a at the boundaries $x = \pm L$ is such that the pressure disappears. The argument of the trigonometric functions runs between $-1.05 > -\frac{\pi}{2}$ and $1.05 < \frac{\pi}{2}$, the cosine is always different from zero and the singularity does not arise. Besides, at both boundaries, the derivatives of the scale factors are different from zero. Hence, in both half-spaces outside the slab, this solution should be matched with the Weyl–Levi-Civita solutions. Let us stress once again that this symmetric solution is a very particular one, arising at some special value of kL, while generally we have a pair of solutions, each of which is not symmetric with respect to the reflection $x \to -x$, instead this reflection transforms one solution into another and vice versa. One can trace here an analogy with a well-known case of two-well potential, which is often considered at the introducing of the spontaneous symmetry breaking phenomenon in quantum field theory (see e.g. [21])

$$V(\phi) = (\phi^2 - \phi_0^2)^2,$$

which is symmetric with respect to $\phi \to -\phi$, while its minimum values $\phi = \pm\phi_0$ are not symmetric.

4 Solution with Vanishing Tangential Pressure

In the preceding section, we have considered a situation where the tangential pressure coincides with the transversal pressure, just like in the internal Schwarzschild solution [2]. In the case of the Schwarzschild spherically symmetric geometry, such a choice is obligatory because otherwise the pressure becomes infinite in the centre of the ball and a non-singular internal solution does not exist (unless it is assumed that radial pressure is identically zero and tangential pressure does not vanish at the boundary; see [22]). However, it is not obvious that in the case of the plane symmetry the situation is the same. Let us consider a more general energy–momentum tensor

$$T_t^t = \rho,\ T_x^x = -p_x,\ T_y^y = -p_y,\ T_z^z = -p_z. \tag{61}$$

Then, the energy–momentum tensor conservation condition (14) takes the following form:

$$p_x' + A(\rho + p_x) + B(p_x - p_y) + C(p_x - p_z) = 0. \tag{62}$$

In our case, $B = C$ and, hence, $p_y = p_z$. We shall consider the case, where the tangential pressure $p_y = p_z = 0$. Now, Eq. (62) becomes

$$p' + A(\rho + p) + 2Bp = 0, \tag{63}$$

where $p \equiv p_x$. We have two unknown functions: p and A. However, it is not convenient to try to find the relation between these functions using Eq. (63). It is better to take Eq. (25) with the vanishing right-hand side:

$$A' + A^2 + B' + B^2 + AB = 0. \tag{64}$$

The function B still satisfies (27) and (28); using (28), we can rewrite (64) in terms of the function a:

$$a'' - \frac{2}{3}\tan k(x + x_0)a' + \left(\frac{4}{3}k^2 \tan^2 k(x + x_0) - \frac{2}{3}\frac{k^2}{\cos^2 k(x + x_0)}\right)a = 0. \tag{65}$$

Looking for the solution of these second-order linear differential equation in the form

$$a(x)(\cos k(x + x_0))^{\alpha}(\sin k(x + x_0))^{\beta}e^{k\gamma(x+x_0)}, \tag{66}$$

we find two sets of the parameters giving the solution of Eq. (65):

$$\alpha = -\frac{1}{3},\ \beta = 0,\ \gamma = \frac{1}{\sqrt{3}},$$
$$\alpha = -\frac{1}{3},\ \beta = 0,\ \gamma = -\frac{1}{\sqrt{3}}. \tag{67}$$

Thus, the general solution of Eq. (65) is

$$a(x) = (\cos k(x + x_0))^{-1/3}(a_4 e^{\frac{1}{\sqrt{3}}k(x+x_0)} + a_5 e^{-\frac{1}{\sqrt{3}}k(x+x_0)}). \tag{68}$$

Now, we find

$$A = \frac{a'}{a} = \frac{k}{3}\tan k(x + x_0) + \frac{k}{\sqrt{3}}\frac{a_4 e^{\frac{1}{\sqrt{3}}k(x+x_0)} - a_5 e^{-\frac{1}{\sqrt{3}}k(x+x_0)}}{a_4 e^{\frac{1}{\sqrt{3}}k(x+x_0)} + a_5 e^{-\frac{1}{\sqrt{3}}k(x+x_0)}}. \tag{69}$$

Substituting this expression into Eq. (23), we find the transversal pressure

$$p = -\frac{4k^2}{3\sqrt{3}}\tan k(x + x_0)\frac{a_4 e^{\frac{1}{\sqrt{3}}k(x+x_0)} - a_5 e^{-\frac{1}{\sqrt{3}}k(x+x_0)}}{a_4 e^{\frac{1}{\sqrt{3}}k(x+x_0)} + a_5 e^{-\frac{1}{\sqrt{3}}k(x+x_0)}}. \tag{70}$$

In order to have the pressure vanishing at $x = -L$, we can again choose $x_0 = L$. Then, fixing

$$a_5 = a_4 e^{\frac{4kL}{\sqrt{3}}}, \tag{71}$$

we have the pressure vanishing also at $x = L$. Finally, we have

$$p = \frac{4k^2}{3\sqrt{3}} \tan k(x+L) \tanh \frac{k}{\sqrt{3}}(L-x), \tag{72}$$

and

$$a(x) = a_6(\cos k(x+L))^{-1/3} \cosh \frac{1}{\sqrt{3}} k(x-L). \tag{73}$$

For $x > L$, this solution should be matched with the Weyl–Levi-Civita solution with the same value of the parameter x_{WLC} as in the previous section. For $x < -L$, the obtained solution is matched with the Rindler solution with

$$x_R = -L + \frac{\sqrt{3}\coth \frac{2kL}{\sqrt{3}}}{k}. \tag{74}$$

It is easy to see that as long as $2kL < \pi/2$ the internal metric is regular and the pressure (72) is finite everywhere in the slab. Thus, in contrast to the case of the Schwarzschild geometry, we have here a non-singular internal solution with an anisotropic pressure, namely with the pressure whose tangential components are identically equal to zero.

5 Concluding Remarks

We have found two static solutions for an infinite slab of finite thickness immersed in the spacetime with plane symmetry. How are these solutions related to the solutions of a matter source localised on an infinitely thin plane? First of all, let us note that our solutions are non-singular inside the slab if the condition (46) is satisfied. If we introduce the notion of the energy of the unit square of the slab M:

$$M = 2\rho L = \frac{8k^2 L}{3}, \tag{75}$$

then the condition (46) becomes

$$L < \frac{\pi^2}{12M}. \tag{76}$$

Thus, if we fix the value of M and begin squeezing the slab, diminishing L, we do not encounter anything similar to the Buchdahl limit for spherically symmetric configurations [5]. In other words, if the relation (76) is satisfied at some value of L_0, it remains satisfied at all finite values of $L < L_0$. On the other hand, if we start increasing the thickness of the slab, then at the value $L = \frac{\pi^2}{12M}$ a singularity arises inside the slab. Moreover, in the case considered in Sect. 4, the pressure also becomes infinite.

What happens when $L \to 0$? Obviously, the energy density will tend to the delta function

$$\rho_{L\to 0} \to M\delta(x). \tag{77}$$

As was discussed in paper [9], the tangential pressure should also tend to infinity to maintain the validity of the energy–momentum conservation equation (14). In our solution presented in Sect. 4, the tangential pressure is identically zero. One can show, using Eqs. (16) and (42), that in the solution with an isotropic pressure presented in Sect. 3, the pressure in the slab is limited by the value $p \approx M^2$ when $L \to 0$. Thus, while both of these solutions are well defined at any arbitrary small, but finite value of the thickness parameter L, there is not a smooth transition to an infinitely thin plane configuration for these two solutions. However, these solutions represent some particular configurations acceptable from a physical point of view. Let us emphasise once again that we did not fix some particular equation of state for the matter filling our slab. We simply required that the energy density on the slab is constant and that the pressure disappears at the boundaries of the slab. These conditions are the same used in the Schwarzschild internal solution [2]. Then, we considered two particular additional conditions: one of them requires the isotropy of the pressure, just like in the Schwarzschild solution [2], and another requires the disappearance of the tangential pressure in all the slab. For both these requirements, we have found exact solutions. In principle, one can imagine the existence of a solution where the transversal and tangential pressures are different functions of the coordinate x, vanishing on the borders of the slab. Then, one cannot exclude that for some solutions of this kind a smooth transition to the localised matter configurations is possible.

There is also another problem here: it would be interesting to find matter distributions, which imply the existence of solutions of the Einstein equations which are matched in the empty regions of the space with the general spatial Kasner solutions (2) and (3) with $p_2 \neq p_3$. We hope to attack these problems in a future work [23].

Acknowledgments We are grateful to R. Casadio, J. Ovalle and G. Venturi for useful discussions.

References

1. K. Schwarzschild, Sitzungsber. Preuss. Akad. Wiss. Berlin (Math. Phys.) **1916**, 189 (1916)
2. K. Schwarzschild, Sitzungsber. Preuss. Akad. Wiss. Berlin (Math. Phys.) **1916**, 424 (1916)
3. R.C. Tolman, Phys. Rev. **55**, 364 (1939)
4. J.R. Oppenheimer, G.M. Volkoff, Phys. Rev. **55**, 374 (1939)
5. H.A. Buchdahl, Phys. Rev. **116**, 1027 (1959)

6. C.S. Trendafilova, S.A. Fulling, Eur. J. Phys. **32**, 1663 (2011)
7. R.P. Geroch, J.H. Traschen, Phys. Rev. D **36**, 1017 (1987)
8. P.A. Amundsen, O. Gron, Phys. Rev. D **27**, 1731 (1983)
9. S.A. Fulling, J.D. Bouas, H.B. Carter, Phys. Scripta **90**(8), 088006 (2015)
10. E. Kasner, Am. J. Math. **43**, 217 (1921)
11. H. Weyl, Annalen Physik **54**, 117 (1917)
12. T. Levi-Civita, Atti Accad. Naz. Rend. **27**, 240 (1918)
13. A.Y. Kamenshchik, T. Vardanyan, Phys. Lett. B **792**, 430 (2019)
14. A.H. Taub, Annals Math. **53**, 472 (1951)
15. O. Heckmann, E. Schucking, Handbuch der Physik **53**, 489 (1959)
16. E.M. Lifshitz, I.M. Khalatnikov, Adv. Phys. **12**, 185 (1963)
17. V.A. Belinsky, I.M. Khalatnikov, E. M. Lifshitz, Adv. Phys. **19**, 525 (1970)
18. C.W. Misner, Phys. Rev. Lett. **22**, 1071 (1969)
19. T. Damour, M. Henneaux, H. Nicolai, Class. Quant. Grav. **20**, R145 (2003)
20. W. Rindler, Am. J. Phys. **34**, 1174 (1966)
21. N.N. Bogolyubov, D.V. Shirkov, *Quantum Fields* (Benjamin/Cummings, Reading, 1983)
22. P.S. Florides, Proc. R. Soc. Lond. A **337**, 529 (1974)
23. A.Y. Kamenshchik, T. Vardanyan, JETP Lett. **111**(6), 306–310 (2020)

Emergence of Classicality from an Inhomogeneous Universe

Adamantia Zampeli

Contents

1 Introduction

More than a century now, we know that the world is fundamentally quantum mechanical; yet our everyday experience fools us with a classical world. How do these two pictures reconcile? Indeed, if one accepts the thesis that the world is fundamentally quantum mechanical, the natural question to ask is how classical world we experience in our everyday life emerges from the quantum structures. There must be a mechanism for this to happen; but more importantly, what are the requirements for a system to be considered classical? Here, we will adopt the position that this transition from quantum to classical happens through the mechanism of decoherence, which destroys the interference terms between different quantum states. Usually, this happens through the interaction of the system with an environment. In quantum cosmology, on which we focus our considerations here, the universe is a closed system, and the role of the environment is played by inho-

Prepared for the proceedings of the 2nd Domoschool, Domodossola.

A. Zampeli (✉)
Institute of Theoretical Physics, Faculty of Mathematics and Physics, Charles University, Prague, Czech Republic
e-mail: azampeli@phys.uoa.gr

S. L. Cacciatori, A. Kamenshchik (eds.), *Einstein Equations: Local Energy, Self-Force, and Fields in General Relativity*, Tutorials, Schools, and Workshops in the Mathematical Sciences, https://doi.org/10.1007/978-3-031-21845-3_12

mogeneous degrees of freedom acting as perturbations in an overall homogeneous background (e.g. [1, 2]). We are interested to explore a different path, by starting with a genuinely inhomogeneous spacetime and investigating the emergence of classicality due to the presence of symmetries.

Our starting point is the Szekeres spacetime, which is a dust-type D exact solution of general relativity. This classical model has attracted the interest since it is an alternative not only to the Friedman–Lemaitre–Robertson–Walker (FLRW) model but also of the Bianchi I (Kasner). The first currently serves as the "standard" cosmological model, while the latter as a model for the dynamics close to the singularity [3]. It is therefore clear that the class of models described by the Szekeres spacetimes can provide useful information for many properties of physically interesting models and possible new effects which might appear due to quantum gravity.

There are two criteria we consider to examine the emergence of classicality: (i) Hartle's criterion that states that predictions in quantum cosmology are possible when there is a peak on the configuration space, which accordingly indicates a correlation between conjugate momenta on the phase space [4, 5] and (ii) decoherence, which is quantified in the condition that the sum of the non-diagonal terms of the reduced density matrix is much smaller than the sum of its diagonal terms [1, 6]. To this end, we define the reduced density matrix by tracing out the constant of motion related to the classical symmetry in question.

In the following sections, we first summarise the previous results regarding the classical symmetries of a reduced Lagrangian for the Szekeres spacetime and the solution of the quantum equations. Then, we proceed to check whether the first but mainly the second criterion holds for a reduced density matrix defined as described above. In the last section, we discuss the results and the connection with the current cosmological observations about the homogeneity of the universe.

2 Preliminary Results

The general spacetime element for the Szekeres solution is [7]

$$ds^2 = -dt^2 + e^{2A(t,x,y,z)}dx^2 + e^{2B(t,z,y,z)}(dy^2 + dz^2), \tag{1}$$

where the functions $A(t, x, y, z)$ and $B(t, x, y, z)$ can be specified by the solution of the Einstein equations with energy–momentum tensor of the dust,

$$G_{\mu\nu} = T^{(D)}_{\mu\nu}. \tag{2}$$

Instead of the metric variables, we choose the physical variables, since we can take advantage of the fact that in these solutions the two components of the electric part of the Weyl tensor and the two components of the shear for the observer denoted by a time-like 4$-$vector u^μ are equal, respectively. In these variables, the evolution equations take the form [8]

$$\dot{\rho} + \theta\rho = 0, \tag{3a}$$

$$\dot{\theta} + \frac{\theta^2}{3} + 6\sigma^2 + \frac{1}{2}\rho = 0, \tag{3b}$$

$$\dot{\sigma} - \sigma^2 + \frac{2}{3}\theta\sigma + E = 0, \tag{3c}$$

$$\dot{E} + 3E\sigma + \theta E + \frac{1}{2}\rho\sigma = 0, \tag{3d}$$

where $\dot{} = u^\mu \nabla_\mu$ and the energy density is defined as $\rho = T^{\mu\nu}u_\mu u_\nu$. The constraint equation is

$$\frac{\theta^2}{3} - 3\sigma^2 + \frac{{}^{(3)}R}{2} = \rho. \tag{4}$$

We can find a second-order differential system by solving equation (3a) with respect to θ and Eq. (3d) with respect to σ and replacing the results to the other two. The new system, which we omit to write, can be further simplified through the transformation

$$\rho = \frac{6}{u^3(1 - \frac{v}{u})}, \quad E = \frac{v}{u^4\left(\frac{v}{u} - 1\right)}, \tag{5}$$

thus taking the simplified form

$$\ddot{v} - \frac{2v}{u^3} = 0, \tag{6}$$

$$\ddot{u} + \frac{1}{u^2} = 0. \tag{7}$$

These correspondingly can be seen as the Euler–Lagrange equations of the following Lagrangian function:

$$L = \dot{u}\dot{v} - \frac{v}{u^2} \tag{8}$$

It is interesting to note the initial spacetime, despite the degeneracy between the different components of the Weyl tensor and the shear has no Killing vector field and it can only be said that these solutions are locally axisymmetric [9]. However, this reduced Lagrangian possess generalised symmetries which facilitates the quantisation of this system [10]. The symmetries of this Lagrangian are

$$h = \dot{u}\dot{v} - \frac{v}{u^2}, \tag{9a}$$

$$I_0 = \dot{u}^2 - \frac{2}{u}, \tag{9b}$$

and in Appendix, it is shown that their presence is due to the existence of two trivial Killing tensor fields on the configuration space of the (u, v) variables. It is clear that the first equation is the Hamiltonian function, and thus it plays the role of "energy" of the reduced system. The stability analysis of this system [8] showed that there are two exact solutions when $h = 0$ and $I_0 = 0$ of the form $u(t) = u_0 z^{-1}$, $v(t) = v_0 t^{2/3}$ and $u(t) = u_0 t^{2/3}$, $v(t) = v_0 t^{2/3}$. The first solution corresponds to an unstable critical point for the dynamical system (6) while the latter to a stable one [8].

3 Quantum Dynamics and Classical Emergence

We now turn to the quantum dynamics which arises by turning to quantum operators the fundamental variables on the phase space and the classical observables to self-adjoint operators. Then, we find the following eigen equations [10]:

$$\left(-\partial_{uv} + \frac{v}{u^2}\right)\Psi = h\Psi, \tag{10}$$

$$\left(\partial_{vv} + \frac{2}{u}\right)\Psi = -I_0\Psi, \tag{11}$$

which are the quantum analogues of (9). We note in passing that, contrary to what happens for gravitational systems, which are constrained due to the presence of the arbitrary functions (the lapse function and the shift vector), here the dynamics of the reduced system is not constrained. We now limit ourselves to the case $h = 0$, in which the wave function takes the form

$$\Psi\left(I_0, u, v\right) = \frac{\sqrt{u}}{\sqrt{2 + I_0 u}}\left(\Psi_1 \cos\left(\sqrt{\frac{2 + I_0 u}{u}}v\right) + \Psi_2 \sin\left(\sqrt{\frac{2 + I_0 u}{u}}v\right)\right). \tag{12}$$

If we select the constants such that $\Psi_2 = i\Psi_1 = C$, the wave function is written in polar form as

$$\Psi(I_0, u, v) = \frac{C\sqrt{u}}{\sqrt{2 + I_0 u}} \exp\left(i\sqrt{\frac{2 + I_0 u}{u}}v\right), \tag{13}$$

where C is a constant. In [10], we performed the Bohmian analysis for this wave function as well as for general values of the constant h. It was shown that the quantum potential, which is given by

$$\mathcal{Q}(q^i) = -\frac{\Box\Omega(q^i)}{2\Omega(q^i)}, \tag{14}$$

where $\Omega(q^i)$ is the amplitude of the wave function (13), q^i are the variables of the configuration space, in this case (u, v) and $\Box$ the Laplacian for this space, becomes zero. Since the quantum potential appears in the Hamilton–Jacobi equation as an additional term arising due to quantum effects, its value becoming zero means that the classical dynamics emerges from these quantum solutions. This can also be attributed to the fact that the variables $q_i = u, v$ and their conjugate momenta which are defined as $p_i = \nabla_i S$ are highly correlated. Indeed, following the analysis of [11], where it was shown that for the case of WKB-type wave functions strong correlations between the variables and their conjugate momenta on the phase space lead to strong peaks of the wave function and to classicality. We can conclude that the first criterion for the emergence of classicality as introduced by Hartle is satisfied [4, 5].

We are now interested to check whether the second criterion holds, which is decoherence; this is the destruction of the interference terms between different systems due to correlations [12] and happens due to the interaction between subsystems. One plays the role of environment, which has infinite degrees of freedom and is of no interest in the analysis. Therefore, it is traced out, keeping only the degrees of freedom of the system under physical interest. In cosmology, the environment is usually inhomogeneous degrees of freedom of some scalar field. In our case, though, we are interested to examine decoherence in relation to the existence of a symmetry. The induction of decoherence due to symmetries has been discussed before elsewhere, e.g. [13, 14]. In order to quantify this effect, we will define the reduced density matrix as

$$\rho^{red}(u_i, v_j, u_k, v_l) = \left|\Psi(u_i, v_j)\right\rangle \langle\Psi(u_k, v_l)| = \int_0^\alpha DI_0 \Psi^*(u_i, v_j, I_0)\Psi(u_k, v_l, I_0), \tag{15}$$

i.e. by tracing out the symmetry constant I_0. If we insert the polar form of the wave function, it becomes

$$\rho_{ijkl}^{red} = \int_0^\alpha dI_0 \Omega^*(u_i, I_0)\Omega(u_k, I_0)e^{-i(S(u_i,v_j,I_0)-S(u_k,v_l,I_0))}. \tag{16}$$

The condition for decoherence is that the sum of the real part of the non-diagonal elements of this matrix should be much smaller than the sum of the diagonal elements [6]

$$|\sum_{i\neq j} \mathrm{Re}\,\rho^{red}(u_i, u_j)| < \epsilon \sum_{i=j} \rho^{red}(u_i, u_j). \tag{17}$$

The diagonal elements are the ones with

$$\rho^{red}_{ijij} = \int_0^{\alpha} dI_0 \Psi^*(u_i, v_j)\Psi(u_i, v_j) = \int_0^{\alpha} dI_0 |\Omega(u_i, I_0)|^2, \tag{18}$$

which, after substituting the explicit form of the solution, become

$$\rho^{red}_{diag} = \int_0^{\alpha} dI_0 \frac{|C|^2 u}{2 + I_0 u} = |C|^2 \ln(1 + \frac{\alpha u}{2}) \tag{19}$$

and depend only on u. For the non-diagonal elements, we are interested in the behaviour of their real part. These are given by $i \neq k$ and/or $j \neq l$. Their real part is given by

$$\mathrm{Re}\,\rho^{red}(u_i, v_j, u_k, v_l, I_0) = \int_0^{\alpha} dI_0 \Omega^*(u_i, I_0)\Omega(u_k, I_0) \left(\cos(S_{ij} - S_{kl})\right). \tag{20}$$

This expression is always bounded, i.e., it always satisfies the relationship

$$|\,\mathrm{Re}\,\rho^{red}(u_i, v_j, u_k, v_l, I_0)| \leq \int_0^{\alpha} dI_0 \Omega^*(u_i, I_0)\Omega(u_k, I_0) \tag{21}$$

with the equality holding for the diagonal elements when $S_{ij} = S_{kl}$. The right-hand side can be calculated, and it is equal to

$$|C|^2 \ln\left(\frac{u_i\left(\alpha u_j + 1\right) + \sqrt{u_i u_j\left(\alpha u_i + 2\right)\left(\alpha u_j + 2\right)} + u_j}{2\sqrt{u_i u_j} + u_i + u_j}\right) \tag{22}$$

from which we recover the relation (18) for $i = j$. The relation we wish to show that holds for all the range of values of u is the sum of the corresponding term, i.e.

$$|2\,\mathrm{Re}\,\rho^{red}(u_1, u_2)| < \epsilon(\rho^{red}(u_1, u_1) + \rho^{red}(u_2, u_2)) \tag{23}$$

since $\mathrm{Re}\,\rho^{red}(u_1, u_2) = Re\rho^{red}(u_2, u_1/)$, which is written in our case as

$$2\ln\left(\frac{\sqrt{u_i u_j\left(\alpha u_i + 2\right)\left(k u_j + 2\right)} + u_i\left(\alpha u_j + 1\right) + u_j}{2\sqrt{u_i u_j} + u_i + u_j}\right) < \epsilon\left(\ln\left(\frac{\alpha u_i}{2} + 1\right) + \ln\left(\frac{\alpha u_j}{2} + 1\right)\right). \tag{24}$$

This relation is true for every $\alpha > 0$ and positive values of the configuration variables, which is of our interest, and therefore we do have decoherence for this case induced by the presence of symmetry.

4 Discussion

We studied the quantum solution of an inhomogeneous gravitational model which is an exact classical solution. We showed that the presence of symmetry can satisfy two criteria for the emergence of classicality for the particular case of $h = 0$. Instead of separating our system to environment and subsystem, we traced out over the classical constant of motion I_0. We examined the possibility that the interference terms are destroyed due to the existence of symmetries, and we found that this can indeed be the case. It is a known fact that symmetries can lead to decoherence and this can also be manifest formally through the existence of superselection rules. These are rules prohibiting the existence of pure states which are superpositions of states that belong to different coherent subspaces of the Hilbert space [14].

These considerations strengthen the results in [10] and give further motivation to study possible quantum effects at the low-energy limit coming from inhomogeneous spacetimes.

Appendix: The Killing Tensors of the Lagrangian

The metric on the configuration space is

$$G_{\mu\nu} = \begin{pmatrix} 0 & 1 \\ 1 & 0 \end{pmatrix}. \tag{A.1}$$

The Killing fields are

$$\xi_1 = \partial_u, \ \xi_2 = \partial_v, \ \xi_3 = u\partial_u - v\partial_v. \tag{A.2}$$

The (trivial) Killing tensors constructed by these Killing vector fields are found by the relation

$$K = \frac{1}{2}\left(\xi_i \otimes \xi_j + \xi_j \otimes \xi_i\right) \tag{A.3}$$

and have the form

$$K_1 = \begin{pmatrix} 1 & 0 \\ 0 & 0 \end{pmatrix} \tag{A.4}$$

and

$$K_2 = \begin{pmatrix} 0 & 1 \\ 1 & 0 \end{pmatrix}. \tag{A.5}$$

The conserved quantities are given by $K_i = K_i^{\mu\nu} p_\mu p_\nu$, and one can see that K_1 corresponds to Eq. (7), while K_2 to the energy, and thus can be associated with the constants of motion considered in the main text as $K_1 \to I_0$ and $K_2 \to h$.

Acknowledgments I would like to thank the organisers of the 2nd Domoschool for the high level of lectures and their kind hospitality. During the school, I was benefited from discussions with Profs. Sergio Cacciatori, Vittorio Gorini and Alexander Kamenshchik. I also thank Drs. Andronikos Paliathanasis, Georgios Pavlou and Otakar Svitek for suggestions and corrections of the manuscript. Finally, I acknowledge the financial support from the Albert Einstein Center.

References

1. J.J. Halliwell, Decoherence in quantum cosmology. Phys. Rev. **D39**, 2912 (1989). https://doi.org/10.1103/PhysRevD.39.2912
2. A.O. Barvinsky, A.Yu. Kamenshchik, C. Kiefer, I.V. Mishakov, Decoherence in quantum cosmology at the onset of inflation. Nucl. Phys. **B551**, 374–396 (1999). https://doi.org/10.1016/S0550-3213(99)00208-4, gr-qc/9812043
3. V.A. Belinsky, I.M. Khalatnikov, E.M. Lifshitz, Oscillatory approach to a singular point in the relativistic cosmology. Adv. Phys. **19**, 525–573 (1970). https://doi.org/10.1080/00018737000101171
4. J.B. Hartle, Prediction in quantum cosmology. NATO Sci. Ser. B **156**, 329–360 (1987). https://doi.org/10.1007/978-1-4613-1897-2_12
5. R. Geroch, The Everett interpretation. Noûs **18**, 617–633 (1984) .
6. H.F. Dowker, J.J. Halliwell, The Quantum mechanics of history: the Decoherence functional in quantum mechanics. Phys. Rev. **D46**, 1580–1609 (1992). https://doi.org/10.1103/PhysRevD.46.1580
7. P. Szekeres, A class of inhomogeneous cosmological models. Commun. Math. Phys. **41**, 55 (1975). https://doi.org/10.1007/BF01608547
8. A. Paliathanasis, P.G.L. Leach, Symmetries and singularities of the Szekeres system. Phys. Lett. **A381**, 1277–1280 (2017). https://doi.org/10.1016/j.physleta.2017.02.009, 1702.01593
9. M. Bruni, S. Matarrese, O. Pantano, Dynamics of silent universes. Astrophys. J. **445**, 958–977 (1995) . https://doi.org/10.1086/175755, astro-ph/9406068
10. A. Paliathanasis, A. Zampeli, T. Christodoulakis, M.T. Mustafa, Quantization of the Szekeres system. Class. Quant. Grav. **35**, 125005 (2018). https://doi.org/10.1088/1361-6382/aac227, 1801.01276
11. J.J. Halliwell, Correlations in the wave function of the universe. Phys. Rev. **D36**, 3626–3640 (1987). https://doi.org/10.1103/PhysRevD.36.3626
12. W.H. Zurek, Decoherence and the transition from quantum to classical — revisited, in *Quantum Decoherence* (2006), pp. 1–31. https://doi.org/10.1007/978-3-7643-7808-0_1
13. D. Giulini, C. Kiefer, H.D. Zeh, Symmetries, superselection rules, and decoherence. Phys. Lett. **A199**, 291–298 (1995). https://doi.org/10.1016/0375-9601(95)00128-P, gr-qc/9410029
14. D. Giulini, Superselection rules and symmetries, in *Decoherence and the Appearance of a Classical World in Quantum Theory* (1996), pp. 187–222. https://doi.org/10.1007/978-3-662-03263-3_6

Printed in the United States
by Baker & Taylor Publisher Services